Fluorine

Chemistry, Analysis, Function and Effects

Food and Nutritional Components in Focus

Series Editor:
Professor Victor R Preedy, *School of Medicine, King's College London, UK*

Titles in the Series:
1: Vitamin A and Carotenoids: Chemistry, Analysis, Function and Effects
2: Caffeine: Chemistry, Analysis, Function and Effects
3: Dietary Sugars: Chemistry, Analysis, Function and Effects
4: B Vitamins and Folate: Chemistry, Analysis, Function and Effects
5: Isoflavones: Chemistry, Analysis, Function and Effects
6: Fluorine: Chemistry, Analysis, Function and Effects

How to obtain future titles on publication:
A standing order plan is available for this series. A standing order will bring delivery of each new volume immediately on publication.

For further information please contact:
Book Sales Department, Royal Society of Chemistry, Thomas Graham House, Science Park, Milton Road, Cambridge, CB4 0WF, UK
Telephone: +44 (0)1223 420066, Fax: +44 (0)1223 420247
Email: booksales@rsc.org
Visit our website at www.rsc.org/books

Fluorine

Chemistry, Analysis, Function and Effects

Edited by

Victor R Preedy
School of Medicine, King's College London, UK
Email: editor@publicationeditor.org.uk

ROYAL SOCIETY
OF CHEMISTRY

THE QUEEN'S AWARDS
FOR ENTERPRISE:
INTERNATIONAL TRADE
2013

Food and Nutritional Components in Focus No. 6

Print ISBN: 978-1-84973-888-0
PDF eISBN: 978-1-78262-850-7
ISSN: 2045-1695

A catalogue record for this book is available from the British Library

Published by The Royal Society of Chemistry,
Thomas Graham House, Science Park, Milton Road,
Cambridge CB4 0WF, UK

Registered Charity Number 207890

For further information see our web site at www.rsc.org

Preface

Recently there have been major advances in our understanding of the chemistry and function of nutritional components. This has been enhanced by rapid developments in analytical techniques and instrumentation. Chemists, food scientists and nutritionists are, however, separated by divergent skills, and professional disciplines. Hitherto, this transdisciplinary divide has been difficult to bridge.

The series *Food and Nutritional Components in Focus* aims to cover in a single volume the chemistry, analysis, function and effects of single components in the diet or its food matrix. Its aim is to link scientific disciplines so that information becomes more meaningful and applicable to health in general.

The series *Food and Nutritional Components in Focus* covers the latest knowledge base and has a structured format with major subsections covering

- Compounds in context
- Chemistry
- Analysis
- Function and Effects

In some books the section on Chemistry is also linked with Biochemistry. Each chapter has a novel cohort of features namely by containing:

- *Key Facts* (areas of focus explained for the lay person)
- *Definitions of words and terms*
- *Summary points*

The series covers numerous classes of dietary components including, for example, minerals, vitamins, food additives, and so on. The chapters are written by national or international experts, specialists and leaders in the field.

Food and Nutritional Components in Focus No. 6
Fluorine: Chemistry, Analysis, Function and Effects
Edited by Victor R Preedy

Published by the Royal Society of Chemistry, www.rsc.org

Fluorine contains the following material: In Section 1 Fluorine In Context, there are chapters on the environment, and dental fluorosis. Section 2 covers chemistry and metabolism. Section 3 Analysis, includes milk, tea, water, high-resolution, high-temperature molecular absorption spectrometry, blood, plasma and bone. Section 4 Function and Effects, has contributions on crops and vegetables, herbal and tea infusions, bioavailability, saliva, health, teeth–saliva migration of fluoride, bone metabolism, remodelling, dietary calcium, PET bone scans, oxidative damage, hippocampal cells, the liver, insulin, selenium and caspases.

Fluorine is specifically designed for chemists, analytical scientists, forensic scientists, food scientists, dieticians, nutritionists, and research academics. The series is suitable for lecturers and teachers in food and nutritional sciences. Importantly, the series will be a valuable resource for college or university libraries as a reference guide.

Professor Victor R Preedy
King's College London

Contents

Fluorine in Context

Food and Nutritional Components in Focus No. 6
Fluorine: Chemistry, Analysis, Function and Effects
Edited by Victor R Preedy

Published by the Royal Society of Chemistry, www.rsc.org

Chemistry and Biochemistry

Analysis

Section I
Fluorine in Context

CHAPTER 1

Fluoride in the Context of the Environment

MARÍA GABRIELA GARCÍA*[a] AND LAURA BORGNINO[a]

[a]Centro de Investigaciones en Ciencias de la Tierra (CICTERRA). CONICET, Universidad Nacional de Córdoba, Córdoba, Argentina
*E-mail: ggarcia@efn.uncor.edu

1.1 Fluoride in the Environment and Health Implications

Fluoride has interesting properties related to human health, particularly in preventing dental caries. However, when it is concentrated in drinking water at levels higher than 1 mg l^{-1}, a clinical condition called dental fluorosis may appear. This consists of a dental enamel hypomineralization that manifests through greater surface and subsurface porosity than in normal enamel, a condition that develops as a result of excessive fluoride intake (Burt and Eklund, 1992). Typical symptoms of dental fluorosis are fine white stripes to dark stains in the teeth surface. Endemic fluorosis is known to affect many millions of people around the world due to chronic exposure to high levels of fluoride in drinking water. The incidence of this illness is also closely correlated with climatic conditions, eating habits, and the social status of the population.

The effects of fluoride are best predicted by the dose and the duration of exposure. For example, the F requirements set for drinking water in temperate climates are not directly applicable to hot humid regions, where significantly more water is consumed. However, most epidemiological studies concerning the effects of fluoride on teeth and bone have correlated the effects with the concentration of fluoride in drinking water (Fawell *et al.*, 2006). The World

Food and Nutritional Components in Focus No. 6
Fluorine: Chemistry, Analysis, Function and Effects
Edited by Victor R Preedy

Published by the Royal Society of Chemistry, www.rsc.org

Health Organization has established a maximum concentration of 1.5 mg l^{-1} F as a guideline value for drinking water (Fawell *et al.*, 2006) and most countries have adopted this value in their corresponding national regulations.

Long-term intake of high doses of fluoride can also produce alterations in the musculoskeletal, reproductive, developmental, renal, endocrine, neurological, and genotoxic systems. Bones and teeth are the target organs of fluoride, as they tend to accumulate it with age (Ayoob and Gupta, 2006). Hydroxyapatite is the mineral deposited in and around the collagen fibrils of skeletal tissues to form bone (Ozsvath, 2009). In the presence of F-rich solutions, hydroxyapatite is transformed into the more stable and less soluble fluoroapatite, *via* a process of ion exchange. When the exposure to fluoride is high, a dose-related disruption of enamel mineralization is produced, resulting in anomalously large gaps in its crystalline structure, excessive retention of enamel proteins, and increased porosity (Aoba and Fejerskov, 2002).

Fluorine (F) is a univalent poisonous gaseous halogen that occurs under standard conditions (*i.e.*, room temperature and normal vapor pressure) as a light yellow-green, pungent, acrid gas. Fluorine has the atomic number 9 and a molecular weight of 18.998 g mol^{-1}. It is the most electronegative element in the Periodic Table and the most chemically reactive of all elements. Fluorine readily forms compounds with most other elements, including the noble gases krypton, xenon and radon. The superb reactivity (*i.e.*, oxidizing capacity) of fluorine is due to its high electronegativity, its unusually low dissociation energy in relation to other halogens, and the great bond strength of its compounds. Fluorine compounds are characterized by an oxidation state −1. In soil, water, plants, foods and even in air, fluorine is present in its ionic form. During weathering, fluorine can be leached out and dissolved into water, where its natural concentration depends on the geological, chemical and physical characteristics of the reservoir.

Fluorosis, along with other diseases produced by the intake of fluoride-rich waters, is a major health concern. Understanding the causes of elevated fluoride concentrations in natural waters is therefore essential in order to define appropriate protocols of exploitation and management of the water resources, as well as to design adequate methodologies of remediation. To achieve this goal, it is necessary to identify the sources of fluorine and the mechanisms that control its mobility and biodisponibility in natural media.

1.2 Natural Sources of Fluoride

In pristine areas, the source of solutes can be mostly attributed to lithology. When rocks are exposed to the weather agents, a series of chemical processes start. These chemical processes lead to the alteration of the rock minerals and the release of dissolved components.

Minerals are characterized by their definite chemical composition and distinctive crystalline structure. Some elements dominate the chemical composition of a mineral, while some others may be incorporated into lattices as impurities, which neither affect the chemical or physical properties

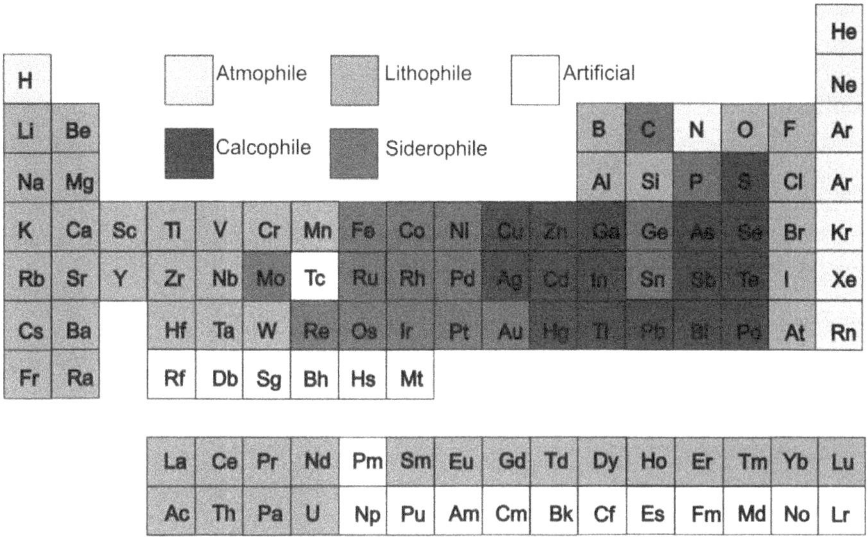

Figure 1.1 Periodic table showing the Goldschmidt's classification of elements.

of the system as a whole or to a significant extent. This process, known as substitution, depends on two physical properties such as ionic radius and electronegativity. Based on these two properties, the prestigious geochemist Goldschmidt recognized four categories of elements: atmophile, lithophile, chalcophile, and siderophile (Figure 1.1). Atmophile elements are, in general, extremely volatile and concentrated in the atmosphere and hydrosphere. Lithophile elements are those that show an affinity for silicate phases and are concentrated in the silicate portion – crust and mantle – of the Earth. Siderophile elements have an affinity for a metallic liquid phase. They are depleted in the silicate portion of the earth and presumably concentrated in the core. Chalcophile elements have an affinity for a sulfide liquid phase, so they too are depleted in the crust and may be concentrated in the mantle and core.

Fluorine is an incompatible lithophile element that preferentially partitions into silicate melts as magmatic crystallization proceeds. The geochemical behavior of F determines its accumulation in the Upper Continental Crust, where its average abundance is 611 mg kg^{-1} (Wedepohl, 1995).

1.2.1 F-Bearing Minerals

A large number of minerals contain F in their chemical formulas as a primary constituent or include it as an impurity. Among primary minerals, biotite and muscovite may contain about 1 wt% of F, while contents are higher in accessory minerals, such as fluorapatite (~3.8 wt%), topaz (~11.5 wt%), and fluorite (~48 wt%). Rare minerals such as cryolite may contain up to 54 wt% F.

As fluoride ions have the same charge and nearly the same radius as hydroxyl ions, they may replace each other in the octahedral sheet of mineral structures (Brigatti and Guggenheim, 2002). This kind of substitution is a common process in micas lattices, where the halogen is found at the same level as the apical oxygens of the tetrahedral sheets where they are bonded to octahedral cations (Figure 1.2).

Minerals of the apatite group are some of the other important F-bearing minerals associated with igneous rocks. The apatite-group minerals, which have a general chemical formula, $Ca_{10}(PO_4)_6X_2$, include three accepted species: fluorapatite (where X = F), chlorapatite (X = Cl), and hydroxylapatite (X = OH). Fluorapatite, in particular, is a ubiquitous accessory phase in igneous, metamorphic, and sedimentary rocks (Pan and Fleet, 2002). Along with hydroxylapatite and their carbonate-bearing varieties, it is an important mineral component of bones, teeth and fossils (Ivanova *et al.*, 2001).

Fluorite (CaF_2) is a common mineral that forms in hydrothermal systems usually with a cubic shape but can also form into octahedral or other suboctahedral shapes, depending on the temperature and composition of the solution where the crystals grow (Zidarova, 2010). It is also found as a cement in

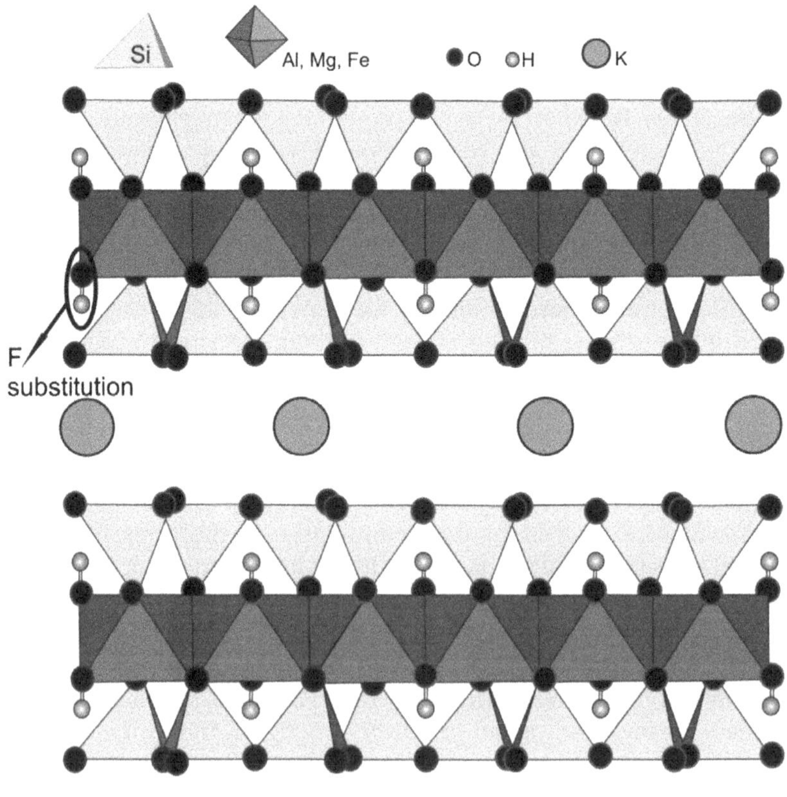

Figure 1.2 Crystalline structure of micas showing the substitution of OH^- by F^-.

carbonate rocks (limestones and dolomites), and less often as an accessory mineral in pegmatites and granites or as a precipitated salt associated with hot springs.

Other F-bearing minerals that are much less abundant include amphiboles (*e.g.*, hornblende, tremolite), topaz ($Al_2SiO_4(OH,F)_2$), and cryolite (Na_3AlF_6). Villiamite (NaF), an almost infinitely soluble mineral, may contribute considerably to fluoride concentrations in groundwater associated with certain peralkaline intrusive bodies, such as the Lovozero Massif in Russia (Kraynov *et al.*, 1969).

1.2.2 Volcanic Sources

Fluoride is also commonly associated with volcanic activity and fumarolic gases. As magmas ascend and decompress their volatile species exsolve into a vapor phase. Trace elements diffuse through the liquid and partition into the vapor, to be liberated during degassing from volcanoes, fumaroles and solfataras.

Active volcanoes emit a variety of gases, including H_2O, CO_2, SO_2, HCl, NH_3, H_2S, HF and a few other minor constituents. It is estimated that the annual global emissions of gaseous fluoride compounds from volcanic sources range from 60 000 to 6 million tons (Symonds *et al.*, 1988). The emitted gases interact rapidly with the ash particles of the volcanic plume and form extremely thin salt coatings. This material is composed of relatively soluble sulfate and halide salts mixed with sparingly soluble fluorine compounds (Delmelle *et al.*, 2007), such as CaF_2, AlF_3 and $Ca_5(PO_4)_3F$). As a consequence, the water in contact with volcanic ash deposits usually contains high concentrations of fluoride (*e.g.*, Wolff-Boenisch *et al.*, 2004; Ruggeri *et al.*, 2010).

Additionally, the (calc-) alkaline volcanoes, typical of continental rift (East Africa), hot spot, continental margin (Andes) or island arcs (Japan), produce relatively fluorine-rich lavas (Rosi *et al.*, 2003).

1.2.3 Geothermal Sources

Geothermal waters are associated with active areas of volcanic activity. Geothermal fluids that rise to the surface include hot water and steam both containing dissolved solutes and gases. They may be discharged from hot springs, fumaroles, hot-water wells and wet- and dry-steam wells. Although a small amount of geothermal water may be derived from gases that were originally dissolved in magmas, most correspond to rainwater that descends through faults to considerable depths, and then is heated by the magma attaining high temperatures (>250 °C) and pressures.

The most common types of geothermal water are alkali-chloride solutions with near-neutral pH values (Edmunds and Smedley, 2013), although a great variability in the chemical composition and in the range of pH can be observed worldwide. Features that are common to most high-temperature waters include unusually high proportions of Li, Rb, Cs, Si, B, As, F and NH_4^+.

The concentration of fluoride in geothermal waters is closely linked to the solubility equilibria of fluorite (Nordstrom and Jenne, 1977), which in turn, also depends on the fluid temperature.

1.2.4 Seawater

The average abundance of fluorine in seawater is 1.3 mg l^{-1}. The aqueous speciation of marine waters reveals that fluorine is mainly found as F^- and MgF^+ ions; these account for about 51 and 47%, respectively, of the total F concentration in seawater. Fluorine's removal from seawater is produced as aerosol sea spray and through the incorporation into marine sediments, either as part of marine organisms, such as in shells or fish bones (Carpenter, 1969) or by the precipitation of authigenic minerals, especially carbonate fluorapatite (Van Cappellen and Berner, 1988). This is the most abundant authigenic marine phosphate phase and consists of a highly substituted apatite with a formula $Ca_5(PO_4)_{2.5}(CO_3)_{0.5}F$ that contains ~3.90 wt% F.

1.2.5 Natural and Anthropogenic Sources in the Atmosphere

Fluoride contributions from natural sources like volcanic eruptions, rock dust or the marine environment are relatively small in the atmosphere. Most of these derive from anthropogenic sources, which include aluminum smelters, fertilizer factories, and industrial activities such as brick and ceramic industries, iron and steel production, fossil fuel burning, cement works, and glass manufacture. These pollution sources release fluorine to the environment as gaseous (*e.g.*, HF, SiF_4, F_2, and H_2SiF_4) or particulate (*e.g.*, CaF_2, NaF, and Na_2SiF_6) species (Ozsvath, 2009). Atmospheric chlorofluorocarbons also contribute fluoride to the atmosphere, however, the measured concentrations in rainwater are insignificant when compared to natural background levels. Fluoride concentrations in precipitation are generally low, typically <50 μg l^{-1}, although levels can be higher where atmospheric deposition of fluoride is enhanced by sources of industrial pollution, volcanic activity or in coastal areas.

Several cases of fluorosis due to indoor pollution have been reported in many regions of SW China, where they were attributed to coal-burning practices (*e.g.*, Liu *et al.*, 2007). Although the F-bearing phase in coal has not yet been identified, many authors suggest that it likely occurs in minerals that are commonly present in this material, such as apatites, fluorites, amphiboles, clays and mica.

1.3 Controls on Fluoride Mobilization

The occurrence, distribution, and mobility of fluorine in aqueous reservoirs depend on the interplay of a number of geochemical processes that determine its removal or release into the solution. Environmental conditions, such as water pH, the chemical composition of the system (*e.g.*, water chemical composition, lithology, *etc.*), and the reaction kinetics strongly influence all these processes.

1.3.1 Adsorption

Colloids in the environment play a fairly important role in controlling pollutant cycling and transport, as they may retain or release ions through adsorption and desorption processes. Adsorption may be defined as the transfer of ions (adsorbate) from the solution phase to the solid phase (adsorbent) *via* various mechanisms, such as physical and chemical adsorption and surface precipitation. Adsorption is described in terms of a set of reactions between solutes and functional groups existing at the solids surfaces. The variety of natural colloid compositions leads to the occurrence of a broad spectrum of surface functional groups that form during the water–solid interaction.

The metal ions on the surface layer have a reduced coordination number, and therefore behave as Lewis acids. In the presence of water, the surface metal ions may first tend to coordinate with water molecules (Figure 1.3(a)) and the dissociative chemisorption of water leads to the formation of surface of hydroxyl groups (Figure 1.3(b)). As they can be either deprotonated or exchanged readily for an anion in aqueous solution, these functional groups are highly reactive (Stumm, 1992).

These kinds of OH groups are found on metal (hydr)oxides, clays minerals and amorphous silicates, where they may form two types of complexes with ions at the solid/water interface (Figure 1.4):

a) Inner sphere complexes (specific adsorption): complexation occurs through short range bonding between the solute complex and surface oxygens. This type of interaction is also called chemisorption.
b) Outer-sphere complexes (nonspecific adsorption): complexation occurs between aqueous ions and the surface functional group throughout electrostatic interactions. This type of interaction is also called physisorption.

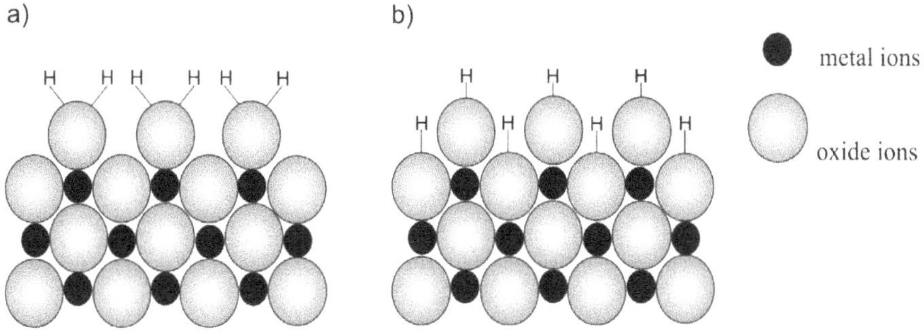

Figure 1.3 Surface functional groups normally present at the water/solid interface of metal oxides, oxyhydroxides, hydroxides, clays minerals and on amorphous silicates. a) Coordination of surface metal ions with water molecules; b) formation of surface hydroxyl groups by dissociative chemisorption of water.

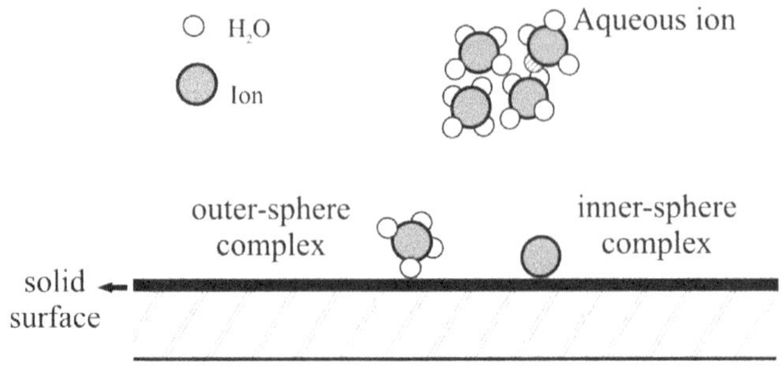

Figure 1.4 Inner- and outer-surface complexes on solid surfaces.

It is important to note that even if the ion and surface have the same charge, the chemical affinity of an ion for the surface can override the electrostatic repulsion (*i.e.*, if there is a sufficiently strong chemical interaction, a negatively charged ion can adsorb onto a negatively charged surface). Ions retained strictly by electrostatic forces are more likely to be easily displaced by ions of a similar charge and thus, are termed exchangeable. Exchangeable ions are essential in the maintenance of plant nutrient levels, but are not strong enough to immobilize environmental pollutants. Chemically retained ions form very strong associations with solids that are often considered to be irreversible. As a result, chemically bound ions will have a diminished potential for release and should therefore pose a lower risk than ions held by only electrostatic forces.

A number of studies have been carried out in order to characterize the interactions of fluoride with different mineral surfaces. The identification of the factors that control the adsorption and desorption of fluoride from mineral surfaces is crucial to understanding its mobility in aquatic environments, where the most reactive minerals are metal (hydr)oxides, carbonates and clays.

1.3.1.1 Adsorption Onto Metal (Hydr)Oxides and Calcite

The adsorption of fluoride onto Fe (hydr)oxides has been studied extensively (*e.g.*, Hiemstra and Van Riemsdijk, 2000; Sujana and Anand, 2010). The kinetics of adsorption is relatively fast (10 min to 10 h, depending on the oxide crystallinity) and the process is pH dependent, as higher adsorption occurs at lower pH values. Optimum removal is determined in the range from pH 4 to 8. The interaction between surface groups and fluoride is by inner sphere surface complexes and involved a replacement of the fluoride with the surface hydroxyl groups (Hiemstra and Van Riemsdijk, 2000) and the consequent release of OH^- into the solution as show in eqn (1.1).

$$\equiv FeOH^{-1/2} + F^- = \equiv FeF^{-1/2} + OH^- \quad (1.1)$$

Farrah *et al.* (1987) first investigated the interaction of fluoride with aluminum (hydr)oxides over a range of pH from 3.0 to 8.0. The obtained results indicated that under strongly acid conditions, the oxides dissolve and form F–Al complexes, while maximum adsorption occurs in the pH range from 5.5 to 6.5 (~170 mg g^{-1}). As the adsorption of fluoride onto Al oxides is low (~0.2–0.4 mg g^{-1}), this material needs to be activated by heating in order to become an effective adsorbent.

Calcite is another important fluoride sink in soils and sediments, where this mineral may be present as either dentritic grains or cement. Due to the strong affinity of fluoride for calcite sites, this mineral has largely been used as an adsorbent in wastewater-treatment technologies (*e.g.*, Turner *et al.*, 2005).

1.3.1.2 Adsorption Onto Clay Minerals and Soils

The electrical charge in natural clay minerals is generally negative, which makes these materials unsuitable for anion removal. However, this condition is altered in soils and sedimentary deposits, where clays are usually associated with metal (hydr)oxide coatings, particularly Fe (hydr)oxides (Zhuang and Yu, 2002). Thus, the presence of these coatings increases their capacity for fluoride adsorption (Bia *et al.*, 2012). The formation of Al–F complexes also enhances the adsorption of fluoride in soil systems (Harrington *et al.*, 2003; Zhu *et al.*, 2004). Harrington *et al.*, (2003) indicated that fluoride complexation to surficial Al and Fe *via* ligand exchange with surficial OH^- groups and water molecules appears to be a dominant mechanism of adsorption.

Competitive adsorption is a common mechanism in soils and sedimentary environments. Fluoride usually competes with other anions such as phosphate, arsenate, chloride, sulfate and carbonate for surface sites in aquatic environments (*e.g.*, Zhu *et al.*, 2007; Bia *et al.*, 2012).

1.3.2 Precipitation

The main authigenic F-bearing minerals that may precipitate from saturated solutions in the aquatic environments are fluorite, fluorapatite and, depending on the alkalinity of the media, carbonate fluorapatite. Table 1.1 shows the solubility constant of some of these minerals.

Table 1.1 Solubility constants at 25 °C for most common F-bearing minerals.

Mineral	**Chemical Formula**	**log k_{ps}**
Fluorite	CaF_2	−10.0370
Fluorapatite	$Ca_5(PO_4)_3F$	−24.9940
Carbonate-fluorapatite	$Ca_{9\cdot 316}Na_{0.36}Mg_{0.144}(PO_4)_{4.8}(CO_3)_{1.2}F_{2.48}$	−114.4

1.3.3 Dissolution and Desorption

As mentioned above, the main F-bearing minerals in the natural environment are amphiboles, micas, fluorite and apatite, and, to a lesser extent, topaz and cryolite. All these phases show marked differences in regards to their solubility in water, as shown in Figure 1.5. Cryolite and fluorite are much more soluble than fluorapatite and micas in the range of pH of natural water (4–9) and their solubilities are almost constant across the entire pH range. On the contrary, the solubilities of fluorapatite and micas are highly pH dependent in the range from pH 2.5 to 12.5.

Fluorapatite is more soluble under acidic pH, but is still one to two orders of magnitude less soluble than fluorite and cryolite, respectively. Chaïrat *et al.* (2007) determined that fluorapatite dissolution rates decreased at $pH \geq 3$, remained constant between $7 \leq pH \leq 10$ and decreased again at $pH \geq 10$.

Despite their low solubilities, the dissolution of biotite and muscovite in granitic terrains has been considered an important source of fluoride in water that is in contact with such rocks. Leaching experiments performed with suspensions of biotite in pure water at pH 7 and 25 °C revealed that fluoride is released after 200 h of reaction reaching concentrations in water of about 6–10 mg l^{-1} (Chae *et al.*, 2006)

It is commonly accepted that dissolution of fluorite is the main cause of elevated concentrations of fluoride in groundwater in many parts of the world (*e.g.*, Currel *et al.*, 2011; Edmunds and Smedley, 2013). The dissolution of this mineral is usually favored by the co-occurrence of mechanisms that promote the scavenging of Ca^{2+}, such as calcite precipitation and cation exchange. For

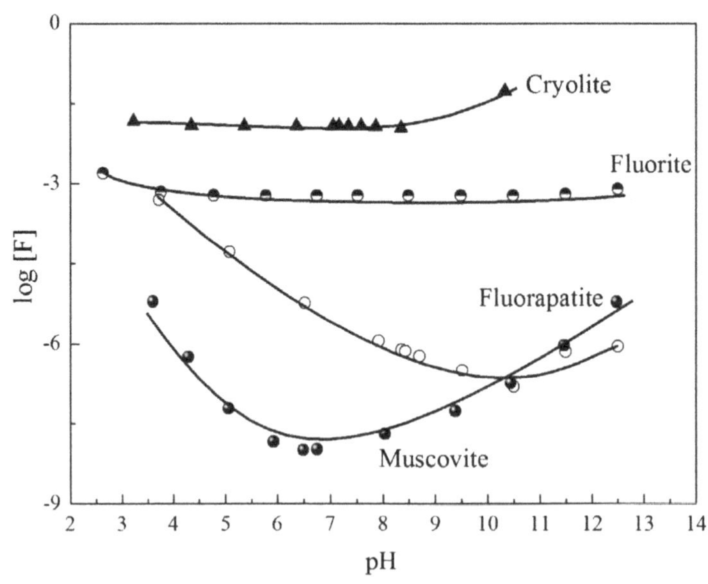

Figure 1.5 Solubility diagram of most common F-bearing minerals.

this reason, in such reservoirs, it is usual to observe inverse trends between Ca^{2+} and fluoride concentrations along the flow path.

Desorption from Fe, Al and Mn (oxy)hydroxides has usually been considered a secondary source of fluoride in groundwater reservoirs where alkaline conditions predominate (*e.g.*, Currel *et al.*, 2011; Borgnino *et al.*, 2013).

1.4 Fluoride in Natural Waters

1.4.1 Speciation of Fluoride in Natural Waters

Fluoride, as a strong ligand in water, may form soluble complexes with polyvalent cations, such as Mg^{2+}, Fe^{3+}, Al^{3+}, and Ca^{2+} depending upon the water pH (Nordstrom and Jenne, 1977), as well as with trace elements present in water, such as B, Be, Si, U, V, and rare earth elements. Free fluoride, however, is the predominant species in most groundwater reservoirs worldwide, while metal complexes with Ca, Mg, and Na, along with the neutral species HF, are only present in low concentrations under acidic conditions (*i.e.*, pH < 3.5). Table 1.2 summarizes the thermodynamic data and the aqueous speciation of some fluoride species commonly found in natural aquatic reservoirs.

1.4.2 Concentration and Speciation of Fluoride in Natural Environments

1.4.2.1 Igneous and Metamorphic Terrains

Many of the world's high-fluoride districts (Figure 1.6) are underlain by crystalline igneous and metamorphic rocks (*i.e.*, parts of India, Sri Lanka, Senegal, Ghana, Cameroon, Tanzania, South Africa, the Pampean ranges in southern South America, and Scandinavia). The average fluorine concentration in crystalline rocks can range from 100 mg kg^{-1} up to over 1000 mg kg^{-1}, because these contain a large proportion of F-bearing minerals. In affected areas, fluoride concentrations in natural water have been reported to be higher than 0.5 mg l^{-1} and up to 40 mg l^{-1} (*e.g.*, Chae *et al.*, 2007; Fantong *et al.*, 2010; Reddy *et al.*, 2010; Berger *et al.*, 2012; Garcia *et al.*, 2012; Padhi and Muralidh, 2012; He *et al.*, 2013; Vikas *et al.*, 2013). Most of these works attribute the elevated fluoride concentrations in water to the dissolution and weathering of F-bearing minerals present in the dominant rocks. Among them, fluorite is one of the more soluble, but it is just an occasional accessory mineral in these crystalline rocks. On the other hand, apatite, micas, and amphiboles are much less soluble than fluorite (Figure 1.5) but they are considered to be the more likely source of fluoride in natural water in granitic terrains. Recently, on the basis of experimental results, Borgnino *et al.*, (2013) determined that fluoride is preferentially released under acidic pH and low Ca^{2+} concentrations from fluvial sediments of the granitic Pampean ranges of Argentina. When equilibrium conditions are reached, water typically shows high pH and alkalinity, which in turn promotes the precipitation of calcite and further enhances the dissolution of Ca- and F-bearing minerals present in the rocks (*i.e.*, fluorapatite). The combination of these processes results in water with high fluoride concentrations along with low levels of Ca, Mg and total hardness.

1.4.2.2 Volcanic and Geothermal Domains

In volcanic areas, drinking-water sources often have inputs from geothermal activity related to volcanic and/or hydrothermal processes. Consequentially, aqueous reservoirs in the Pacific volcanic belt, the largest volcanic area in the world, are affected by high fluoride concentrations (Figure 1.6). For example, in the Andean cordillera, concentrations as high as 8.5 mg l^{-1} have been measured in thermal springs from the Puna/Altiplano region (*e.g.*, Hudson-Edwards and Archer, 2012) In the western USA, fluoride

Table 1.2 Thermodynamic data for aqueous speciation of fluoride species.

	Aqueous species	Reaction	log *k*
Al	AlF^{+2}	$F^- + Al^{3+} = AlF^{2+}$	7.0
	AlF_2^+	$2F^- + Al^{3+} = AlF_2^+$	12.6
	AlF3	$3F^- + Al^{3+} = AlF_3$	16.7
	AlF_4^-	$4F^- + Al^{3+} = AlF_4^-$	19.1
B	$BF_2(OH)_2^-$	$F^- + H^+ + B(OH)_3 = BF2(OH)_2^- + H_2O$	6.62
	BF_3OH^-	$3F^- + H^+ + B(OH)_3 = BF_3OH^- + H_2O$	13.2
	BF_4^-	$4F^- + 3H^+ + B(OH)_3 = BF_4^- + 3H_2O$	18.0
Ba	BaF^+	$F^- + Ba^{2+} = BaF^+$	−0.18
Cu	CuF^+	$F^- + Cu^{2+} = CuF^+$	1.2
Eu	EuF^+	$F^- + Eu^{2+} = EuF^+$	−1.3
	EuF^{+2}	$F^- + Eu^{3+} = EuF^{2+}$	4.4
	EuF_2	$2F^- + Eu^{2+} = EuF_2$	−2.04
	EuF_2^+	$2F^- + Eu^{3+} = EuF_2^+$	4.0
	EuF_3	$3F^- + Eu^{3+} = EuF_3$	10.2
	EuF_3^-	$3F^- + Eu^{2+} = EuF_3^-$	−2.5
	EuF_4^-	$4F^- + Eu^{3+} = EuF_4^-$	12.1
	EuF_4^{-2}	$4F^- + Eu^{2+} = EuF_4^{2-}$	−2.8
Fe	FeF^+	$Fe^{2+} + F^- = FeF^+$	1.3
	FeF^{2+}	$Fe^{3+} + F^- = FeF^{2+}$	4.1
	FeF_2^+	$2F^- + Fe^{3+} = FeF_2^+$	8.3
H	HF_2^-	$H^+ + 2F^- = HF^-$	3.76
	H_2F_2	$2H^+ + 2F^- = H_2F_2$	6.8
Mn	MnF^+	$Mn^{2+} + F^- = MnF^+$	1.4
Pb	PbF^+	$Pb^{2+} + F^- = PbF^+$	0.82
	PbF_2	$2F^- + Pb^{2+} = PbF_2$	1.61
Sm	SmF^{2+}	$Sm^{3+} + F^- = SmF^{2+}$	4.36
	SmF_2^+	$2F^- + Sm^{3+} = SmF_2^+$	7.63
	SmF_3	$3F^- + Sm^{3+} = SmF_3$	10.02
	SmF_4^-	$4F^- + Sm^{3+} = SmF_4^-$	11.97
Sn	SnF^+	$Sn^{2+} + F^- = SnF^+$	4.08
	SnF_2	$2F^- + Sn^{2+} = SnF_2$	6.68
	SnF_3^-	$3F^- + Sn^{2+} = SnF_3^-$	9.46

concentrations up to 50 mg l^{-1} were determined in thermal waters collected from Yellowstone National Park (*i.e.*, Deng *et al.*, 2011), while in New Zealand and Taiwan, deep geothermal waters have shown concentrations as high as 12 mg l^{-1} (Cronin *et al.*, 2003; Bernal *et al.*, 2014)

The speciation of fluorine in high-temperature geothermal waters is usually quite different from that found in cold groundwater, since several complexation reactions may occur between F and other constituents that are usually present in geothermal water, such as Al, Fe and B (Guo, 2012). In these systems, fluoride concentrations are typically in the range of 1–10 mg l^{-1}, although concentrations as high as 1000 mg l^{-1} can be found under acidic conditions. In such solutions, HF, HF_2^- and SiF_6^{2-} species are present and may escape into the air as HF and SiF_4 gases at atmospheric pressures (Nordstrom and Jenne, 1977).

The region occupied by the East African Rift is a large tectonic depression with volcanic activity that encompasses several countries in East Africa, including Sudan, Ethiopia, Uganda, Kenya and Tanzania. Natural waters here are characterized by anomalous concentrations of fluoride that mostly exceed the World Health Organization guideline value for drinking water. This is the cause of prevalent endemic fluorosis disease in the region (*e.g.*, Rango *et al.*, 2009).

A thick layer (up to 1500 m) of basaltic rocks covers an extensive area over the Parana River basin, in southern Brazil, eastern Paraguay, northern Uruguay and Argentina (Figure 1.6). These rocks, known as Serra Geral Formation, overlay the formations of the Guarani Aquifer System, one of the most important reservoirs of fresh water in the world. Thermal waters extracted from deep boreholes located along the Uruguay River show fluoride concentrations of up to 13 mg l^{-1}. Despite these elevated concentrations, fluorosis has not been identified in the region, most likely because water use is generally limited to recreational purposes.

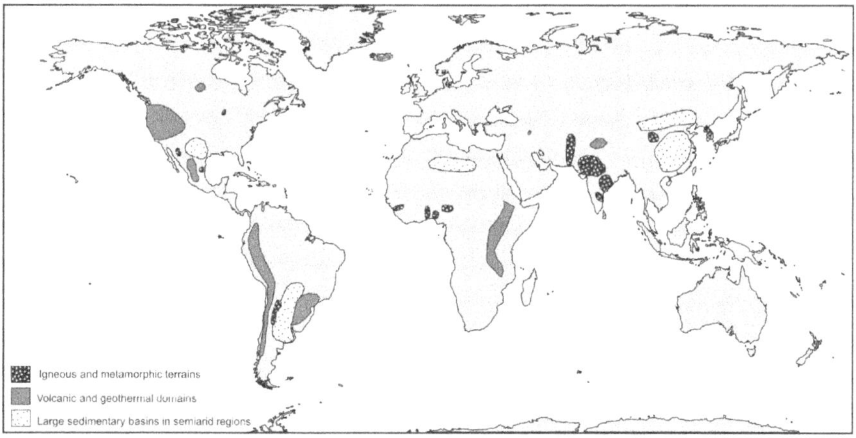

Figure 1.6 Worldwide distribution of fluoride-rich districts.

1.4.2.3 Large Sedimentary Basins in Semiarid Regions

In large sedimentary basins located in semiarid regions of the world, such as in the Chacopampean Plain in South America, the Yellow River Basin in China, and the arid and semiarid regions of Southern USA and Northern Mexico, groundwater constitutes one of the main drinking-water sources. Common features in these aquifers are the predominance of oxidizing conditions, alkaline pH, and a chemical composition of water dominated by Na^+ and HCO_3^- ions. Elevated concentrations of fluoride are typically found, along with other hazardous elements such as As, Mo, V, and U (*e.g.*, Currel *et al.*, 2011; Valenzuela Vásquez *et al.*, 2006; Nicolli *et al.*, 2012).

It is estimated that about 1.2 million inhabitants in the Chaco Pampean Plain drink groundwater with fluoride levels that exceed the guideline value of 1.5 mg l^{-1} for drinking water. The primary sources of fluoride have been traditionally attributed to the presence of volcanic shards and some scarce fluorapatite grains spread within the loessic sediments that are in contact with these water reservoirs. However, the association of fluoride in the volcanic glass still remains unknown. The concentration of this ion in groundwater is considered to be the result of the release from these primary sources, as well as from desorption from Fe and Mn (hydr)oxide coatings (*e.g.*, Borgnino *et al.*, 2013).

High concentrations of fluoride and arsenic are also found in groundwater that is in contact with loess sediments accumulated in arid and semiarid river basins in northern China (*e.g.*, Guo *et al.*, 2007; Currel *et al.*, 2011). The mechanism that mainly controls the mobility of fluoride is attributed to desorption from Fe, Al and Mn oxides present in the loess sediments that compose the aquifer (Currel *et al.*, 2011).

In the desert area along the border between Mexico and USA, groundwater also shows high contents of As and F (*e.g.*, Scanlon *et al.*, 2009; Vinson *et al.*, 2011). The sources of both natural pollutants has been attributed to lithology, mainly composed of alluvial sediments that contain fragments of volcanic rocks such as rhyolites, tuff, and volcanic glass (Reyes Gomez *et al.*, 2013).

Summary Points

- More than 70 million people suffer from fluorosis worldwide.
- Although anthropogenic sources account for some important contributions, the sources of this element in water are primarily geogenic.
- Minerals such as micas, apatites and fluorite are responsible for the release of elevated concentrations of fluoride to natural waters.
- Fluoride is associated with volcanic activity.
- The dynamics of fluoride in natural environments depend on the dissolution and precipitation of F-bearing minerals, along with the adsorption/desorption from metal (hydr)oxides and clay minerals.
- Most of the world's high-fluoride districts coincide with areas affected by volcanic activity, regions underlain by crystalline igneous and metamorphic rocks, and large sedimentary basins in arid and semiarid conditions.

Key Facts

Key Features of Weathering

1. Weathering is the process by which rocks originally formed at higher temperatures and pressure deep in the Earth come to equilibrium with water at temperatures prevailing at the surface of the Earth.
2. Weathering produces two kinds of products: secondary minerals, formed by the chemical alteration of primary minerals, and dissolved components.
3. The global reaction that resumes the chemical weathering can be expressed as: rock + H_2O + $CO_{2\ atm}$ → secondary minerals (altered rock) + solutes.
4. During weathering, rock minerals may completely dissolve without leaving a solid residue. For example, calcite ($CaCO_3$) and dolomite ($CaMg(CO_3)_2$), dissolve congruently in slightly acidified water, while silicate minerals are usually altered, forming a solid residue, generally a clay mineral and/or amorphous Fe or Mn (hydr)oxides.

Key Terms

Accessory minerals: Any mineral in an igneous rock not essential to the naming of the rock. They usually constitute less than 1% of a rock by volume.

Authigenic minerals: Minerals that form *in situ* within the depositional site in response to geochemical processes. Typical examples are calcite, halite, and gypsum.

Basaltic rocks: Dark colored, fine-grained, igneous rock, with low contents of silica and comparatively rich in iron and magnesium. It most commonly forms as an extrusive rock, such as a lava flow.

Cratonic area: Tectonically stable areas composed of crystalline rocks that make up the oldest part of a continent.

Igneous rocks: Rocks formed from the solidification of magma, which is a hot (600 to 1300 °C) molten or partially molten rock material. Rocks formed from lava at the Earth surface, are extrusive or volcanic rocks, while rocks formed from magma at depth are intrusive, or plutonic rocks.

Incompatible element: Elements that tend to partition into the melt phase when melting or crystallization occurs.

Lewis acids: Elements or compounds that behave as electron-pair acceptors.

Primary minerals: Minerals that form by solidification from magma under conditions different from those at the surface of the Earth.

Secondary minerals: Minerals that form through processes such as weathering or by hydrothermal alteration.

Rift: Wide valley formed at a divergent plate boundary.

References

Aoba, T. and Fejerskov, O., 2002. Dental fluorosis: chemistry and biology. *Critical Reviews in Oral Biology and Medicine.* **13**: 155–170.

Ayoob, S. and Gupta, A. K., 2006. Fluoride in drinking water: a review on the status and stress effects. *Critical Reviews in Environmental Science and Technology.* **36**: 433–487.

Berger, T., Peltola, P., Drake, H. and Åström, M., 2012. Impact of a fluorine-rich granite intrusion on levels and distribution of fluoride in a small boreal catchment. *Aquatic Geochemistry.* **18**: 77–94.

Bernal, N. F., Gleeson, S. A., Dean, A. S., Liu, X.-M. and Hoskin, P., 2014. The source of halogens in geothermal fluids from the Taupo Volcanic Zone, North Island, New Zealand. *Geochimica et Cosmochimica Acta.* **126**: 265–283.

Bia G., De Pauli C. P. and Borgnino L., 2012. The role of Fe(III) modified montmorillonite on fluoride mobility: adsorption experiments and competition with phosphate. *Journal of Environmental Management.* **100** : 1–9.

Borgnino, L., Garcia, M. G., Bia, G., Stupar, Y., Le Coustumer, Ph. and Depetris, P. J., 2013. Mechanisms of fluoride release in sediments of Argentina's central region. *Science of the Total Environment.* **443**: 245–255.

Brigatti, M. F. and Guggenheim, S., 2002. Mica crystal chemistry and the influence of pressure, temperature, and solid solution on atomistic models. In: Mottana, A., Sassi, F. P., Thompson, Jr. J. B. and Guggenheim, S. (ed.) *Micas: Crystal Chemistry & Metamorphic Petrology*. Mineralogical Society of America, pp. 1–98.

Burt, B. A. and Eklund, S. A., 1992. Dentistry, Dental Practice and the Community. *WB Saunders*, Philadelphia, p. 147.

Carpenter, R., 1969. Factors controlling the marine geochemistry of fluorine. *Geochimica et Cosmochimica Acta.* **33**: 1156–1167.

Chae, G.-T., Yun, S.-T., Kwon, M.-J., Kim, Y.-S. and Mayer, B., 2006. Batch dissolution of granite and biotite in water: Implication for fluorine geochemistry in groundwater. *Geochemical Journal.* **40**: 95–102.

Chae, G.-T., Yun, S.-T., Mayer, B., Kim, K.-H., Kim, S.-Y., Kwon, J.-S., Kim, K. and Koh, Y.-K., 2007. Fluorine geochemistry in bedrock groundwater of South Korea. *Science of the Total Environment.* **385**: 272–283.

Chaïrat, C., Schott, J., Oelkers, E. H., Lartigue, J.-E. and Harouiya, N., 2007. Kinetics and mechanism of natural fluorapatite dissolution at 25° C and pH from 3 to 12. *Geochimica et Cosmochimica Acta.* **71**: 5901–5912.

Cronin, S. J., Neall, V. E., Lecointre, J. A., Hedley, M. J. and Loganathan, P., 2003. Environmental hazards of fluoride in volcanic ash: a case study from Ruapehu volcano, New Zealand. *Journal of Volcanology and Geothermal Research.* **121**: 271–291.

Currel, M., Cartwright I., Raveggi M. and Han S., 2011. Controls on elevated fluoride and arsenic concentrations in groundwater from the Yuncheng Basin, China. *Applied Geochemistry.* **26**: 540–552.

Delmelle, P., Lambert, M., Dufrene, Y., Gerin, P. and Oskarsson, N., 2007. Gas/aerosol–ash interaction in volcanic plumes: new insights from surface analyses of fine ash particles. *Earth and Planetary Science Letters.* **259**: 159–170.

Deng, Y., Nordstrom D. K. and McCleskey, R. B., 2011. Fluoride geochemistry of thermal waters in Yellowstone National Park: I. Aqueous fluoride speciation. *Geochimica et Cosmochimica Acta.* **75**: 4476–4489.

Edmunds, W. M. and Smedley, P. L., 2013. Fluoride in natural waters. In: Selinus, O. (ed.). Essentials of Medical Geology: Revised Edition. British Geological Survey. pp. 311–336.

Fantong, WY., Satake, H., Ayonghe, S. N., Suh, E. C., Adelana, S. M. A., Fantong, E. B. S., Banseka, H. S., Gwanfogbe, C. D., Woincham, L. N., Uehara, Y. and Zhang, J., 2010. Geochemical provenance and spatial distribution of fluoride in groundwater of Mayo Tsanaga River Basin, Far North Region, Cameroon: implications for incidence of fluorosis and optimal consumption dose. *Environmental Geochemistry and Health.* **32**: 147–163.

Farrah, H., Slavek, J. and Pickering, W. F., 1987. Fluoride interactions with hydrous aluminum oxides and alumina. *Australian Journal of Soil Research.* **25**: 55–69.

Fawell, J., Bailey, K., Chilton, J. and Dahi, E., 2006. *Fluoride in Drinking-Water.* World Health Organization, Cornwall, UK, p. 144.

García, M. G., Lecomte, K. L., Stupar, Y., Formica, S. M., Barrionuevo, M., Vesco, M., Gallará, R. and Ponce, R., 2012. Geochemistry and health aspects of F-rich montainous streams and groundwaters from Sierras Chicas de Córdoba, Argentina. *Environmental Earth Sciences.* **65**: 535–545.

Guo, Q., 2012. Hydrogeochemistry of high-temperature geothermal systems in China: a review. *Applied Geochemistry.* **27**: 1887–1898.

Guo, Q., Wang, Y., Ma, T. and Ma, R., 2007. Geochemical processes controlling the elevated fluoride concentrations in groundwaters of the Taiyuan Basin, Northern China. *Journal of Geochemical Exploration.* **93**: 1–12.

Harrington, L. F., Cooper, E. M. and Vasudevan, D. J., 2003. Fluoride sorption and associated aluminum release in variable charge soils. *Journal of Colloid and Interface Science.* **267**: 302–313.

He, J., An, Y. and Zhang, F., 2013. Geochemical characteristics and fluoride distribution in the groundwater of the Zhangye Basin in Northwestern China. *Journal of Geochemical Exploration.* (in press).

Hiemstra, T. and Van Riemsdijk, W. H., 2000. Fluoride adsorption on goethite in relation to different types of surface sites. *Journal of Colloid and Interface Science.* **225**: 94–104.

Hudson-Edwards, K. A. and Archer, J., 2012. Geochemistry of As-, F- and B-bearing waters in and around San Antonio de los Cobres, Argentina, and implications for drinking and irrigation water quality. *Journal of Geochemical Exploration.* **112**: 276–284.

Ivanova, T. I., Frank-Kamenetskaya, O. V., Kol'tsov, V. and Ugolkov, L., 2001. Crystal structure of calcium-deficient carbonated hydroxyapatite. Thermal decomposition. *Journal of Solid State Chemistry.* **160**: 340–349.

Kraynov, S. R., Merkov, A. N., Petrova, N. G., Baturinskaya, I. V. and Zharikova, V. M., 1969. Highly alkaline (pH 12) fluosilicate waters in the deeper zone of the Lovozero Massif. *Geochemistry International.* **6**: 635–640.

Liu, G., Zheng, L., Qi, C. and Zhang, Y., 2007. Environmental geochemistry and health of fluorine in Chinese coals. *Environmental Geology.* **52**: 1307–1313.

Nicolli, HB., Bundschuh J., Blanco, M. del C., Tujchneider O. C., Panarello, H. O., Dapeña, C. and Rusansky, J. E., 2012. Arsenic and associated trace-elements in groundwater from the Chaco-Pampean plain, Argentina: results from 100 years of research. *Science of the Total Environment.* **429**: 36–56.

Nordstrom, D. K. and Jenne, E. A., 1977. Fluorite solubility equilibria in selected geothermal waters. *Geochimica et Cosmochimica Acta.* **41**: 175–188.

Ozsvath, D. L., 2009. Fluoride and environmental health: a review. *Reviews in Environmental Science and Biotechnology.* **8**: 59–79.

Padhi, S. and Muralidh, D., 2012. Fluoride occurrence and mobilization in geo-environment of semi-arid Granite watershed in southern peninsular India. *Environmental Earth Sciences.* **66**: 471–479.

Pan, Y. and Fleet, M. E., 2002. Compositions of the Apatite-group Minerals: Substitution Mechanisms and Controlling Factors. In: Kohn, M. L., Rakovan, J. and Hughes, J. M. (ed.) *Phosphates*. Mineralogical Society of America: pp. 13–50.

Rango T., Bianchini, G., Beccaluva, L., Ayenew, T. and Colombani, N., 2009. Hydrogeochemical study in the Main Ethiopian Rift: new insights to the source and enrichment mechanism of fluoride. *Environmental Geology.* **58**: 109–118.

Reddy, D. V., Nagabhushanam, P., Sukhija, B. S., Reddy, A. G. S. and Smedley P. L., 2010. Fluoride dynamics in the granitic aquifer of the Wailapally watershed, Nalgonda district, India. *Chemical Geology.* **269**: 278–289.

Reyes-Gómez, V. M., Alarcón-Herrera, M. T., Gutiérrez, M., Núñez López, D., 2013. Fluoride and arsenic in an alluvial aquifer system in Chihuahua, Mexico: contaminant levels, potential sources, and co-occurrence. *Water Air and Soil Pollution.* **224**: 1433–1448.

Rosi, M., Papale, P., Lupi, L. and Stoppato, M., 2003. *Volcanoes.* Firefly Books Ltd., Spain.

Ruggeri, F., Saavedra J., Fernandez-Turiel J. L., Gimeno D. and Garcia-Valles, M., 2010. Environmental geochemistry of ancient volcanic ashes. *Journal of Hazardous Materials.* **183**: 353–365.

Scanlon, B. R., Nicot, J. P., Reedy, R. C., Kurtzman, D., Mukherjee, A. and Nordstrom, D. K., 2009. Elevated naturally occurring arsenic in a semiarid oxidizing system, Southern High Plains aquifer, Texas, USA. *Applied Geochemistry.* **24**: 2061–2071.

Stumm, W., 1992. *Chemistry of the Solid-Water Interface.* Wiley Interscience Publication.

Sujana, M. G. and Anand, S., 2010. Iron and aluminium based mixed hydroxides: a novel sorbent for fluoride removal from aqueous solutions *Applied Surface Science.* **256**: 6956–6962.

Symonds, R. B., Rose, W. I. and Reed, M. H., 1988. Contribution of Cl– and F– bearing gases to the atmosphere by volcanoes. *Nature.* **334**: 415–419.
Turner, B. D., Binning, P., Stipp, S. L. S., 2005. Fluoride removal by calcite: evidence for fluoride precipitation and surface adsorption. *Environmental Science and Technology.* **39**: 9561–9568.
Valenzuela Vásquez, L., Ramírez Hernándcz, J., Reyes López, J., Sol Uribe, A. and Lázaro Mancilla, O., 2006. The origin of fluoride in groundwater supply to Hermosillo City, Sonora, México. *Environmental Geology.* **51**: 17–27.
Van Cappellen, P. and Berner, R., 1988. A mathematical model for the early diagenesis of phosphorous and fluorine in marine sediments: apatite precipitation. *American Journal of Science.* **288**: 289–333.
Vikas, C., Kushwaha, R., Ahmad, W., Prasannakumar, V. and Reghunath, R., 2013. Genesis and geochemistry of high fluoride bearing groundwater from a semi-arid terrain of NW India. *Environmental Earth Sciences.* **68**: 289–305.
Vinson, D. S., McIntosh, J. C., Dwyer, G. S. and Vengosh, A., 2011. Arsenic and other oxyanion-forming trace elements in an alluvial basin aquifer: evaluating sources and mobilization by isotopic tracers (Sr, B, S, O, H, Ra). *Applied Geochemistry.* **26**: 1364–1376.
Wedepohl, K. H., 1995. The composition of the continental crust. *Geochimica et Cosmochimica Acta.* **59**: 1217–1232.
Wolff-Boenisch, D., Gislason, S. R. and Oelkers, E. H., 2004. The effect of fluoride on the dissolution rates of natural glasses at pH 4 and 25 °C. *Geochimica et Cosmochimica Acta.* **68**: 4571–4582.
Zhu, M.-X., Ding, K.-Y., Jiang, X. and Wang, H.-H., 2007. Investigation on co-sorption and desorption of fluoride and phosphate in a red soil of China. *Water Air and Soil Pollution.* **183**: 455–465.
Zhu, M.-X., Jiang, X. and Ji. G.-L., 2004. Interactions between variable-charge soils and acidic solutions containing fluoride: an investigation using repetitive extractions. *Journal of Colloid and Interface Science.* **276**: 159–166.
Zhuang, J. I. E. and Yu, G.-R., 2002. Effects of surface coatings on electrochemical properties and contaminant sorption of clay minerals. *Chemosphere.* **49**: 619–628.
Zidarova, B., 2010. Hydrothermal fluorite-forming processes in the Mikhalkovo deposit (Central Rhodopes, Bulgaria) field observation and experimental confirmation. *Neues Jahrbuch für Mineralogie.* **187**: 133–157.

CHAPTER 2

Fluoride Intake in the Context of Dental Fluorosis

ALINE DE LIMA LEITE[a], CAMILA PERES BUZALAF[a], AND MARÍLIA AFONSO RABELO BUZALAF*[a]

[a]Department of Biological Sciences, Bauru Dental School, University of São Paulo, Al. Octávio Pinheiro Brisolla, 9-75, 17012-901, Bauru, São Paulo, Brazil
*E-mail: mbuzalaf@fob.usp.br

2.1 Introduction

Fluoride was introduced in dentistry in the 1940 s and since then dental caries prevalence has declined significantly. In the beginning, it was believed that to have a cariostatic effect fluoride should be incorporated into dental enamel and fluorosis would be expected in areas that received fluoride in drinking water (Buzalaf and Levy, 2011). Recently, simultaneously to the substantial decline of dental caries there was a significant increase of dental fluorosis both in fluoridated and nonfluoridated regions (CDC, 2001; Whelton *et al.*, 2004; Khan *et al.*, 2005). This high incidence of fluorosis was due mostly to the excessive consumption of fluoride from different sources during tooth formation (Warren and Levy, 1999). Currently, it is known that the major mechanism of fluoride action is during the de- and remineralization cycle, *i.e.*, the effect is essentially topical (Rodrigues *et al.*, 2009). This fact drew attention to the issue of the control the amount of fluoride intake because it is possible to

Food and Nutritional Components in Focus No. 6
Fluorine: Chemistry, Analysis, Function and Effects
Edited by Victor R Preedy

Published by the Royal Society of Chemistry, www.rsc.org

achieve a maximum protective effect without the necessity of ingesting fluoride.

It is important to recall that dental fluorosis occurs only when the amount of fluoride intake is higher than recommended at the period of dental formation. Therefore, it is important to know the sources of fluoride intake, as well as the critical periods of tooth formation, because excessive fluoride intake during these periods can promote a mineralization disturbance, resulting in dental fluorosis. For this reason, the propose of this chapter is to make clear what are the safe levels of fluoride intake, what are the main sources of this element and which are the critical periods of tooth formation when there is risk of occurrence of dental fluorosis if fluoride is ingested in excess.

2.2 Too Much or Too Little?

What is the safe fluoride level? The first discussion about this level was introduced by (McClure, 1943) in the 1940s. He estimated that the fluoride intake by children aged 1–12 years would be between 1.0 and 1.5 mg of fluoride to provide about 0.05 mg fluoride per kilogram body weight (mgF kg^{-1} b.w.) per day for those children. During the 1970s, the first recommendation was suggested by Farkas and Farkas (1974), who cited several sources that reported 0.06 mgF kg^{-1} b.w. "...is generally regarded as optimum". A few years later, this level started to be used as the optimal amount of fluoride intake (Ophaug *et al.*, 1980). More recently, the range between 0.05 and 0.07 mgF kg^{-1} b.w. has been regarded as the *optimal* amount of fluoride intake (Burt, 1992). However, it is not totally clear if this range can be associated with *optimal* to caries prevention, with fluorosis prevention or both. Some authors have reported that fluoride ingestion above 0.1 mg kg^{-1} b.w. can enhance the risk of dental fluorosis (Forsman, 1977). On the other hand, there are reports showing the occurrence of fluorosis with a daily fluoride intake of less than 0.03 mg kg^{-1} b.w. (Manji *et al.*, 1986; Baelum *et al.*, 1987). Other factors such as residence at high altitude, renal insufficiency, malnutrition and genetics are associated with an increased susceptibility to dental fluorosis (for review, see Buzalaf *et al.*, 2011).

In 2009, results of the prospective Iowa fluoride study regarding the effect of total fluoride exposure on tooth health after 9 years of a child's life were reported (Warren *et al.*, 2009). According to this study, the mean daily of fluoride intake for children with no history of caries or fluorosis at age 9 years was around or below 0.05 mg kg^{-1} b.w. or below during the first 4 years of life and this level declined after this period. However, there was considerable individual variation and overlap in mean fluoride intake among caries/fluorosis groups, which makes it difficult to strictly recommend an "optimal" fluoride intake. Thus, the range of fluoride intake between 0.05 and 0.07 mg kg^{-1} b.w. should still be used.

2.3 Critical Period for Fluorosis Development

As mentioned above, dental fluorosis occurs as a result of excess fluoride ingestion during the critical period of tooth formation. Thus, it is important to know which is the most susceptible period for fluorosis development, in order that fluoride intake during this period can be monitored.

The literature has reported that the critical age for fluorosis development considering the whole permanent dentition (except the third molars) are the first 6–8 years of life. Several epidemiological studies of dental fluorosis in humans show that the permanent maxillary central incisors, which are the most important teeth from the aesthetic point of view, are most susceptible for fluorosis during the first 2–3 years of postnatal life (Pendrys and Katz, 1989; Evans and Stamm, 1991a; Evans and Stamm, 1991b; Jackson *et al.*, 1999). The same applies for the permanent first molars (Bardsen and Bjorvatn, 1998; Bardsen, 1999). Tables 2.1 and 2.2 show the development timeline of deciduous and permanent human teeth.

Enamel formation is a complex and orchestrated process that consists in secretion of organic matrix, rich in protein and water, and subsequent mineralization of this matrix. These processes are divided into presecretory, secretory, transition and maturation stages, each one with unique properties that affect fluoride susceptibility. The presecretory stage is characterized by differentiation of pre-secretory ameloblasts into secretory ameloblasts (Denbesten and Li, 2011). At this stage, there is no evidence that fluoride exposure could affect tooth development, once there is no change in tooth size and form, even in teeth with severe fluorosis. At the secretory stage, ameloblasts deposit matrix proteins and establish a thin layer of aprismatic enamel, which is deposited against dentin. Several studies reported that chronic exposure to fluoride in drinking water can reduce the enamel thickness by about 10% (Kerley and

Table 2.1 Chronology of human deciduous teeth.

Tooth	Mineralization begins (weeks *in utero*)	Enamel completed (months)	Eruption (mean age in months	Root completed (years)
		Maxillary teeth		
I1	14	1.5	7.5	1.5
I2	16	2.5	9	2
C	17	9	18	3.25
M1	12.5–15.5	6	14	2.5
M2	12.5–19	11	24	3
		Mandibular teeth		
I1	18	2.5	6	1.5
I2	18	3	7	1.5
C	20	9	16	3.25
M1	12.5–15.5	5.5	12	2.25
M2	12.5–18	10	20	3

I1, Central incisor; I2, lateral incisor; C, canine; M1, first molar; M2, second molar. Adapted from Ten Cate's Oral Histology.

Kollar, 1977; Levenson, 1980; Bronckers *et al.*, 1984; Smith *et al.*, 1993; Kierdorf and Kierdorf, 1997) and the exposure to high doses fluoride in this stage has been associated to enhanced cell apoptosis (Bartlett *et al.*, 2005). In transitional stage, there is loss of most organelles involved in protein synthesis and the ameloblasts morphology changes (they become similar to epithelial cells). At this stage, ion transport occurs, as well as a reduction in the number of ameloblasts through apoptosis. The ameloblasts appear to be more sensitive to fluoride at this stage than secretory ameloblasts (Richards *et al.*, 1986; Lyaruu *et al.*, 1990; Lyaruu *et al.*, 2006; Lyaruu *et al.*, 2012). Finally, the maturation stage is marked by cyclic ameloblasts modulation from cells with a smooth-ended to a ruffle-ended distal membrane that result in the removal of the proteins, followed by a gradual increase of enamel mineralization. Denbesten *et al.* (1985) reported that modulation between smooth-ended/ruffle-ended frequencies are decreased after chronic fluoride exposure. Other alterations include decrease in numbers of lysosomes and phagosomes (Ribeiro *et al.*, 2006), reduced lysosomal activity (Smid *et al.*, 1990), and increase of stress proteins in endoplasmatic reticle (Sharma *et al.*, 2008). This suggests that the maturation stage seems to be most sensitive to the effects of excessive fluoride ingestion.

While there is general consensus that the early maturation stage of enamel development is more critical for fluorosis than the secretory stage (Larsen *et al.*, 1985; Baelum *et al.*, 1987; Pendrys and Katz, 1989; Pendrys and Morse, 1990; Evans and Stamm, 1991b; Bronckers *et al.*, 2009), the evidence is not

Table 2.2 Chronology of human permanent teeth.

Tooth	Mineralization begins	Enamel completed (yr) male/female	Eruption (yr) male/female	Root completed (yr) male/female
		Maxillary teeth		
I1	3–4 mo[a]	3.7/3.3	8.3/7.4	10.6/9.3
I2	10–12 mo	4.0/3.8	9.1/8.1	11.1/9.7
C	4–5 mo	4.9/4.1	11/9.4	13.7/11.9
PM1	1.5–1.75 yr[a]	5.8/5.1	11.1/9.7	13.5/11.8
PM2	2.2–2.5 yr	6.3/5.9	11.6/10.6	13.8/12.6
M1	At birth	2.7/2.6	7.8/7.2	10.1/9.2
M2	2.5–3 yr	6.7/6.3	12.4/11.8	14.6/13.6
M3	7–9 yr	13.3/12.7	17.4/17.8	18.2/18.8
		Mandibular teeth		
I1	3–4 mo	3.6/3.3	7.3/6.7	9.2/8.1
I2	3–4 mo	4.0/3.7	8.1/7.3	9.9/8.8
C	4–5 mo	4.8/4.1	10.9/9.2	13.5/11.4
PM1	1.75–2 yr	5.6/5.0	11.2/9.9	13.3/11.9
PM2	2.25–2.5 yr	6.3/5.9	11.9/10.6	14/12.8
M1	At birth	2.7/2.6	7.8/7.2	10/9.2
M2	2.5–3 yr	6.7/6.3	12.5/11.8	14.8/13.8
M3	8–10 yr	13.3/12.8	17.4/17.7	18.5/18.3

I1, Central incisor; I2, lateral incisor; C, canine; P1, first premolar; P2, second premolar; M1, first molar; M2, second molar; M3, third molar. Adapted from Ten Cate's Oral Histology.
[a]mo, months; yr, years.

completely conclusive regarding the age at which maxillary central incisors are most susceptible to dental fluorosis. A study conducted by (Ismail and Messer, 1996) reported that the first year of life was the most critical period for developing fluorosis in the permanent central maxillary incisors. On the other hand, three studies concluded that the first 3 years are critical (Burt *et al.*, 2000; Burt *et al.*, 2003; Hong *et al.*, 2006) and another one described the period between 35 and 42 months as critical (Ishii and Suckling, 1986). Most of the studies recognize the first two years of life as a crucial period to fluorosis development. The *meta*-analysis conducted by Bårdsen (1999) concluded that there is no specific period of enamel formation that could be classified as critical for dental fluorosis development; the duration of fluoride exposure would be more important, *i.e.*, long or short periods of exposure seem to explain the occurrence of dental fluorosis. These observations were reinforced in a more recent publication (Hong *et al.*, 2006) showing that ingestion of high levels of fluoride for longer periods leads to higher fluorosis prevalence. Thus, it seems important to monitor the fluoride intake by children in the first three years of life, as well as the total time of exposure.

2.4 Sources of Fluoride Intake

Concerns with the increase in prevalence of dental fluorosis have led many researchers to evaluate the fluoride concentration in potential sources and estimate total intake from all sources, especially those consumed by young children. The main sources of fluoride intake, which constitute the main risk factors for dental fluorosis, are described in the following sections.

2.4.1 Fluoride in Drinking Water

Community water fluoridation is considered one of the top ten greatest public health achievements in the world in the last century. It is a method of great scope that benefits all socioeconomic groups and has an excellent cost–benefit relationship. The mean estimated costs are very small. They are estimated at around US$ 0.50 per person per year in large communities to US$ 3.00 per person per year in small communities. For every US$ 1.00 invested in this preventive measure, approximately US$ 38.00 are saved in dental treatment costs (ADA, 2005), reinforcing the cost–benefit relationship. The main concern of the epidemiologists was to find the fluoride concentration that would be sufficient to give effective protection against caries with a minimum risk of dental fluorosis. Since then, plethora of studies have been conducted to demonstrate the beneficial effects of water fluoridation.

Dean *et al.* (1942) were the first to suggest that 1 ppm fluoride was the optimal concentration in community drinking water, since this concentration would be able to considerably reduce the prevalence of dental caries and fluorosis prevalence would be kept low (*i.e.*, 10–12%). Currently, concentrations ranging between 0.7 and 1.2 ppm fluoride in the drinking water are regarded as optimal, depending on the mean maximum daily temperature that influences the amount of water intake. Thus, in hotter areas the fluoride concentration should be lower than in areas with lower temperatures (Galagan and Vermillion, 1957).

After the introduction of water fluoridation, several studies were conducted to determine the prevalence of dental fluorosis. In the 1980s and 1990s many studies concluded that the prevalence of dental fluorosis in areas with fluoridation at 0.8 ppm increases four-fold when compared with nonfluoridated areas (Szpunar and Burt, 1988; Riordan and Banks, 1991; Riordan, 1993). A systematic review analyzed 214 studies, of which 88 involving dental fluorosis, and the authors found a significant dose–response association between the fluoride concentration in the drinking water and the prevalence of dental fluorosis. The results showed that the prevalence of fluorosis was 48% in communities containing fluoride at 1 ppm in drinking water, but no other adverse effects could be attributed to the presence of fluoride in the drinking water (McDonagh *et al.*, 2000). These findings are higher than those reported by Dean and coworkers in 1942.

An interesting way to analyze the impact of water fluoridation in the prevalence of dental fluorosis are the studies where water fluoridation was interrupted for some period. Results from these studies show that the prevalence of dental fluorosis is affected by changes in fluoride exposure from drinking water. (Buzalaf *et al.*, 2004b) conducted a study evaluating the effects of interruption in water fluoridation for 7 years. The prevalence of dental fluorosis was lower in children who were 36, 27 and 18 months old when water fluoridation ceased when compared with children who were born 18 months after fluoridation was interrupted. These results suggest that currently water fluoridation seems not to be as important in the occurrence of dental fluorosis as it was in the past, mainly due to different sources of fluoride intake now available. Recent studies have shown that other sources such as dentifrice and reconstituted milk are important contributors for the daily fluoride intake (de Almeida *et al.*, 2007). Thus, fluoridated water *per se* seems to have a small role on dental fluorosis, but has a great impact when used to prepare foods and drinks that are consumed by infants and small children. Thus, to minimize the possible impact of water fluoridation on dental fluorosis, dentists should instruct parents about components of the diet that could have a high fluoride content.

2.4.2 Fluoridated Dentifrices

For many years, fluoridated water was regarded as the main risk factor for dental fluorosis. In the last decades, it has been observed that the prevalence of dental fluorosis has increased more in nonfluoridated areas than in fluoridated areas, which turned the attention of researchers to search for other sources of fluoride intake such as fluoridated dentifrices (Buzalaf and Levy, 2011). It has been reported that fluoridated dentifrices are an important source of fluoride intake during the first years of life, since an inverse relationship can be observed between the age of the child and the mean percentage ingestion of toothpaste. This is because small children are not able to completely control swallowing (Richards and Banting, 1996; de Almeida *et al.*, 2007; Kobayashi *et al.*, 2011). Besides, behaviors associated with toothbrushing, such as the amount of toothpaste used, are associated with occurrence of dental fluorosis in optimally fluoridated and nonfluoridated areas (Pendrys, 2000; Davies *et al.*, 2003). To minimize the ingestion of fluoride from dentifrices, researchers have suggested the use of "drop" or "transverse" techniques that reduce the amount of dentifrice applied to the toothbrush, which could reduce the risk of dental fluorosis for young children (Villena, 2000; Vilhena *et al.*, 2008)

A systematic review and *meta*-analysis examined 25 studies between 1988 and 2006 with the purpose to establish the relationship between the use of fluoridated dentifrices and dental fluorosis. Among these, 2 randomized controlled trials (RCTs), 1 cohort study, 6 case-control studies and 16 cross-sectional surveys were included (Walsh *et al.*, 2010). The main findings of this systematic review with *meta*-analysis were: (a) a significant reduction in the risk of dental fluorosis was found if toothbrushing with fluoride toothpaste did not start until the age of 12 months, but the evidence for starting toothbrushing with fluoride toothpaste before the age of 24 months was inconsistent (data from case-control and cross-sectional studies); (b) no significant association was found between the frequency of toothbrushing or amount of toothpaste used and fluorosis (data from cross-sectional surveys); and (c) using toothpaste with a higher concentration of fluoride increased the risk of dental fluorosis (data from two RCTs; evidence from cross-sectional studies was inconsistent). From the available evidence, the authors concluded that decisions involving the use of topical fluorides (including toothpastes) should balance their benefits in caries prevention and the risk of causing dental fluorosis. The authors suggest that to minimize the influence of dentifrice on dental fluorosis development, the fluoride levels in young children toothpaste should be lower than 1000 ppm" (Wong *et al.*, 2010). However, it is generally agreed that only dentifrices containing 1000 ppm fluoride or higher concentrations are able to significantly reduce caries when compared with a placebo, since the number of studies published so far that compared low-fluoride dentifrices (around 500 ppm) with a placebo is very reduced, especially in the deciduous dentition (Wong *et al.*, 2011).

Due to the uncertainty around the caries preventive effect of low-fluoride dentifrices due to the lack of clinical trials, some authors have attempted to improve the anticaries efficacy of these dentifrices by reducing their pH. Vilhena and coworkers (Vilhena *et al.*, 2010) performed a RCT to evaluate the anticaries effectiveness of low-fluoride (550 ppm) acidic (pH 4.5) liquid dentifrice in 4 year-old children in fluoridated area. The results showed that the caries progression rate was quite similar to that seen for the conventional 1100 ppm F dentifrice, after 20 months of evaluation. These results were confirmed in a more recent clinical trial conducted with 2–4 year-old children (de Almeida Baldini Cardoso *et al.*, 2013). The low-fluoride (550 ppm) acidic (pH 4.5) dentifrice led to the same caries progression and net increment as the conventional dentifrice (1100 ppm fluoride, pH 7, 0) and both performed significantly better than the low-fluoride (550 ppm) neutral dentifrice for caries-active children. However, for the caries-inactive children, no significant difference among the tested dentifrices was found. It has been shown that the reduction of the dentifrice pH is able to increase biofilm fluoride concentrations (Buzalaf *et al.*, 2009), which could, at least in part, explain the better anticaries performance of the acidic dentifrices when compared to their neutral counterparts (Vilhena *et al.*, 2010; de Almeida Baldini Cardoso *et al.*, 2013). In addition, it has been reported that the use of low-fluoride dentifrices (550 ppm) leads to fingernail fluoride concentrations significantly lower when compared with conventional (1100 ppm fluoride dentifrices) (Buzalaf *et al.*, 2009; de Almeida Baldini Cardoso *et al.*, 2013), indicating that the first ones are able to reduce fluoride intake from dentifrices by small children.

In conclusion, based on the available evidence and taking into account risks (dental fluorosis increase) and benefits (caries reduction), if there is concern with the occurrence of dental fluorosis, low-fluoride dentifrices (around 500 ppm) could be indicated for low-caries risk children with 3 years of age or less (when there is risk to develop fluorosis in the central incisors), especially if they live in a fluoridated area. In all other situations, conventional dentifrices (around 1000 ppm fluoride) should be used. Recent studies have reported that acidic low-fluoride dentifrices have the same anticaries potential as the conventional (1000 ppm fluoride, neutral pH) dentifrices, but additional studies are still necessary before they can be definitively indicated as substitutes for the conventional dentifrices. In addition, despite there being no clear evidence regarding the association between the amount of fluoride loaded onto the toothbrush and the risk of dental fluorosis development (Wong *et al.*, 2010), it is appropriate to recommend the use of a small amount of dentifrice by young children by using the "transversal" or "drop" techniques. Moreover, it is essential that small children brush under the supervision of an adult and be instructed not to swallow the foam.

2.4.3 Infant Formulas

Exclusive breastfeeding is recommended up to 6 months of age. In many cases, however, this practice is not possible because some mothers do not have enough milk or some babies present lactose intolerance. For this reason, infant formulas were developed. They were designed to be roughly based on human mother's milk (Lonnerdal, 2014).

Generally, these formulas can be purchased as powder or liquid concentrate that have to be reconstituted with water before use, or as a ready-to-feed formulation. Unlike human milk and cow's milk that present low fluoride concentrations (between 0.01 and 0.10 ppm), infant formulas can present high concentrations of fluoride, due to the manufacturing procedures or to the use of fluoridated water to reconstitute them (Silva and Reynolds, 1996; Buzalaf *et al.*, 2001). The concern about infant formulas arises from the fact that their consumption can lead infants to exceed the tolerable levels of fluoride intake established at 0.7 mg day^{-1} for the first 6 months and 0.9 mg day^{-1} for 7–12 months (Hujoel *et al.*, 2009). Worldwide, many studies have shown that the fluoride content of infant formulas undergoes large variations when analyzed in different countries (for a review, see Buzalaf and Levy, 2011).

A systematic review compiled the results of 19 studies that associated infant formula consumption from birth to the age of 24 months and dental fluorosis development (Hujoel *et al.*, 2009). Data came from one prospective cohort study, five retrospective cohort studies, case-control studies, four cross-sectional studies and three historical-control studies. In summary, the authors concluded that infant-formula consumption during infancy increases the risk of developing dental fluorosis in the permanent teeth (odds ratio [OR] 1.8, 95% CI 1.4–2.3). However, there was significant heterogeneity in the magnitude of the odd ratios among the studies, which should be interpreted with caution. The authors also observed that the association between dental fluorosis risk and the use of infant formula depended on the level of fluoride in the water supply, as mentioned before. Formulas prepared with 1 ppm water fluoridated increase the risk of fluorosis by 67% (OR 1.67, 95 percent CI 1.18–2.36).

Recently a panel of experts was commissioned by the American Dental Association in order to develop evidence-based clinical recommendations on the ingestion of fluoride from infant formulas, from birth to 12 months of age, and its potential relationship with the occurrence of dental fluorosis in the permanent dentition (Berg *et al.*, 2011). The experts, based on the systematic review by Hujoel *et al.* (2009) and on two other clinical trials, concluded that the dentist can recommend to parents and caregivers of infants that consume infant formulas as the main source of nutrition, to reconstitute them with fluoridated water, provided that they are informed about the potential risk of dental fluorosis development. If parents express their concern with the occurrence of dental fluorosis, water containing less than 0.5 ppm fluoride should be recommended to reconstitute the formulas.

Beyond infant formulas, other ailments, called beikost, also present a wide variation in fluoride content (between 0.01 and 8.30 ppm). Generally, chicken-based foods have the highest values due to the inclusion of bones in the manufacturing process (Forsman, 1977). In some studies, fish-based products also have been reported to have high fluoride content (Levy *et al.*, 2001). In general, the fluoride concentrations of most beikost is usually low (Fomon and Ekstrand, 1999). However, some cereals commonly added to milk that are usually consumed by infants in Brazil have been shown to have high fluoride concentrations. This is the case for Mucilon and Neston (Nestlé), which have fluoride concentrations of 2.4 and 6.2 ppm, respectively (Buzalaf *et al.*, 2002; Buzalaf *et al.*, 2004a). A relatively high fluoride concentration was also found in ready-to-drink chocolate milk (1.2 ppm, Toddynho, Quaker) (Buzalaf *et al.*, 2002; Buzalaf *et al.*, 2004a).

In summary, we can conclude that infant formulas *per se* are not able to increase the fluorosis risk, but when they are reconstituted with fluoridated water the risk increases significantly. Thus, if there is concern with the occurrence of dental fluorosis, water containing less than 0.5 ppm fluoride should be used to dilute infant formulas.

2.4.4 Fluoride Supplements

Dietary fluoride supplements were originally introduced to help prevent dental caries in children living in low-fluoride areas. The recommended daily dose was based on the age of the child and fluoride concentration in the drinking water. However, the controversy about the use of fluoride supplements led community dentistry to suspend this recommendation (Levy, 2003).

Ismail and coworkers (Ismail and Bandekar, 1999) published a systematic review that evaluated 10 cross-sectional/case studies to assess the association between the use of fluoride supplements and dental fluorosis. Data showed a strong association between these variables, suggesting that children who used fluoride supplements had a significantly higher risk to develop dental fluorosis. The OR of dental fluorosis in children living in nonfluoridated areas who use fluoride supplements was 2.5 compared with nonusers. Almost 10 years after publication, the authors updated this systematic review including 4 new studies, which reinforced the existence of a positive association between the use of fluoride supplements and occurrence of dental fluorosis, since data showed that the odds ratio for dental fluorosis increases 84% for each year of use of fluoride supplements between the ages of 6–7 years, especially for the first 3 years of life (Ismail and Hasson, 2008). The authors also evaluated if the fluoride supplements could prevent dental caries effectively, and they reported weak and inconsistent evidence showing that fluoride supplements are effective to prevent caries in the primary dentition. However, they have an effective role in preventing of caries in the permanent teeth (children older than 6 years) when they are chewed instead of being swallowed.

Based on the available evidence, it has been suggested that fluoride supplements should not be prescribed in optimally fluoridated areas, for children younger than 6 months or for those that present low risk of developing dental caries. However, certain countries have adopted different rules for recommendation for the appropriate use of supplements. In 2006, Australia established that supplements should not be used (Health, 2006). The American Dental Association (ADA) recommends 0.25 mg fluoride per day from 6 months–3 years for children living in areas containing less than 0.3 ppm fluoride in the drinking water; 0.50 and 0.25 mg per day for children aged 3–6 years, living in areas with less than 0.3 ppm fluoride and with 0.3 to 0.6 ppm fluoride in the drinking water, respectively, while double the dose is recommended from 6–16 years. In Canada, the recommendation is only for children who present high risk of caries or for those whose fluoride intake is less than 0.07 mg kg^{-1} b.w. However, it should be noted that is quite difficult to estimate the total daily fluoride intake (Swan, 2000). Finally, a panel of Europeans recommended supplements containing 0.5 mg day^{-1} only for children aged 3 years or more considering the individual risks for caries (Clarkson, 1992).

Based on the exposed, it is important to emphasize that fluoride supplements might help caries prevention, but they should be prescribed with caution, based on the individual caries risk. However, populations that do not receive fluoride from other sources and that do not have access to other caries-preventive measures could benefit from the use of supplements.

Summary Points

- This chapter focuses on fluoride intake and fluorosis.
- In recent years dental fluorosis increased substantially.
- Fluoride intake has a crucial role in dental fluorosis appearance.
- The critical age for fluorosis development are the first 6–8 years of life.
- Dentifrice and infant formula provide almost all ingested fluoride.

Key Facts

- Fluoride is the thirteenth chemical element most abundant element in the Earth's crust and has important role in oral health.
- Dental fluorosis is a change in tooth enamel characterized by staining and pitting of the teeth, and in more severe cases all the enamel may be damaged.
- The ameloblasts are responsible for enamel formation and their life cycle is marked by several changes in morphology during amelogenesis.
- Dental fluorosis on permanent dentition can be observed when children are exposed to fluoride during enamel formation.
- Infant formulas simulate human milk and are used to feed babies and infants under 12 months of age.

Dictionary of Terms

Dental fluorosis: defects on dental enamel caused by excessive fluoride exposure during tooth development.

Ameloblasts: group of cells responsible for enamel formation.

Aprismatic enamel: layer composed of parallel crystallites of enamel.

Prismatic enamel: layer composed of prisms that are parallel to each other.

Beikost: infant first nonliquid food, such as mashed vegetables or applesauce, given to supplement or replace breast milk.

ppm: parts per million.

References

ADA., 2005. Fluoridation facts. Chicago, Illinois, American Dental Association.

de Almeida Baldini Cardoso, C., Mangueira, D. F., Olympio, K. P., Magalhaes, A. C., Rios, D., Honorio, H. M., Vilhena, F. V., Sampaio, F. C. and Buzalaf, M. A., 2013. The effect of pH and fluoride concentration of liquid dentifrices on caries progression. *Clinical Oral Investigations.*

de Almeida, B. S., da Silva Cardoso, V. E. and Buzalaf, M. A., 2007. Fluoride ingestion from toothpaste and diet in 1- to 3-year-old Brazilian children. *Community Dentistry and Oral Epidemiology.* **35**(1): 53–63.

Baelum, V., Fejerskov, O., Manji, F. and Larsen, M. J., 1987. Daily dose of fluoride and dental fluorosis. *Tandlaegebladet.* **91**(10): 452–456.

Bardsen, A., 1999. "Risk periods" associated with the development of dental fluorosis in maxillary permanent central incisors: a *meta*-analysis. *Acta Odontologica Scandinavica.* **57**(5): 247–256.

Bardsen, A. and Bjorvatn, K., 1998. Risk periods in the development of dental fluorosis. *Clinical Oral Investigations.* **2**(4): 155–160.

Bartlett, J. D., Dwyer, S. E., Beniash, E., Skobe, Z. and Payne-Ferreira, T. L., 2005. Fluorosis: a new model and new insights. *Journal of Dental Research.* **84**(9): 832–836.

Berg, J., Gerweck, C., Hujoel, P. P., King, R., Krol, D. M., Kumar, J., Levy, S., Pollick, H., Whitford, G. M., Strock, S., Aravamudhan, K., Frantsve-Hawley, J., and Meyer, D. M., American Dental Association Council on Scientific Affairs Expert Panel on Fluoride Intake From Infant, F. and Fluorosis., 2011. Evidence-based clinical recommendations regarding fluoride intake from reconstituted infant formula and enamel fluorosis: a report of the American Dental Association Council on Scientific Affairs. *Journal of the American Dental Association.* **142**(1): 79–87.

Bronckers, A. L., Jansen, L. L. and Woltgens, J. H., 1984. Long-term (8 days) effects of exposure to low concentrations of fluoride on enamel formation in hamster tooth-germs in organ culture in vitro. *Archives of Oral Biology.* **29**(10): 811–819.

Bronckers, A. L., Lyaruu, D. M. and DenBesten, P. K., 2009. The impact of fluoride on ameloblasts and the mechanisms of enamel fluorosis. *Journal of Dental Research.* **88**(10): 877–893.

Burt, B. A., 1992. The changing patterns of systemic fluoride intake. *Journal of Dental Research.* **71**(5): 1228–1237.

Burt, B. A., Keels, M. A. and Heller, K. E., 2000. The effects of a break in water fluoridation on the development of dental caries and fluorosis. *Journal of Dental Research.* **79**(2): 761–769.

Burt, B. A., Keels, M. A. and Heller, K. E., 2003. Fluorosis development in seven age cohorts after an 11-month break in water fluoridation. *Journal of Dental Research.* **82**(1): 64–68.

Buzalaf, M. A., de Almeida, B. S., Cardoso, V. E., Olympio, K. P. and Furlani Tde, A. (2004a). Total and acid-soluble fluoride content of infant cereals, beverages and biscuits from Brazil. *Food Additives and Contaminants.* **21**(3): 210–215.

Buzalaf, M. A., de Almeida, B. S., Olympio, K. P., da, S. C. V. E. and de, C. S. P. S. H. (2004b). Enamel fluorosis prevalence after a 7-year interruption in water fluoridation in Jau, Sao Paulo, Brazil. *Journal of Public Health Dentistry.* **64**(4): 205–208.

Buzalaf, M. A., Granjeiro, J. M., Damante, C. A. and de Ornelas, F., 2001. Fluoride content of infant formulas prepared with deionized, bottled mineral and fluoridated drinking water. *ASDC Journal of Dentistry for Children.* **68**(1): 37–41, 10.

Buzalaf, M. A., Granjeiro, J. M., Duarte, J. L. and Taga, M. L., 2002. Fluoride content of infant foods in Brazil and risk of dental fluorosis. *ASDC Journal of Dentistry for Children.* **69**(2): 196–200, 125–196.

Buzalaf, M. A. and Levy, S. M., 2011. Fluoride intake of children: considerations for dental caries and dental fluorosis. *Monographs in Oral Science.* **22**: 1–19.

Buzalaf, M. A., Rodrigues, M. H., Pessan, J. P., Leite, A. L., Arana, A., Villena, R. S., Forte, F. D. and Sampaio, F. C., 2011. Biomarkers of fluoride in children exposed to different sources of systemic fluoride. *Journal of Dental Research.* **90**(2): 215–219.

Buzalaf, M. A., Vilhena, F. V., Iano, F. G., Grizzo, L., Pessan, J. P., Sampaio, F. C. and Oliveira, R. C., 2009. The effect of different fluoride concentrations and pH of dentifrices on plaque and nail fluoride levels in young children. *Caries Research.* **43**(2): 142–146.

CDC., 2001. Recommendations for using fluoride to prevent and control dental caries in the United States. *MMWR.* **50**: 26.

Clarkson, J., 1992. A European view of fluoride supplementation. *British Dental Journal.* **172**(9): 357.

Davies, R. M., Ellwood, R. P. and Davies, G. M., 2003. The rational use of fluoride toothpaste. *International Journal of Dental Hygiene.* **1**(1): 3–8.

Dean, H. T., Arnold, F. A. and Elvolve, E., 1942. Additional studies of the relation of fluoride domestic waters to dental caries experience in 4,425 white children aged 12–14 years in 13 cities in 4 states. *Public Health Reports.* **57**: 1155–1179.

Denbesten, P. K., Crenshaw, M. A. and Wilson, M. H., 1985. Changes in the fluoride-induced modulation of maturation stage ameloblasts of rats. *Journal of Dental Research.* **64**(12): 1365–1370.

Denbesten, P. and Li, W., 2011. Chronic fluoride toxicity: dental fluorosis. *Monographs in Oral Science.* **22**: 81–96.

Evans, R. W. and Stamm, J. W. (1991a). Dental fluorosis following downward adjustment of fluoride in drinking water. *Journal of Public Health Dentistry.* **51**(2): 91–98.

Evans, R. W. and Stamm, J. W. (1991b). An epidemiologic estimate of the critical period during which human maxillary central incisors are most susceptible to fluorosis. *Journal of Public Health Dentistry.* **51**(4): 251–259.

Farkas, C. S. and Farkas, E. J., 1974. Potential effect of food processing on the fluoride content of infant foods. *Science of the Total Environment.* **2**(4): 399–405.

Fomon, S. J. and Ekstrand, J., 1999. Fluoride intake by infants. *Journal of Public Health Dentistry.* **59**(4): 229–234.

Forsman, B., 1977. Early supply of fluoride and enamel fluorosis. *Scandinavian Journal of Dental Research.* **85**(1): 22–30.

Galagan, D. J. and Vermillion, J. R., 1957. Determining optimum fluoride concentrations. *Public Health Reports.* **72**(6): 491–493.

Health, A. R. C. P. O., 2006. The use of fluorides in Australia: guidelines *Australian Dental Journal.* **51**: 195–199.

Hong, L., Levy, S. M., Warren, J. J., Broffitt, B. and Cavanaugh, J., 2006. Fluoride intake levels in relation to fluorosis development in permanent maxillary central incisors and first molars. *Caries Research.* **40**(6): 494–500.

Hujoel, P. P., Zina, L. G., Moimaz, S. A. and Cunha-Cruz, J., 2009. Infant formula and enamel fluorosis: a systematic review. *Journal of the American Dental Association.* **140**(7): 841–854.

Ishii, T. and Suckling, G., 1986. The appearance of tooth enamel in children ingesting water with a high fluoride content for a limited period during early tooth development. *Journal of Dental Research.* **65**(7): 974–977.

Ismail, A. I. and Bandekar, R. R., 1999. Fluoride supplements and fluorosis: a meta-analysis. *Community Dentistry and Oral Epidemiology.* **27**(1): 48–56.

Ismail, A. I. and Hasson, H., 2008. Fluoride supplements, dental caries and fluorosis: a systematic review. *Journal of the American Dental Association.* **139**(11): 1457–1468.

Ismail, A. I. and Messer, J. G., 1996. The risk of fluorosis in students exposed to a higher than optimal concentration of fluoride in well water. *Journal of Public Health Dentistry.* **56**(1): 22–27.

Jackson, R. D., Kelly, S. A., Katz, B., Brizendine, E. and Stookey, G. K., 1999. Dental fluorosis in children residing in communities with different water fluoride levels: 33-month follow-up. *Pediatric Dentistry.* **21**(4): 248–254.

Kerley, M. A. and Kollar, E. J., 1977. Regeneration of tooth development in vitro following sodium fluoride treatment. *American Journal of Anatomy.* **149**(2): 181–195.

Khan, A., Moola, M. H. and Cleaton-Jones, P., 2005. Global trends in dental fluorosis from 1980 to 2000: a systematic review. *SADJ.* **60**(10): 418–421.

Kierdorf, H. and Kierdorf, U., 1997. Disturbances of the secretory stage of amelogenesis in fluorosed deer teeth: a scanning electron-microscopic study. *Cell and Tissue Research.* **289**(1): 125–135.

Kobayashi, C. A., Belini, M. R., Italiani Fde, M., Pauleto, A. R., Araujo, J. J., Tessarolli, V., Grizzo, L. T., Pessan, J. P., Machado, M. A. and Buzalaf, M. A., 2011. Factors influencing fluoride ingestion from dentifrice by children. *Community Dentistry and Oral Epidemiology.* **39**(5): 426–432.

Larsen, M. J., Richards, A. and Fejerskov, O., 1985. Development of dental fluorosis according to age at start of fluoride administration. *Caries Research.* **19**(6): 519–527.

Levenson, G. E., 1980. The effect of fluoride on ameloblasts of mouse molar tooth germs "in vitro". *Journal de Biologie Buccale.* **8**(3): 255–263.

Levy, S. M., 2003. An update on fluorides and fluorosis. *Journal of the Canadian Dental Association.* **69**(5): 286–291.

Levy, S. M., Warren, J. J., Davis, C. S., Kirchner, H. L., Kanellis, M. J. and Wefel, J. S., 2001. Patterns of fluoride intake from birth to 36 months. *Journal of Public Health Dentistry.* **61**(2): 70–77.

Lonnerdal, B., 2014. Infant formula and infant nutrition: bioactive proteins of human milk and implications for composition of infant formulas. *The American Journal of Clinical Nutrition.* **99**(3): 712S–717S.

Lyaruu, D. M., Bervoets, T. J. and Bronckers, A. L., 2006. Short exposure to high levels of fluoride induces stage-dependent structural changes in ameloblasts and enamel mineralization. *European Journal of Oral Sciences.* **114** Suppl. 1: 111–115; discussion 127–119, 380.

Lyaruu, D. M., Tros, G. H., Bronckers, A. L. and Woltgens, J. H., 1990. Micro-PIXE (proton-induced X-ray emission) study of the effects of fluoride on mineral distribution patterns in enamel and dentin in the developing hamster tooth germ. *Scanning Microscopy.* **4**(2): 315–322.

Lyaruu, D. M., Vermeulen, L., Stienen, N., Bervoets, T. J., Denbesten, P. K. and Bronckers, A. L., 2012. Enamel pits in hamster molars, formed by a single high fluoride dose, are associated with a perturbation of transitional stage ameloblasts. *Caries Research.* **46**(6): 575–580.

Manji, F., Baelum, V., Fejerskov, O. and Gemert, W., 1986. Enamel changes in two low-fluoride areas of Kenya. *Caries Research.* **20**(4): 371–380.

McClure, F. J., 1943. Ingestion fo fluoride and dental caries. Quantitative relations based on food and water requirements for children 1 to 12 years old. *American Journal of Diseases of Children.* **66**: 8.

McDonagh, M. S., Whiting, P. F., Wilson, P. M., Sutton, A. J., Chestnutt, I., Cooper, J., Misso, K., Bradley, M., Treasure, E. and Kleijnen, J., 2000. Systematic review of water fluoridation. *British Medical Journal.* **321**(7265): 855–859.

Ophaug, R. H., Singer, L. and Harland, B. F., 1980. Estimated fluoride intake of average two-year-old children in four dietary regions of the United States. *Journal of Dental Research.* **59**(5): 777–781.

Pendrys, D. G., 2000. Risk of enamel fluorosis in nonfluoridated and optimally fluoridated populations: considerations for the dental professional. *Journal of the American Dental Association.* **131**(6): 746–755.

Pendrys, D. G. and Katz, R. V., 1989. Risk of enamel fluorosis associated with fluoride supplementation, infant formula, and fluoride dentifrice use. *American Journal of Epidemiology.* **130**(6): 1199–1208.

Pendrys, D. G. and Morse, D. E., 1990. Use of fluoride supplementation by children living in fluoridated communities. *ASDC Journal of Dentistry for Children.* **57**(5): 343–347.

Ribeiro, D. A., Hirota, L., Cestari, T. M., Ceolin, D. S., Taga, R. and Assis, G. F., 2006. Ultrastructural morphometric analysis of ameloblasts exposed to fluoride during tooth development. *Journal of Molecular Histology.* **37**(8–9): 361–367.

Richards, A. and Banting, D. W., 1996. Fluoride toothpastes. Fluoride in Dentistry. Fejerskov, O., Ekstrand, J. and Burt, B.A. Copenhagen, Munksgaard, Wiley, John & Sons: 328–344.

Richards, A., Kragstrup, J., Josephsen, K. and Fejerskov, O., 1986. Dental fluorosis developed in post-secretory enamel. *Journal of Dental Research.* **65**(12): 1406–1409.

Riordan, P. J., 1993. Dental fluorosis, dental caries and fluoride exposure among 7-year-olds. *Caries Research.* **27**(1): 71–77.

Riordan, P. J. and Banks, J. A., 1991. Dental fluorosis and fluoride exposure in Western Australia. *Journal of Dental Research.* **70**(7): 1022–1028.

Rodrigues, M. H., Leite, A. L., Arana, A., Villena, R. S., Forte, F. D., Sampaio, F. C. and Buzalaf, M. A., 2009. Dietary fluoride intake by children receiving different sources of systemic fluoride. *Journal of Dental Research.* **88**(2): 142–145.

Sharma, R., Tsuchiya, M. and Bartlett, J. D., 2008. Fluoride induces endoplasmic reticulum stress and inhibits protein synthesis and secretion. *Environmental Health Perspectives.* **116**(9): 1142–1146.

Silva, M. and Reynolds, E. C., 1996. Fluoride content of infant formulae in Australia. *Australian Dental Journal.* **41**(1): 37–42.

Smid, J. R., Monsour, P. A., Harbrow, D. J. and Young, W. G., 1990. A histochemical study of the effects of high doses of sodium fluoride on dipeptidyl peptidase II activity in the rat incisor ameloblast. *Archives of Oral Biology.* **35**(8): 671–675.

Smith, C. E., Nanci, A. and Denbesten, P. K., 1993. Effects of chronic fluoride exposure on morphometric parameters defining the stages of amelogenesis and ameloblast modulation in rat incisors. *Anatomical Record.* **237**(2): 243–258.

Swan, E., 2000. Dietary fluoride supplement protocol for the new millennium. *Journal of Canadian Dental Association.* **66**(7): 362; discussion 363.

Szpunar, S. M. and Burt, B. A., 1988. Dental caries, fluorosis, and fluoride exposure in Michigan school children. *Journal of Dental Research.* **67**(5): 802–806.

Vilhena, F. V., Olympio, K. P., Lauris, J. R., Delbem, A. C. and Buzalaf, M. A., 2010. Low-fluoride acidic dentifrice: a randomized clinical trial in a fluoridated area. *Caries Research.* **44**(5): 478–484.

Vilhena, F. V., Silva, H. M., Peres, S. H., Caldana Mde, L. and Buzalaf, M. A., 2008. The drop technique: a method to control the amount of fluoride dentifrice used by young children. *Oral Health and Preventive Dentistry.* **6**(1): 61–65.

Villena, R. S., 2000. An investigation of the transverse technique of dentifrice application to reduce the amount of fluoride dentifrice for young children. *Pediatric Dentistry.* **22**(4): 312–317.

Walsh, T., Worthington, H. V., Glenny, A. M., Appelbe, P., Marinho, V. C. and Shi, X., 2010. Fluoride toothpastes of different concentrations for preventing dental caries in children and adolescents. *The Cochrane Database of Systematic Reviews.* (**1**): CD007868.

Warren, J. J. and Levy, S. M., 1999. Systemic fluoride. Sources, amounts, and effects of ingestion. *Dental Clinics of North America.* **43**(4): 695–711.

Warren, J. J., Levy, S. M., Broffitt, B., Cavanaugh, J. E., Kanellis, M. J. and Weber-Gasparoni, K., 2009. Considerations on optimal fluoride intake using dental fluorosis and dental caries outcomes–a longitudinal study. *Journal of Public Health Dentistry.* **69**(2): 111–115.

Whelton, H. P., Ketley, C. E., McSweeney, F. and O'Mullane, D. M., 2004. A review of fluorosis in the European Union: prevalence, risk factors and aesthetic issues. *Community Dentistry and Oral Epidemiology.* **32** Suppl. 1: 9–18.

Wong, M. C., Clarkson, J., Glenny, A. M., Lo, E. C., Marinho, V. C., Tsang, B. W., Walsh, T. and Worthington, H. V., 2011. Cochrane reviews on the benefits/risks of fluoride toothpastes. *Journal of Dental Research.* **90**(5): 573–579.

Wong, M. C., Glenny, A. M., Tsang, B. W., Lo, E. C., Worthington, H. V. and Marinho, V. C., 2010. Topical fluoride as a cause of dental fluorosis in children. *The Cochrane Database of Systematic Reviews.* (1): CD007693.

Section II
Chemistry and Biochemistry

CHAPTER 3

The Chemistry of Fluorine

BRENDA LORENA FINA[a] AND ALFREDO RIGALLI*[a]

[a]Bone Biology Laboratory, School of Medicine, Rosario National University, Santa Fe 3100, 2000, Rosario, Argentina
*E-mail: arigalli@conicet.gov.ar

3.1 History

In 1529 a compound having fluorine atoms in its structure was first described and in 1771 it was shown that the treatment of the mentioned compound with sulfuric acid [H_2SO_4] produced a new compound that attacked glass, which was then called hydrofluoric acid [HF]. Fluorine could not be isolated until many years later because of its high reactivity. Finally, it was in 1886 that for the first time fluorine in its elemental state [F_2] was obtained by electrolysis of potassium hydrogen fluoride [KHF_2] by Henri Moissan. In terms of human health and nutrition fluoride began to be considered for both its beneficial and adverse effects in the first half of the twentieth century.

3.2 Fluorine as an Element

Fluorine is the most electronegative element of nature. In the Periodic Table of the elements it is located in the halogen group and in the second period. It is not found in nature as its elemental state and can only be obtained in the laboratory or industry in processes involving strict physicochemical conditions, such as molten salt electrolysis at elevated temperatures. Fluoride is not only difficult to obtain but it also has a high

Food and Nutritional Components in Focus No. 6
Fluorine: Chemistry, Analysis, Function and Effects
Edited by Victor R Preedy

Published by the Royal Society of Chemistry, www.rsc.org

chemical reactivity and can react explosively with the different elements found in nature, including some noble gases. At room temperature, fluorine is a yellow gas and has chlorine and ozone-like odor. Its boiling point is −187 °C and the melting point is −223 °C.

The fluorine atom is found in nature as a single isotope containing 9 protons, 9 electrons and 10 neutrons, with an atomic weight of 19. This stable isotope of fluorine is represented as ${}_9F^{19}$, where 9 is the atomic number and 19 the atomic weight. Its nine electrons are distributed in two energy levels, and its electronic configuration can be written as follows: ${}_9F^{19}$: $1s^2\ 2s^2\ 2px^2\ 2py^2\ 2pz^1$.

As can be seen, level 2 has 7 electrons, which shows that gaining only one electron becomes a stable electronic configuration. Its chemical properties are a consequence of its electronic configuration and high electronegativity: 3.98 Pauling. This high value of electronegativity determines that, being in the presence of any chemical element, fluoride has a great tendency to be reduced, gaining electrons and completing 8 electrons in its valence shell. Because of its high electronegativity, fluorine is not found in nature as its elemental state but it is found as its reduced state fluoride with oxidation number −1, which we denote by the symbol: F^-. Atomic number and atomic weight for simplicity in its graphical presentation are omitted.

Artificial Isotopes with atomic weights 17, 18 and 22 were obtained from fluorine. The isotope of atomic weight 18 [${}_9F^{18}$] is, at this time, widely used in medicine and releases a positive beta-type corpuscular radiation, with a half-life of 110 min.

Fluoride can form covalent bonds with carbon, but the formation of this bond will not be discussed in this chapter. However, destruction and biological effects resulting thereof will be discussed.

3.3 Fluorine in Nature

The percentage of fluorine in the Earth's crust is variable and depends on the measurement technique employed. Fluoride is present in the Earth's crust in 300 part per million (ppm). However, fluoride can be found in variable proportions in different places, the most common amounts found in the surface of the Earth are between 0.01 and 100 ppm (Lupo *et al.*, 2012). In nature, the fluorine is in a state of forming fluoride water-soluble compounds such as sodium fluoride [NaF] or other alkali-metal fluorides such as potassium fluoride [KF]. But it can also be part of some nondissociated salts such as calcium fluoride [CaF_2] also known as fluorite or fluorspar. The difference in water solubility between the CaF_2 and NaF is that the Ca^{++}, having greater nuclear charge than Na^+ and K^+, produces a greater polarization of the ionic bond, increasing the covalent nature of the bond and consequently the compounds become less soluble in water. Another common mineral in nature is cryolite or sodium hexafluoroaluminate [Na_3AlF_6], which is used in the industrial production of aluminum or as a pesticide.

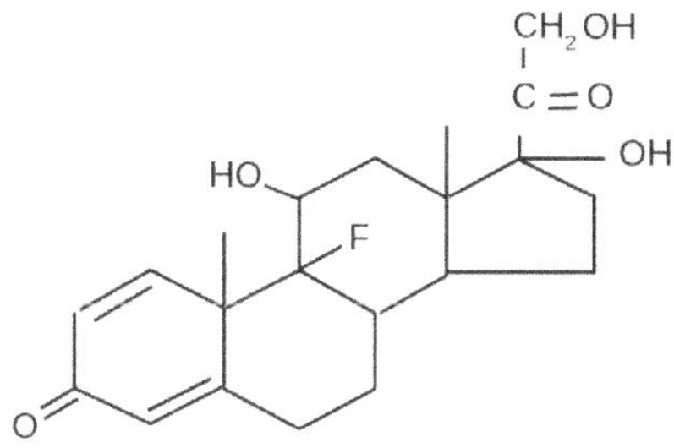

Figure 3.1 Chemical structure of dexamethasone. Fluorine is part of the structure of dexamethasone. The presence of fluorine in the structure produce changes in the biological activity of several compounds.

3.4 Fluorine and Fluoride

Although fluorine is never found in its elemental state in nature, throughout this text the term "fluorine" will be used to refer to the element in its pure state or when referring to the element as part of other chemical compounds. The term "fluoride" will be used when referring to water-soluble salts or to the ion in aqueous solution.

As mentioned above, there are fluorine atoms in different natural compounds. However, it is also common to find fluorine atoms in organic molecules, which is incorporated in order to change its biological activity or promote certain chemical reactions. As example of fluorinated molecules, we can mention dexamethasone which is a well-known steroidal anti-inflammatory (Figure 3.1).

3.5 Fluorine in Medicine and Health

Fluorine is used in medicine in various health-related applications. The natural stable isotope, ${}_9F^{19}$, is used in treatment and prevention of dental cavities (Russof *et al.*, 1962) and was also used for several years in the prevention and treatment of osteoporosis (Briançon and Meunier, 1981).

The radioactive isotope, ${}_9F^{18}$, is associated with various biological molecules and used in imaging technologies (see Chapter 14). Briefly, ${}_9F^{18}$ is chemically bound to the molecule of the 2-deoxy-D-glucose, to form 2-[F^{18}]-fluoro-2-deoxy-D-glucose (FDG) and is used in positron emission tomography (PET), which is an extremely sensitive methodology applied to diagnostic imaging, especially in studies related to cancer. FDG is taken up by cells and the radioactivity can be three-dimensionally located in the body by a PET camera, detecting the precise location of the site of tumor formation. When FDG undergoes radioactive decay of ${}_9F^{18}$, will become ${}_8O^{18}$ emitting positive beta-type corpuscular radiation, and FDG will turn into glucose, which can be metabolized by the cells (Figure 3.2). The radiation produced by the decay of the isotope ${}_9F^{18}$ is proportional to the amount of FDG incorporated and is a

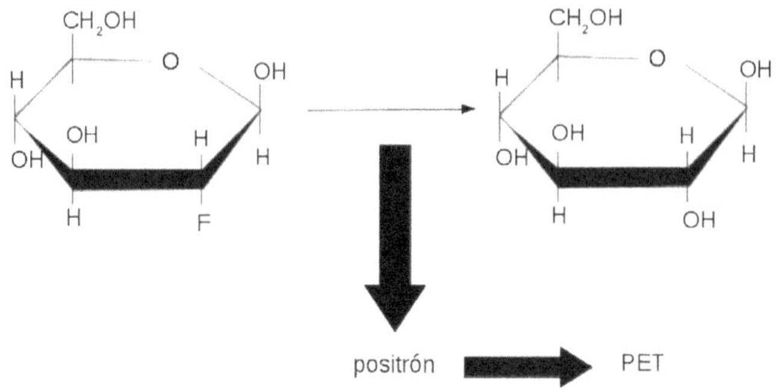

Figure 3.2 2-[^{18}F]-fluoro-2-deoxy-D-glucose (FDG) transformation in D-glucose. FDG inside the cell turn into D-glucose when $_{9}F^{18}$ disintegrates into $_{8}O^{18}$. FDG cannot be metabolized in the cell, and accumulates into the cell. As a consequence, the positron emission is proportional to FDG, which is proportional to the metabolic activity of cells.

$$Ca_{10}(PO_4)_6(OH)_2 \quad + \quad 2\,F^- \longrightarrow Ca_{10}(PO_4)_6F_2 \quad + \quad 2\,OH^-$$

Figure 3.3 Formation of fluorapatite from hydroxyapatite. Fluoride reaction with hydroxyapatite to obtain fluorapatite, a less water soluble compound.

measure of the metabolic activity of the cells. This metabolic activity reports the status of the cells that will be increased in tumor formations.

In the bones of vertebrates that consume fluoride with food or drinking water, fluorine can be found in large amounts as part of the fluorapatite crystal [$Ca_{10}(PO_4)_6F_2$]. This compound is formed by exchange of the hydroxyl group of hydroxyapatite [$Ca_{10}(PO_4)_6(OH)_2$] by fluoride (see Chapter 12). The reaction that produces this exchange is shown in Figure 3.3.

Because fluoride can replace the hydroxyl group in the hydroxyapatite crystal, the isotope $_{9}F^{18}$ can be useful in positron emission tomography. Sodium fluoride, using the radioactive isotope [$Na_{9}F^{18}$], is administered as an intravenous injection for imaging of bone lesions with high bone remodeling. The $_{9}F^{18}$ replaces the hydroxyl group of hydroxyapatite, forming [$_{9}F^{18}$]fluorapatite. The radioactive decay of $_{9}F^{18}$ in $_{8}O^{18}$ generates positrons that interact with the surrounding electrons and generate an annihilation radiation that is detected by positron emission tomography.

[$_{9}F^{18}$]-choline chloride is also used because the high cell division of many tumors produces a high rate of synthesis of phospholipids, such as phosphatidylcholine, used in the structure of cell membranes (Figure 3.4).

3.6 Fluorine Compounds in Bones and Teeth

Hydroxyapatite is a major component of tooth enamel (up to 87%) and the calcified bone tissue (up to 50%). Hydroxyapatite is dissolved in acidic conditions, either as a result of acids produced by bacteria in the mouth or by the

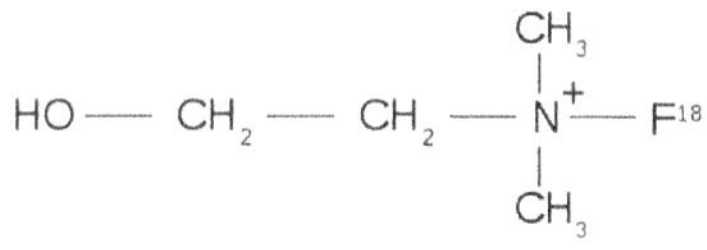

Figure 3.4 [F^{18}]-choline. Chemical structure of fluorcholine, a compound used in positron emission tomography.

$$Ca_{10}(PO_4)_6(OH)_2 \quad + \quad H^+ \longrightarrow Ca_{10}(PO_4)_6(OH)^+ \quad + \quad H_2O$$

Figure 3.5 Dissolution of hydroxyapatite. Hydroxyapatite dissolves in acidic medium by combination of proton with hydroxyl ion.

$$Ca_{10}(PO_4)_6F_2 \quad + \quad H^+ \longrightarrow Ca_{10}(PO_4)_6F^+ \quad + \quad HF$$

Figure 3.6 Dissolution of fluorapatite. Fluorapatite dissolves in acidic medium by combination of proton with fluorine to form hydrofluoric acid [HF].

low pH generated in the Howship's lacunae due to active secretion of protons by osteoclasts.

Although both compounds (fluorapatite and hydroxyapatite) are soluble in acidic medium, the dissolution of hydroxyapatite is faster in acidic media by the larger capacity of the hydroxyl group to join the proton. Hydroxyapatite dissolution in acid is shown in Figure 3.5.

The fluorapatite suffers a similar reaction, but requires a lower pH and also the process is slower (Figure 3.6).

The additional advantage of the presence of fluorapatite in dental enamel is that the generation of hydrofluoric acid has an inhibitory effect of bacterial growth in the oral cavity. HF causes changes in the activity of the electron-transport chain, decreasing production of energy, and also can act as an inhibitor of other enzymes of metabolic pathways involved in the oxidation of nutrients necessary for the bacteria.

3.7 Reaction of Fluorine with Multivalent Metallic Cations

Fluoride may act as a ligand of various complex ions, especially with metals of oxidation number +3, as Fe^{+++}, Al^{+++} and Cr^{+++}. While it may form complexes with one metal atom and 6 fluoride ions, under laboratory conditions similar to those found in living organisms, two fluoride ions complexed only with one metal atom (Rigalli *et al.*, 1994). For example the reaction with aluminum is shown in Figure 3.7.

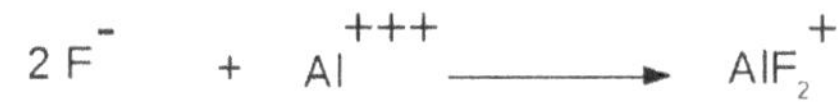

Figure 3.7 Formation of difluoraluminate. Fluoride reaction with aluminum to form a complex ion.

This binding of fluoride to aluminum is strong enough so that it cannot be released from the complex in the digestive tract, precluding fluoride absorption.

3.8 Reaction of Fluorine and its Relation to Storage and Measurement

The reactions of fluorine with silicon compounds are biologically important and should be considered for both storing samples and measuring procedures in biological samples.

In acidic solutions, fluoride is converted to hydrofluoric acid (Figure 3.8). This reaction is important in pH lower than 5 since the hydrofluoric acid is a weak acid having a dissociation constant of 6.8×10^{-4}.

Hydrofluoric acid is capable of reacting with silicon compounds. The glass is composed of silicon dioxide [SiO_2], so it can be attacked by HF and the product, tetrafluorosilane [SiF_4] is a gas evolved from the mixture (Figure 3.9).

The result of the action of HF on the glass produces a frosted surface. This reaction is important to keep in mind when samples containing fluoride are stored. If the sample is stored in a glass container it must be ensured that the pH is greater than 6, so as to avoid the formation of SiF_4, which being volatile will come off the sample and reduce the fluoride content in the solution. Even with neutral pH, storage of solutions containing fluoride is not advisable in glass for extended periods.

Hydrofluoric acid also reacts with hexamethyldisiloxane [$(CH_3)_6OSi_2$] (HMDS), giving trimethylfluorosilane [$(CH_3)_3SiF$] (Figure 3.10). This compound is volatile and when it dissolves in alkaline medium exchanges fluorine for hydroxyl group, giving fluoride and trimethylsilanol. These reactions are those used in the isothermal fluoride distillation with hexamethyldisiloxane (see Chapter 6). This process allows isolation in a few hours of the fluoride from complex samples and retrieval of it from a solution of sodium fluoride.

The distillation process is accelerated in the presence of HMDS as the trimethylfluorosilane has a lower boiling point than the HF, which has a boiling point of 19 °C.

Fluoride can also react with zirconyl chloride [$ZrOCl_2$], giving hexafluorozirconate(IV) [$Zr\ F_6^{2-}$] as product, which is colorless. Zirconyl chloride is colored and hexafluorozirconate (IV) is colorless, then the disappearance of color is proportional to the concentration of fluoride in the sample. This reaction is the basis for the colorimetric method for the detection of fluoride in water.

$$F^{-} + H^{+} \longrightarrow HF$$

Figure 3.8 Formation of hydrofluoric acid [HF]. Hydrofluoric acid [HF] formation by the reaction between fluoride [F^-] and hydrogen ions [H^+].

$$4\,HF + SiO_2 \longrightarrow SiF_4 + 2\,H_2O$$

Figure 3.9 Reaction among hydrofluoric acid [HF] and silicon dioxide [SiO_2]. Hydrofluoric acid react with silicon dioxide to form tetrafluorosilane, which is a covalent volatile compound.

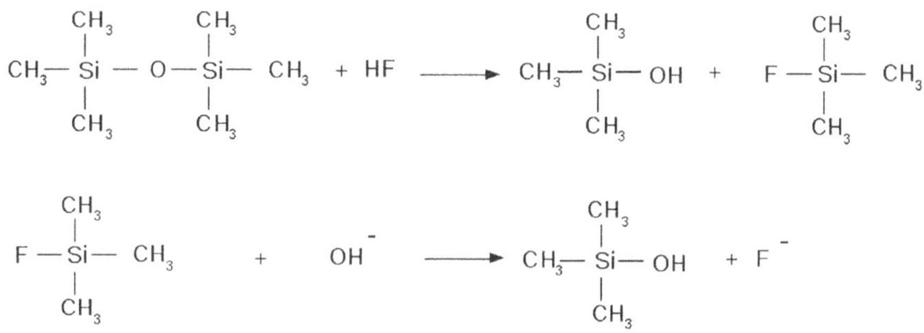

Figure 3.10 Reaction of hydrofluoric acid [HF] with hexamethyldisiloxane. Hydrofluoric acid reacts with hexamenthyldisiloxane to form trimethylfluorosilane, which then reacts in basic solution to form fluoride. This is the reactions that occur during the isothermal distillation for the isolation of fluoride from other compounds in a complex samples.

Lanthanum fluoride [LaF_3] is a crystal at room temperature and is used to construct selective fluoride electrodes for potentiometric measurement of fluoride in water. For the best performance of the fluoride-sensitive membrane, LaF_3 is used doped with europium fluoride [EuF_2] (for additional information see Chapter 6).

3.9 Fluoride Compounds for Use in Medicine

Disodium monofluorophosphate, sodium monofluorophosphate or MFP [$Na_2(FPO_3)$] is used as a source of fluoride in toothpastes and was used as a compound for the treatment of osteoporosis (Resch *et al.*, 1994).

At an industrial level, it is prepared by the reaction of sodium metaphosphate [$NaPO_3$] with sodium fluoride [NaF] at high temperature using sodium hexafluorosilicate [Na_2SiF_6] as flux. The reaction of the process is depicted in Figure 3.11.

This method of synthesis produces MFP having a low percentage of NaF. However, the proportion of NaF in the preparation increases as time passes due to hydrolysis of the P–F link to form NaF (Rigalli *et al.*, 1995).

$$NaPO_3 + NaF \longrightarrow Na_2(FPO_3)$$

Figure 3.11 Synthesis of disodium monofluorophosphate. Disodium monofluorophosphate can be obtained by the reaction between sodium metaphosphate [$NaPO_3$] and sodium fluoride [NaF] at high temperature.

$$H_3PO_4 + 2\,(NH_4)HF_2 \longrightarrow (NH_4)_2(FPO_3) + 3\,HF + H_2O$$

$$(NH_4)_2(FPO_3) + 2\,Ag^+ \longrightarrow Ag_2FPO_3 + 2NH_4^+$$

$$Ag_2FPO_3 + 2Na^+ \longrightarrow Na_2FPO_3 + 2\,Ag^+$$

Figure 3.12 Synthesis of disodium monofluorophosphate. MFP is obtained without NaF contamination by reaction of phosphoric acid [H_3PO_4] with ammonium bifluoride [$(NH_4)HF_2$], precipitation with silver ion and finally by a sodium exchange column.

A method for synthesizing MFP, without NaF as a contaminant, is the reaction of phosphoric acid [H_3PO_4] with ammonium bifluoride [$(NH_4)HF_2$]. Then, MFP is precipitated as a silver salt and finally transformed into MFP in a cation-exchange column. The reactions of this process are depicted in Figure 3.12.

MFP is administered orally and is absorbed at the gastric and intestinal level. For more detail see Chapter 12.

Unlike fluoride that is free in the plasma, the MFP can join some proteins. The chemical structure of the adduct has not yet been studied, but due to the structure of the proteins to which it binds (alpha-macroglobulin and complement component C3) it is suspected that MFP interacts with a thiolester bond present in both proteins and this is necessary to fulfill their biological functions.

Fluorine is also part of some anaesthetics such as halothane (2-chloro-2-bromo-trifluoroethane), methoxyflurane (2, 2-dichloro-1, 1-difluoroethyl methyl ether) and isoflurane (1-chloro-2, 2, 2-trifluoroethyl difluoromethyl ether), among others. Figure 3.13 displays the chemical structure of the above anaesthetics.

In some cases the metabolism of the anaesthetic produces a large release of fluoride that has been shown to have toxic effects in the kidney and other organs. The toxic effects of methoxyflurane have been widely studied, and there is strong evidence that its toxic effects are mediated by the action of cytochrome P450, and the release of fluoride that produces high plasma fluoride concentrations. Fluoride release can be produced by *O*-demethylation or dechlorination. In both cases, hydrogen fluoride is released, which becomes fluoride, and the final product is oxalic acid. Figure 3.14 displays the proposed mechanism for defluorination of methoxiflurane (Kharasch *et al.*, 2006)

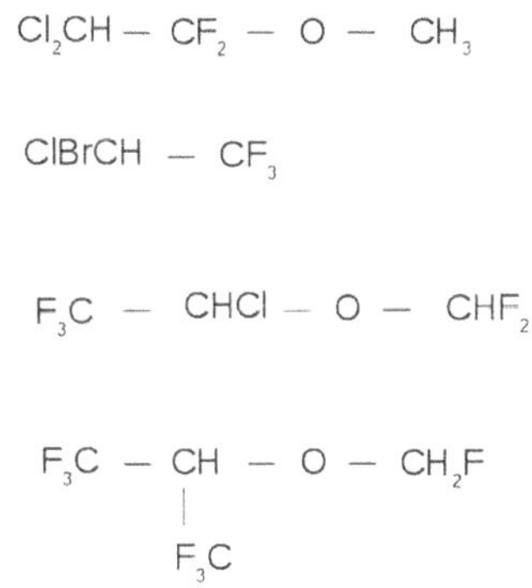

Figure 3.13 Chemical structure of anaesthetic agents. Chemical structure of methoxiflurane, halothane, isofluorane and sevofluorane.

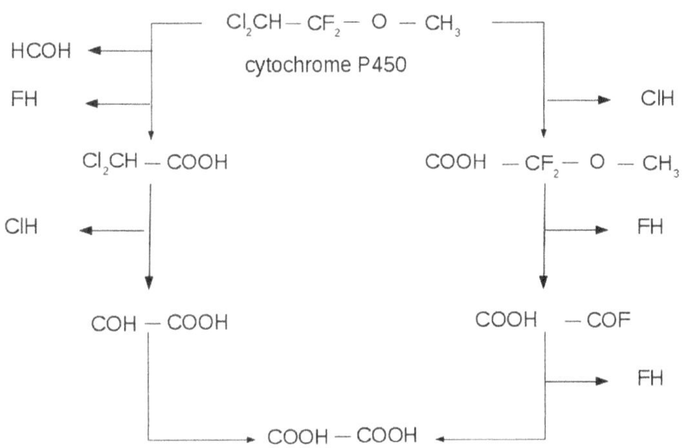

Figure 3.14 Biological defluorination of methoxiflurane. Defluoridation of methoxyflurane can occur by O-demethylation (left) or dechlorination (right) pathways. In both cases oxalic acid is the final product and hydrofluoric acid [HF] is released, which transform into fluoride.

Sevoflurane and enflurane are biotransformed by defluorination and generates fluoride, which has nephrotoxicity (Nuscheler *et al.*, 1998).

3.10 Fluoride Compounds and Industrial Uses that Affect Health

Compounds containing fluorine are important in industry. It has been shown that several compounds containing fluorine have different purposes: industrial, insecticides, medicines, *etc.*

Teflon or polytetrafluoroethylene [$(-F_2C-CF_2-)_x$] is a heat-resistant polymer that has many applications in industry. Its chemical inertness determines that there is no adverse effects of this compound on health and the environment.

Chlorofluorocarbons (CFCs) are compounds having chlorine and fluorine in their structure, are used as refrigerants and propellants in aerosols and are known under the name of Freon. Freon I or dichlorodifluoromethane (CF_2Cl_2) and Freon II or trichloromonofluoromethane [$CFCl_3$] are examples of this series of compounds. These compounds have declined in their use and production for their alleged action on the ozone layer. In the upper layers of the atmosphere, the action of ultraviolet radiation (UV) produces atomic oxygen from dioxygen. The oxygen atoms react with oxygen molecules [O_2] to give ozone [O_3] as a product. Ozone molecules can interact with UV and generate dioxygen and oxygen atoms. Thus, the oxygen and ozone stop UV radiation, which is harmful to human health above certain exposure. Figure 3.15 display the mentioned reactions.

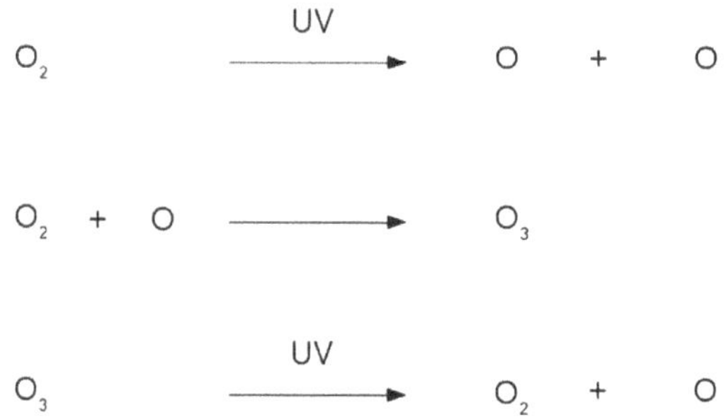

Figure 3.15 Reactions of dioxygen and ozone in the presence of UV radiation. Dioxygen and ozone in the presence of UV radiation generate mono-atomic oxygen that by reaction with dioxygen produce ozone. In this way, dioxygen and ozone stop UV radiation and protect the population from the action of UV radiation.

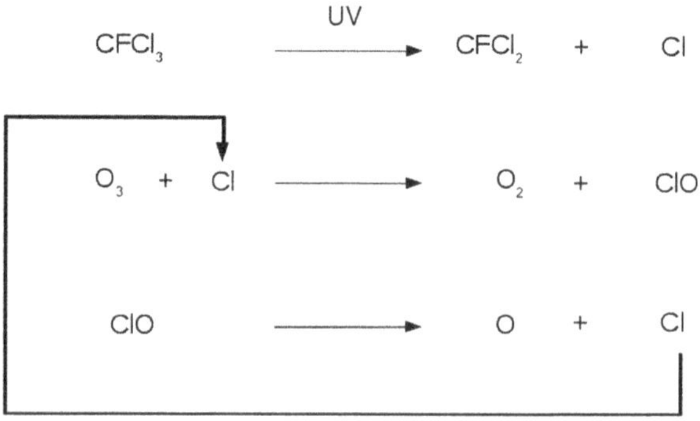

Figure 3.16 Reaction of chlorofluorocarbons [$CFCl_3$] in the ozone layer. Trichloromonofluoromethane [$CFCl_3$], a chlorofluorocarbon compound produces the free radical chlorine [.Cl] which then reacts with the ozone, decreasing its amount in the upper layer of atmosphere.

When UV radiation interacts with CFC it generates chlorine (.Cl) and $CFCl_2$ free radicals. These compounds react with ozone, consuming it, and generates dioxygen hypochlorite free radical. This free radical can be decomposed into oxygen and chlorine free radical that can react with ozone again. It is estimated that a CFC molecule can destroy thousands of ozone molecules, decreasing their concentration in the upper layers of the atmosphere, producing what is known as the ozone hole and the loss of the ability of the atmosphere to stop the UV radiation. Figure 3.16 displays the mentioned reactions.

3.11 Miscellaneous Chemical Compounds Containing Fluorine

Some compounds present in plants are toxic, such as sodium fluoroacetate [$CH_2FCOONa$]. This compound is toxic because it can react with coenzyme A and form fluoroacetyl coenzyme A. Then, by the action of the citrate synthase enzyme, fluoroacetyl coenzyme A is combined with oxaloacetate to form fluorocitrate, a metabolite that is firmly attached to the aconitase enzyme of the Krebs cycle and thus reduces the production of reduced coenzymes by the cycle, depriving cells from energy.

Cryolite [Na_2AlF_6] is used in aluminum metallurgy, acting as a flux, and allowing the electrolysis of aluminum oxide. It is therefore a common fluorine pollution in aluminum industry areas.

Sulfonyl fluoride [$R{-}SO_2F$] is used as an insecticide. Furthermore, fluorine compounds are used as pesticides (fluoraluminate, sulfuryl fluoride and NaF), in floor-polishing products (fluorosilicate) and wood preservation (KF). In the development of industrial products, perfluoro-octane sulfonyl fluoride is used for the preparation of packaging papers, carpets and textiles.

CaF_2 and MgF_2 are used in the glass industry.

Fluoride is also used in the enrichment of uranium since it can react with uranium and form uranium hexafluoride [UF_6]. As UF_6 is a gaseous substance, it can be used to separate the U^{235} from U^{238} because they have different rates of diffusion through membranes.

Summary Points

- Fluorine is the most electronegative element found in nature.
- Fluoride can be found in food and medicines.
- Fluorapatite is part of bones and teeth.
- Fluorapatite is more resistant to dissolution than hydroxyapatite.
- Chlorofluorcarbons are used as refrigerants and may affect the ozone layer.
- The radioactive fluorine isotope ${}_9F^{18}$ is used in medicine for diagnostic purposes.
- Anaesthetics containing fluorine can release hydrofluoric acid and affect human health.

Key Facts of the Chemistry of Fluorine

- Fluorine is an abundant element in the environment.
- Fluorine is part of foods, water and medicines.
- Chlorofluorocarbons destroy the ozone layer.
- Fluorinated anaesthetics can release fluoride and affect health.
- The radioactive isotope of fluorine is used in image diagnosis in medicine.
- Fluoride reacts with trivalent and divalent metals and decreases biological effects.
- Fluoride in acidic conditions reacts with glass.

Definitions of Words and Terms

Chlorofluorocarbons: Compounds containing chlorine and fluorine used as refrigerants, propellants and solvents. They contribute to ozone destruction in the upper layers of atmosphere, called the ozone layer.

Disodium monofluorophosphate: Salt containing fluoride used in the treatment of osteoporosis and as component of dentifrices.

Electronegativity: A property of atoms that indicates the tendency to attract electrons in a chemical bond.

Fluoride: Fluorine monovalent ion.

Fluorine: Element belonging to the group of halogens of atomic number 9 and atomic mass 19.

$_9F^{18}$: Radioactive isotope of fluorine that suffers beta decay.

2-fluor-2-deoxy-D-desoxiglucose: Molecule of glucose with radioactive fluorine on carbon number 2, which is used in positron emission tomography.

Fluorapatite: Mineral containing fluorine that forms part of calcified tissues such as teeth and bones.

Hexamethyldisiloxane: Compound that reacts with hydrofluoric acid and allows isolation of fluoride from biological samples, such as plasma, blood, bones, *etc*.

Isofluorane: Anaesthetic containing fluorine.

Lanthanum fluoride: Solid and non soluble salt of fluoride that is used as part of the sensor in the fluoride-selective electrode.

Methoxyflurane: Anaesthetic containing fluorine.

Ozone layer: The ozone layer is a layer of the atmosphere where oxygen and ozone capture ultraviolet radiation preventing the harmful action of this radiation on the health of the population.

Positron emission tomography: Image diagnostic technique that produces three-dimensional images related to metabolic processes. It uses molecules with the radioactive isotope of fluorine: $_9F^{18}$ bounded to biological molecules.

Teflon: Compound containing fluoride that has many uses in the home and in industry.
Week acid: A substance that partially dissociates into hydrogen ion and anion. Hydrofluoric acid is a weak acid.

List of Abbreviations

CFC	Chlorofluorocarbon
FDG	$[^{18}F]$fluor-2-deoxy-D-glucose
F	Fluorine
$_9F^{19}$	Stable isotope of fluorine
$_9F^{18}$	Radioactive isotope of fluorine that suffers beta decay.
F^-	Fluoride
MFP	Monofluorophosphate
ppm	Parts per million

References

Briançon, D. and Meunier, P. J., 1981. Treatment of osteoporosis with fluoride, calcium and vitamin. *The Orthopedic Clinics of North America.* **12**: 629–648.
Kharasch, E. D., Schroeder, J. L., Liggitt, H. D., Park, S. B., Whittington, D. and Sheffels P., 2006. New insights into the mechanism of methoxyflurane nephrotoxicity and implications for anesthetic development (part 1): identification of the nephrotoxic metabolic pathway. *Anesthesiology.* **105**: 726–736.
Lupo, M., Fina, B. L., Aguirre, M. C., Armendariz, M. N. and Rigalli, A., 2012. Determination of water fluoride concentration and the influence of the geographic coordinate system and time. *Water, Air, and Soil Pollution.* **223**: 5221–5225.
Nuscheler, M., Conzen, P. and Peter, K., 1998. Sevoflurane: metabolism and toxicity. *Der Anaesthesist.* **47** (Suppl. 1): S24–S32.
Resch, H., Libanati, C., Talbot, J., Tabuenca, M., Farley, S., Bettica, P., Tritthart, W. and Baylink, D., 1994. Pharmacokinetic profile of a new fluoride preparation: sustained-release monofluorophosphate. *Calcified Tissue International.* **54**: 7–11.
Rigalli, A., Cabrerizo, M., Beinlich, A. and Puche, R. C., 1994. Gastric and intestinal absorption of monofluorphosphate and fluoride in the rat. *Arzneimittel-Forschung.* **44**: 651–655.
Rigalli, A., Iglesias, A. M. and Puche, R. C., 1995. Long term stability of sodium monofluorophosphate. *Drug Development and Industrial Pharmacy.* **21**: 517–521.
Russof, L. L., Konikoff, B. S., Frye, J. B., Johnston, J. E. and Frye, W. W., 1962. Fluoride addition to milk and its effects on dental caries in school children. *American Journal of Clinical Nutrition.*, **11**: 94–101.

CHAPTER 4

Fluoride Metabolism

CAMILA PERES BUZALAF[a], ALINE DE LIMA LEITE[b], AND MARÍLIA AFONSO RABELO BUZALAF*[b]

[a]Pró-Reitoria de Pesquisa e Pós-Graduação, Universidade do Sagrado Coração, Bauru, São Paulo 17011-160, Brazil; [b]Department of Biological Sciences, Bauru Dental School, University of São Paulo, Bauru, São Paulo 17012-901, Brazil
*E-mail: mbuzalaf@fob.usp.br

4.1 Introduction to Fluoride Metabolism

Fluoride is an inorganic anion of fluorine, which is naturally found in the biosphere with ubiquitous presence in the environment and is the most electronegative and reactive among all elements. This latter characteristic explains why fluorine is mainly found as inorganics instead of the elemental form. Its well-known cariostatic effect led to the implementation of fluoride in systemic (such as water, salt, sugar, milk and supplements) and topic sources (such as toothpastes, gels, mouth rinses and varnishes). The benefits of fluoride include both caries control and bone formation (Bratthall *et al.* 1996; Kobayashi *et al.*, 2014). However, it can negatively affect the quality of the developing mineralized tissues, causing dental and skeletal fluorosis. The differential responses seem to be determined by the dose and time of exposure (Everett, 2011). Thus, the knowledge of fluoride metabolism is crucial not only to understand its biological effects but also to optimize fluoride-driven preventive and therapeutics strategies.

Food and Nutritional Components in Focus No. 6
Fluorine: Chemistry, Analysis, Function and Effects
Edited by Victor R Preedy

Published by the Royal Society of Chemistry, www.rsc.org

4.2 General Aspects of Fluoride Metabolism

Fluoride metabolism includes absorption, distribution and excretion, where each step depends on the pH. Hydrogen fluoride (HF) is a weak acid with a pK_a of 3.4. By definition, at pH 3.4, 50% of fluoride is in the undissociated form (HF), while the remaining 50% is in the ionic form (F^-). As pH decreases from 3.4, the concentration of HF increases, and as pH increases, the concentration of F^- increases (Whitford, 1996) (Figure 4.1). This means that fluoride permeates cell membranes as HF, in response to a pH gradient between adjacent body-fluid compartments, going from the more acidic to the more alkaline compartment (Figure 4.1). After ingestion, plasma fluoride reaches a peak within 20–60 min, due to its pH-dependent gastric absorption, followed by a rapid decline as a result of both uptake in calcified tissues and urine excretion (Figures 4.2 and 4.3). The small intestine also contributes to fluoride absorption in a pH-independent mechanism. Nonabsorbed fluoride is excreted in feces (Figure 4.3). From plasma, fluoride is distributed to both hard and soft tissues followed

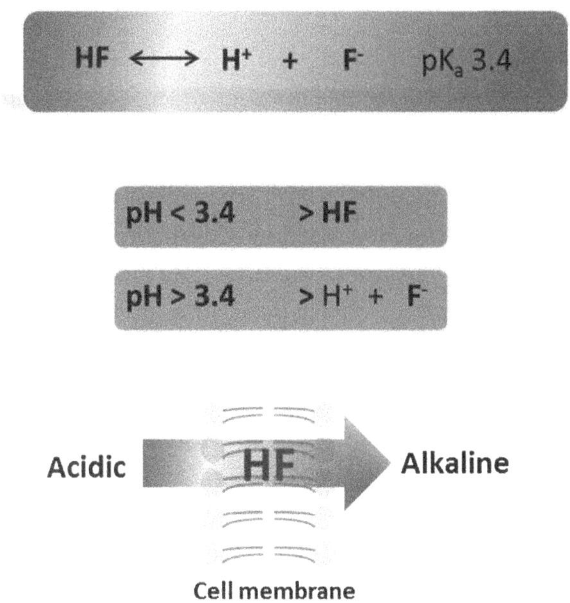

Figure 4.1 Metabolism of fluoride is dependent on pH. HF is a weak acid with pK_a of 3.4, meaning that at pH 3.4, half of fluoride is in the nonionic form (HF) and half is in the ionic form (F^-). As pH decreases from 3.4, the concentration of HF enhances, while as pH increases, the concentration of F^- increases. Besides, the coefficient of permeability of cell membranes to HF is much higher than that of F^-. Thus, fluoride drives through cell membranes as HF, in response to a pH gradient between adjacent compartments, going from more acidic to more alkaline compartments.

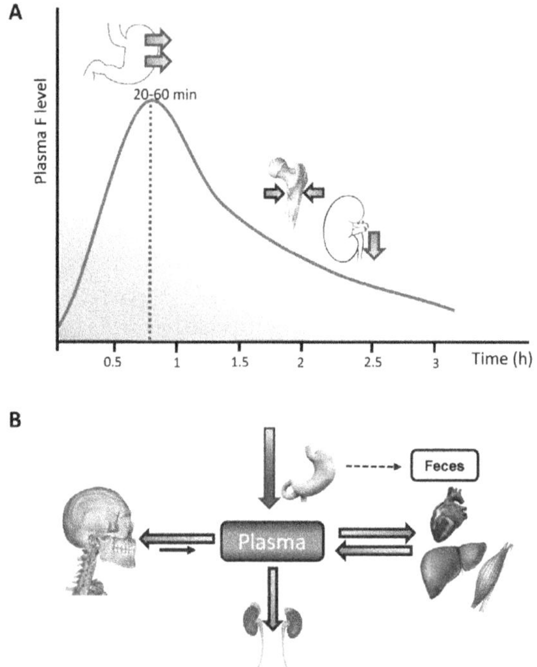

Figure 4.2 Plasma fluoride levels after a small amount of fluoride ingested. (A) After ingestion, plasma fluoride reaches a peak within 20–60 min, due to its pH-dependent gastric absorption, followed by a rapid decline as a result of both uptake in calcified tissues and urine excretion. The fluoride plasma level returns to baseline levels within 3–4 h. The small intestine also contributes to fluoride absorption in a pH-independent mechanism. (B) From plasma, fluoride is distributed to both hard and soft tissues followed by its renal excretion. Plasma fluoride levels are rapidly decreased mainly due to fluoride uptake in hard tissues and renal excretion, while the nonabsorbed fluoride is excreted in feces (A and B).

by its renal excretion (Figure 4.3). A minor portion of absorbed fluoride is found in soft tissues through a steady-state distribution between extracellular and intracellular fluids. However, about 50% of the absorbed fluoride is incorporated in calcified tissues, mainly in bone, where 99% of the fluoride content in the body is found (Whitford, 1994). Its incorporation in calcified tissue is not irreversible, from where it can be released back to plasma compartment. The fluoride excreted in urine is mostly a result of the total absorbed amount subtracted by the incorporated quantity in hard tissues. Each step related to the fluoride metabolism will be described in sequence.

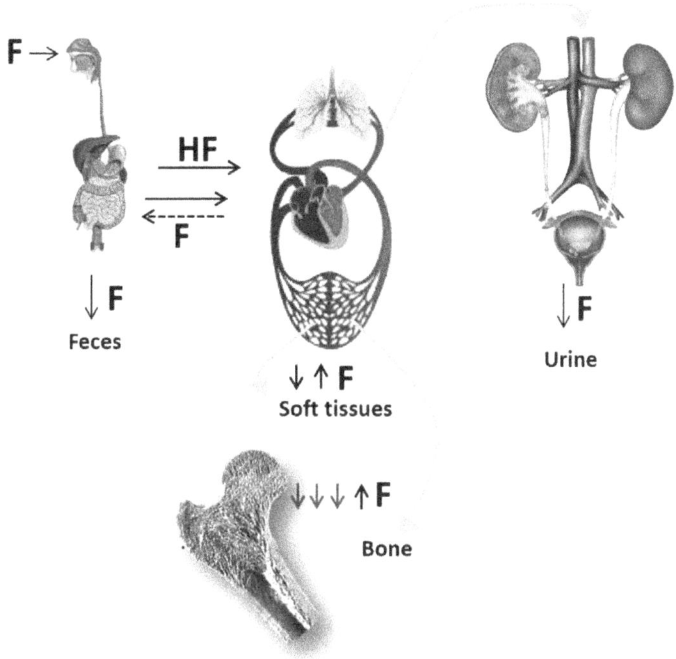

Figure 4.3 General aspects of fluoride metabolism. After infestion, fluoride is absorbed in the stomach and small intestine by a pH-dependent and -independent mechanisms, respectively. Nonabsorbed fluoride is excreted in feces. A minor portion of absorbed fluoride is found in soft tissues through a steady-state distribution between extracellular and intracellular fluids. However, about 50% of the absorbed fluoride is incorporated in calcified tissues, mainly in bone, where 99% of the fluoride content in the body is found. Its incorporation in calcified tissue is not irreversible, from where it can be released back to the plasma compartment. From systemic circulation, fluoride content is followed by renal excretion. The fluoride excreted in urine is mostly a result of the total absorbed amount subtracted by the quantity incorporated in hard tissues.

4.3 Fluoride Absorption

Approximately 80–90% of fluoride ingested is absorbed from the gastrointestinal tract by passive diffusion and does not seem to require specific transporters (Whitford, 1996). However, "Fluc" proteins-based channel was recently purified in phospholipid bilayers with extreme selectivity for F^- over Cl^-, but its expression and role in fluoride absorption in mammalian is still unknown (Stockbridge *et al.*, 2013). Recently, selective bacterial CLC anion-transporting proteins that catalyze transport of fluoride in liposomes were discovered (Stockbridge *et al.*, 2012). In addition, a novel fluoride exporter named *FEX* functioning as fluoride efflux was recently described in eukaryotic cells (Li *et al.*, 2013). This exporter allows organisms such as fungi to evoke from fluoride-mediated toxicity but has not been identified in mammalian cells as yet. Together, these novel findings open new perspectives regarding fluoride-mediated inter- and intracellular transport.

While about 20–25% of the total ingested fluoride is absorbed in the stomach, the remainder is absorbed in the proximal small intestine (Nopakun *et al.*, 1989; Whitford, 1996). Its absorption occurs in a half-life of 30 min, reaching a peak within 20–60 min (Whitford, 1996), independently of the amount ingested. After that, the fluoride concentrations are rapidly decreased and correlate to both uptake in calcified tissues and renal excretion (Whitford, 1996). The gastric absorption is very quick and is determined by acidity and content gastric and fashion of gastric emptying (Whitford and Pashley, 1984). This is explained by the fact that the gastric absorption is inversely related to the pH. *i.e.*, the higher the gastric acidity, the higher the fluoride absorption (Whitford and Pashley, 1984). When ionic fluoride enters the acid gastric lumen environment, it is converted into HF, which is an uncharged molecule that readily crosses cell membranes, including gastric mucosa (Gutknecht and Walter, 1981). Thus, peak plasma fluoride concentration is higher in acidic compared to neutral environments. Thus, the pH dependency of fluoride absorption has a direct implication not only for the treatment in cases of acute fluoride toxicity but also in therapeutic applications. Moreover, delayed gastric emptying may promote both slower and smaller fluoride absorption leading to a decreased plasma concentration (Nopakun *et al.*, 1989).

About 75% of the fluoride not absorbed in the stomach is absorbed predominantly as ionic fluoride by the proximal small intestine in a pH-independent mechanism. In this case, F^- crosses the tight junctions of intestine epithelia arranged between cells and paracellular channels (Nopakun *et al.*, 1989). The greatest absorption in the small intestine may be a consequence of about 100-thousand times more $[F^-]$ compared to [HF] in neutral pH (Messer *et al.*, 1989).

The fluoride absorption is also affected by the diet and other foods with which fluoride interacts. Sodium fluoride (NaF) in water solutions is 100% absorbed (Whitford, 1996). However, when in combination with foods or drinks containing calcium, aluminum and magnesium (bi- and -trivalent cations), fluoride forms insoluble complexes that decrease fluoride absorption (Trautner and Siebert, 1986). This explains the use of calcium-containing solutions in acute fluoride toxicity (Li, 2003).

The absorption is also altered by the type of administered fluoride. So far, there are divergences in the literature about similarities in fluoride absorption when it comes from NaF or disodium monofluorophosphate (Na_2PO_3F, SMFP) (Whitford *et al.*, 1990). The fact is that the absorption of fluoride from SMFP requires phosphatases-mediated enzymatic hydrolysis, leading to lower and delayed peak plasma fluoride levels compared to those seen from NaF. However, the bioavailability of fluoride from both compounds is roughly the same (Buzalaf *et al.*, 2008).

4.4 Fluoride Distribution

After ingestion, plasma fluoride levels rapidly increase until a peak is reached (around 20–60 min). From then on, the levels start declining to baseline levels within 3 to 11 h, depending on the dose that is ingested (Whitford, 1996). Blood plasma is considered the central compartment from which fluoride is distributed to soft and mineralized tissues and specialized body fluids, followed by its elimination or clearance (Whitford, 1996). In soft tissues, fluoride establishes a steady-state distribution between extracellular and intracellular fluids. Variations in the plasma fluoride concentrations lead to proportional alterations in intracellular compartments (Ekstrand, 1996). In mineralized tissues, fluoride is reversibly incorporated mainly in bone, being released back to plasma during bone remodeling. Hereafter, detailed aspects related to each compartment involved in the fluoride distribution will be described.

4.4.1 Plasma Fluoride

Plasma fluoride is found both as inorganic (ionic or free fluoride) and organic (nonionic) forms and together they represent the total of plasma fluoride (Ekstrand, 1996; Whitford, 1996). Inorganic fluoride is not bound to plasmatic macromolecules and can be quantified by a specific electrode. Its biological importance in dentistry, medicine and public health is well established. Instead, organic fluoride is composed chiefly of different types of lipid-like molecules that bind to plasmatic proteins. Although present in greater amounts, the biological role of organic fluoride is still unknown. In contrast to other ions, such as calcium, sodium, phosphorus and chlorine, plasma fluoride is not homeostatically regulated but reflects the balance between the amount and time of intake, absorption, distribution and uptake, and renal filtration and excretion (Ekstrand, 1996; Whitford, 1996). In humans living in areas with fluoridated water (containing 1 ppm F or ~52.6 $\mu mol\ L^{-1}$), plasma fluoride concentrations ranged between 0.5 and 1.5 $\mu mol\ L^{-1}$ (Whitford, 1996). Thus, although plasma fluoride levels are modulated by several physiological factors regardless of fluoride intake, they constitute an import ant biomarker of past and present exposures (Ohmi *et al.*, 2005).

4.4.2 Fluoride in Soft Tissues

The rate of fluoride distribution from plasma to soft tissues is determined by the velocity of blood flow to different tissues and organs. It is influenced by both the magnitude and direction of transmembrane pH gradient. This means that the proportion of fluoride levels in plasma or tissue is dependent on the ratio between the extracellular pH ($_{e}$pH) and intracellular pH ($_{i}$pH). It is important to consider that fluoride accumulates in the more alkaline compartment in response to a pH gradient (Figure 4.1) and that the extracellular fluid is usually more alkaline than the cytosol of mammalian cells. Thus, it is reasonable to expect higher fluoride concentration in plasma and extracellular compared than those found in intracellular fluids (ranging between 10 and 50%) (Whitford *et al.*, 1979). Moreover, as the pH gradient between membranes can be changed by altering extracellular pH, a flux of fluoride into and out of cells may be achieved. When $_{e}$pH increases, fluoride leaves the cell. Inversely, when $_{e}$pH decreases, fluoride goes into the cell. This is the basis for the alkalinization in cases of acute fluoride toxicity, in order to promote releasing of fluoride out of cells (Whitford *et al.*, 1979). Besides pH gradient control, a complex molecular mechanism involved in fluoride toxicity was recently described. It was shown that many species present fluoride RNA-based riboswitches to control the expression of proteins that soften the deleterious effects of this anion (Baker *et al.*, 2012). Thus, differential degrees of fluoride toxicity may be due to differences in expression and/or usage of those fluoride-sensing RNAs, rather than only the pH gradient.

Fluoride concentrations were screened and compared among tissues and organs using ^{18}F, revealing that the kidney presents the greatest ratio in fluoride concentration, followed by liver, lung, spleen and others (Whitford *et al.*, 1979). This may explain why these organs are targets for the fluoride-mediated toxicity (Guo *et al.*, 2003; Pereira *et al.*, 2013). In this context, some compounds have been used to ameliorate the fluoride-driven toxic response by decreasing its accumulation in soft tissues and organs (Blaszczyk *et al.*, 2012). Reduced distribution of fluoride in soft tissue is accompanied of a greater fluoride uptake in bone by uncertain mechanisms (Blaszczyk *et al.*, 2012).

4.4.3 Fluoride in Specialized Body Fluids

Specialized body fluids do not present the same fluoride concentration observed in plasma but changes in fluoride levels are directly modulated according to plasma concentration. From those, cerebrospinal fluid presents 50% less fluoride from those seen in plasma (Whitford, 1996). Moreover, gingival crevicular present fluoride levels in a slightly greater fashion compared to plasma (Whitford, 1994). Instead, concentrations in parotid and submandibular ductal saliva are slightly decreased, showing ductal saliva and plasma fluoride concentrations ratios around 0.8–0.9 (Whitford, 1994). Similarities with plasma allowed them to be used as a biomarker of fluoride exposure aiming to analyze fluoride bioavailability from fluoridated products (Olympio *et al.*, 2007). Fluoride levels in whole saliva are predominantly more variable and higher compared to ductal saliva because they are often contaminated by exogenous fluoride sources such as food, water and fluoridated dental products (Whitford *et al.*, 1999).

4.4.4 Fluoride in Mineralized Tissues

About 99% of total fluoride retained in the organism is found in calcified tissues, mainly in bone. Enamel and dentin also contribute but in a lesser extent (Whitford, 1994). Age seems to be a factor that alters fluoride concentration in bone. About 36% and 55% of the fluoride daily absorbed in incorporated in healthy adults (18–75 years) and children (<7 years), respectively (Villa *et al.*, 2010). This is due to the continuous fluoride intake throughout life (Richards *et al.*, 1994). Chronic intake of fluoride negatively impacts mineralized tissues, promoting dental and skeletal fluorosis dependent on dose, time and duration of exposure. Controversy persists concerning the impact of community water fluoridation on bone health in adults, and few studies have assessed relationships with bone at younger ages. Recently, bone in adolescents measurements were performed after life-long fluoride intake through a prospective cohort study (Levy *et al.*, 2014). Daily fluoride intake was 0.66 mg for females and 0.78 mg for males and no significant relationships between daily fluoride intake and adolescents' bone measures we found. The findings suggest that fluoride exposures at the typical levels for most US adolescents in fluoridated areas do not have significant effects on bone mineral measures (Levy *et al.*, 2014).

Retention of fluoride in bone is not uniform and is more pronounced in the periosteal and endosteal regions and in cancelous rather than contact bone (Richards *et al.*, 1994). Bone fluoride is divided into surface and inner compartments. The first is rapidly exchangeable while the latter is not. In surface and inner bone, fluoride uptake occurs in different steps. First, iso- and heteroionic exchanges on the hydration shells of bone crystallites take place in contact with extracellular fluids. This proximity allows a steady-state relationship between the extracellular fluid (that represents physiological plasma fluoride concentration) and hydration shells of bone regulating fluoride concentration. This characteristic corroborates the use of bone surface as a terminal biomarker of acute fluoride exposure (Buzalaf *et al.*, 2005). Secondly, fluoride incorporates into precursors of hydroxyfluorapatite followed by the apatitic lattice itself (Neuman and Neuman, 1958).

Fluoride associated with the nonexchangeable inner compartment, is irreversibly bound. Instead, it is continuous mobilized during bone remodeling and bone absorption (Rao *et al.*, 1995).

Levels of fluoride in dentin are quite similar to bone and increase after long-term intake. Higher concentration is found in pulp from which it starts to reduce as it approaches the dentin-enamel junction (Hallsworth and Weatherell, 1969). Enamel fluoride content also enhances with age but is comparatively lower and unrelated to dentin (Vieira *et al.*, 2006). Usually, fluoride concentration in tooth enamel reflects fluoride exposure during amelogenesis (Hallsworth and Weatherell, 1969). However, although enamel is the target tissue for dental fluorosis, correlations between fluoride content in enamel and fluorosis occurrence were not found (Vieira *et al.*, 2006).

4.5 Kidney Excretion

Fluoride renal excretion is one of the most important mechanisms for the regulation of fluoride levels in the body (Buzalaf and Whitford, 2011). About 60% and 45% of the daily ingested fluoride is excreted in urine of healthy adults and children, respectively (Villa *et al.*, 2010). Increases in urinary fluoride were observed in children aged 5–8 years living in a nonfluoridated community after using fluoride containing-dental varnishes (Garcia-Hoyos *et al.*, 2012). Thus, plasma and the kidney excretion rate constitutes the physiologic balance determined by fluoride intake, uptake to and removal from bone and the capacity of fluoride clearance by the kidney. Fluoride is freely filtered through the glomerulus and undergoes a variable degree of proximal tubular reabsorption, directly related to glomerular ultrafiltrate pH (Whitford, 1994). After fluoride enters the renal tubules, a variable amount (from 10 to 90%) is reabsorbed to systemic circulation, depending on the urinary pH because transmembrane migration occurs by diffusion of HF (Whitford *et al.*, 1976). When the pH of tubular fluid is relatively high, the amount of fluoride as HF is lower, while it is higher as F^-. Thus, low fluoride concentration is available to be reabsorbed crossing the epithelium of the epithelium renal tubule into the interstitial fluid. Instead, the majority of nonreabsorbed is excreted in urine as F^- (Ekstrand, 1996) (Figure 4.4). In addition, any factor that affects urinary pH will have an impact on the amount of fluoride that is excreted in urine (Buzalaf and Whitford, 2011). Urinary F excretion is also influenced by the glomerular filtration rate since its reduction, as occurs in chronic renal dysfunction as well as in the last decades of life, results in lower excretion and increased plasma F levels (Ekstrand, 1996).

Fluoride causes kidney injury seen by tubular dysfunction at plasma concentrations of fluoride 0.625 μg ml^{-1} and renal cell apoptosis (Bai *et al.*,

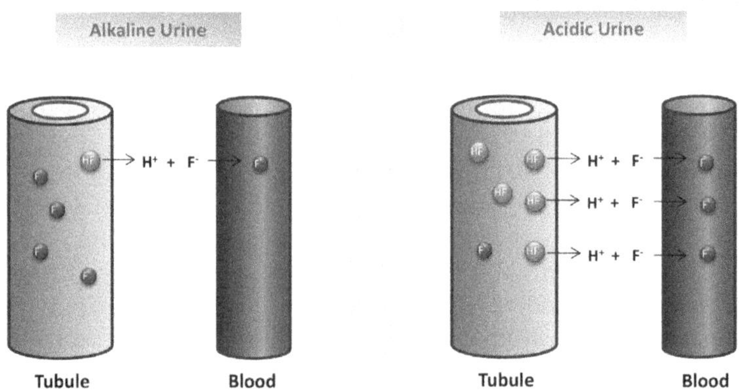

Figure 4.4 Fluoride reabsorption from renal tubule to systemic circulation. The extent of fluoride reabsoption is depedent on urine pH. When the urine is alkaline, the concentration of HF, the most permeable form of fluoride, is lower and most of the fluoride remains in the tubule to be excreted (on the left). Conversely, when the urine is acidic, the concentration of HF is higher that crosses to the renal tubule membrane towards interstitium. Herein, it dissociates forming F^- that diffuses into the peritubular capillaries returning to systemic circulation (on the right).

2010). Several urine markers such as urinary kidney injury molecule (Kim-1), clusterin (Clu), osteopontin, β-2-microglobulin, cystatin-C and heat shock protein (HSP)-72 were shown to be upmodulated in 50 ppm fluoride-treated rats compared to untreated rats (Cardenas-Gonzalez *et al.*, 2013), while a toxicity marker (aflatoxin B1 aldehyde reductase) was up-modulated (Kobayashi *et al.*, 2011).

4.6 Modulators of Fluoride Metabolism and Their Implications

Any systemic, metabolic and genetic alteration can modify fluoride metabolism. They interfere with absorption, excretion and so, with fluoride fate in the body. Imbalances may lead to pathological conditions such as acute and chronic toxicities, dental and skeletal fluorosis. They include acid–base disturbances, physical activity, circadian rhythm and hormones (Whitford, 1996). Other factors such as nutritional status, diet composition, renal impairment and genetic predisposition modify fluoride metabolism. They are described in detail below.

4.6.1 Acid–Base Disturbances

Considering that urinary pH is a limitant factor in determining fluoride reabsorption and/or renal excretion (Figure 4.4), it seems obvious that acid–base disturbances promote variations in fluoride tissue concentration. The acid–base disturbances are usually induced by diet composition. In this sense, vegetarian-based diet turns towards alkaline pH, while a meat-rich diet turns towards acidic pH. Acute respiratory disorders similarly impact renal excretion as well as metabolic disorders (Whitford, 1996).

4.6.2 Physical Activity, Circadian Rhythm and Hormones

Physical activity is associated with either decreased or increased circulating fluoride levels (Whitford, 1996). Prolonged exercise causes enhancement of diffusion of HF from extracellular to intracellular fluid in skeletal muscle cells, a phenomenon caused by reduction in the pH gradient between cell membranes.

A biological rhythmicity of fluoride was found to be opposite to that of calcium and phosphate, discouraging the role of bone as a regulator. The mean fluoride peak is reached around 9 a.m., and decreased around 9 p.m in animals (Whitford, 1996). In humans, plasma fluoride concentrations were positively correlated with urinary fluoride excretion rates and serum parathormone levels, suggesting that both the renal system and hormones regulate fluoride rhythmicity (Cardoso *et al.*, 2008). In agreement with this, Villa *et al.* (2008) demonstrated that fluoride urinary excretion is lower in the diurnal compared to nocturnal period.

4.6.3 Nutritional Status

There is a consensus that malnutrition and dental fluorosis prevalence are related. However, this association still needs to be scientifically proven. The lack of evidence so far is due to difficulties in data interpretations. A relationship between water fluoride concentration, socioeconomic status, nutritional status and prevalence of enamel lesions in child from Saudi Arabia was demonstrated (Rugg-Gunn *et al.*, 1997). In Brazil, using assessment of malnutrition by height-for-age and weight-for-age indexes, dental fluorosis was shown to be dependent on water fluoride concentration but independent of malnutritional status (Correia Sampaio *et al.*, 1999). To clarify this controversy, longitudinal studies should be conducted where variables such as nutritional status, diet composition and fluoride intake are all implicated and controlled during tooth formation.

4.6.4 Diet Composition

Diet that interferes in the acidification and alkalinization of urine promotes significant differences in urinary fluoride clearance and plasma half-lives. As described above, this principle applies for protein-rich and vegetarian diets. Long-term diet-induced changes in urinary pH could decrease (vegetarian-induced alkaline urine) or increase (protein-induced acidic urine) the risk of dental fluorosis (Whitford, 1997). Thus, vegetarianism is inversely related to dental fluorosis prevalence (Awadia *et al.*, 1999).

High dietary concentrations of calcium also reduce fluoride absorption (Whitford, 1994). Its impact over dental fluorosis prevalence was also evidenced (Chen *et al.*, 1997). The rate of dental fluorosis of a milk-drinking group was 7.2%, whereas that of the nonmilk-drinking group was 37.5% (Chen *et al.*, 1997). This and other data support the use of calcium supplementation in endemic fluorosis areas.

Regular diets also seem to regulate fluoride retention in the body. Intake of sorghum was associated with higher fluoride-induced toxicity compared to rice- and wheat-based diets. These data were accompanied by less and more fluoride urinary excretion, respectively (Lakshmaiah and Srikantia, 1977). Also, a multigrain diet enriched with protein was showed to attenuate fluorosis toxicity (Vasant and Amaravadi, 2013).

4.6.5 Renal Impairment

Greater plasma fluoride was observed in patients with renal failure (Schmidt and Leuschke, 1986). However, no differences were observed in plasma insulin level, as a result of fluoride intake, when control and renal-deficient rats were compared (Lupo *et al.*, 2011). Moreover, renal impairment in children has been associated with tooth defects that include enamel pitting and hypoplasia. Intake of fluoride by nephrectomized rats impaired fluoride clearance from plasma and worsened tooth development (Lyaruu *et al.*, 2008). This relationship was also demonstrated in humans (Farge *et al.*, 2006).

4.6.6 Genetic Predisposition

Increasing evidence has led to the affirmative that dental fluorosis is genetically influenced (Polzik *et al.*, 1994). This is clarified through observation that not all children living in the same fluoridated area develop dental fluorosis (Yoder *et al.*, 1998). The genetic character is reinforced when differences in urinary fluoride levels and meal fluoride values are not found among these children. The genetic predisposition to dental fluorosis is also well established in mice (Everett *et al.*, 2002; Everett, 2011). Examination of 12 inbred strains of mice showed differences in susceptibility to dental fluorosis. The A/J strain is "susceptible", with a rapid onset and severe development of dental fluorosis, while the 129P3/J is "resistant", with minimum development of dental fluorosis (Everett *et al.*, 2002). These strains also differ regarding their susceptibilities to the effects of fluoride in bone (Mousny *et al.*, 2006; Everett, 2011). To determine whether such differences were due to differences in F metabolism, a metabolic study was conducted in which total F intake and excretion were measured. The data showed that, compared to A/J mice, 129P3/J mice ingested less water, excreted less urine, had lower urinary fluoride excretion and consequently had higher F retention and plasma and femur fluoride levels (Carvalho *et al.*, 2009). Despite this, the amelogenesis in 129P3/J mice was remarkably unaffected by fluoride (Carvalho *et al.*, 2009). Recently, the molecular mechanisms underlying the renal fluoride metabolism in A/J and 129P3/J mice that may account for their differential metabolic handling of fluoride were determined using a proteomic approach (Carvalho *et al.*, 2013). In F-treated groups, Na(+)/H(+) exchange regulatory cofactor NHE-RF3, a protein involved in the regulation of renal tubular reabsorption capacity was downmodulated in the kidney of 129P3/J mice (Carvalho *et al.*, 2013). Exclusive proteins expressed in A/J or 129P3/J mice exhibited the same profile, regardless of exposure to fluoride, suggesting that genetic background *per se* accounts for such differences between these two strains of mice. In addition, some proteins that are codified by chromosomes 2 and 11, loci previously characterized to determine susceptibility and resistance to dental fluorosis in A/J and 129P3/J mice, respectively (Everett *et al.*, 2009; Everett *et al.*, 2011), were found and may disclose the molecular mechanism of dental fluorosis (Carvalho *et al.*, 2013).

Besides, several gene polymorphisms have been addressed to explore possible correlation with dental fluorosis in humans. They include parathyroid hormone (Wen *et al.*, 2012), myeloperoxidase (Zhang *et al.*, 2013), collagen1A2 (Huang *et al.*, 2008) and estrogen receptor (Ba *et al.*, 2011) gene polymorphisms. Except for parathyroid hormone gene, the others showed an association between polymorphisms and risk and/or development of dental fluorosis.

4.7 Conclusion

Fluoride amounts in the body are tightly regulated and have a direct impact on cellular and systemic levels. It is clear that the beneficial as well as the adverse impact can be attributed to the magnitude and duration of fluoride exposure. In addition to the individual physiological differences in handling fluoride, fluoride levels are also modulated by environmental, biochemical, physiological, pathological and molecular factors. In the past few years, this area of research was aimed to understand the molecular mechanisms involved in fluoride metabolism and its biological effects. Although considerable progress has been achieved, there is still much to be explored.

Summary Points

- Fluoride is a cariostatic agent and is implemented in several systemic and local sources.
- Fluoride presents both positive and negative impacts on hard tissues depend on the dose and time of exposure.
- Fluoride metabolism includes its absorption in the gastrointestinal tract, distribution through soft and hard tissues followed by renal excretion.
- The coefficient of permeability to HF is much higher compared to ionic fluoride.
- Absorption of fluoride in the stomach occurs in a pH-dependent manner and the tubular reabsorption depends on the urinary pH.
- Plasma is an important biomarker of fluoride exposure.
- Most fluoride absorbed is incorporated in bone.
- Several conditions affect fluoride metabolism including acid–base disturbances, physical activity, circadian rhythm and hormones, nutritional status, diet composition, renal impairment and genetic predisposition.

Definitions and Explanation of Key Terms

Systemic toxicity: Requires absorption and distribution of the toxicant to a site distant from its entry point, at which point effects are produced. Most chemicals that produce systemic toxicity do not cause a similar degree of toxicity in all organs, but usually demonstrate major toxicity to one or two organs.

Coefficient of permeability: A rate of particle diffusion through cell membranes that depends on the particle and total area through which it traverses.

pH gradient: A gradient of hydrogen ion concentration between extracellular and intracellular compartments.

Weak acid: An acid that is partially dissociated in an aqueous solution. It is measured by the acid dissociation constant (K_a).

Steady-state distribution: A condition of a physical system or device that does not change over time, or in which any one change is continually balanced by another, such as the stable condition of a system in equilibrium.

Fluc proteins-based channel: Fluoride ion exporters formed by a group of small membrane proteins known as the "crcB" family or Fluc, a protein that is widespread among unicellular organisms.

CLC anion-transporting proteins: A subclass of the widespread CLC superfamily of anion-transport proteins.

Fluoride exporter named FEX: A widespread family of fluoride export proteins that mediated eukaryotic resistance to fluoride toxicity.

Tight junctions: Form the closest contact between adjacent cells known in nature.

Bone remodeling: A lifelong process where mature bone tissue is removed from the skeleton, a process called bone resorption, and new bone tissue is formed, a process called ossification or new bone formation.

RNA-based riboswitches: Riboswitches are specific components of an mRNA molecule that regulates gene expression. The riboswitch is a part of an mRNA molecule that can bind and target small target molecules.

Glomerular filtration rate: The flow rate of filtered fluid through the kidney.

Amelogenesis: The formation ofenamel onteeth and begins when the crown is forming during the bell stage oftooth development after dentinogenesis, which is the formation of dentin. It has three stages: the inductive, the secretory, and the maturation stages.

Dental fluorosis: A developmental disturbance of dental enamel caused by excessive exposure to high concentrations of fluoride during tooth development. It presents different appearances, varying from tiny white streaks or specks in the enamel to discoloration or brown markings in the tooth. The enamel may be pitted and rough.

Skeletal fluorosis: A bone disease caused by excessive consumption of fluoride that causes pain and damage to bones and joints.

List of Abbreviations

HF	hydrogen fluoride
pK_a	the logarithmic form of acid dissociation constant
F^-	fluoride ion
NaF	sodium fluoride
SMFP	disodium monofluorophosphate
$_{e}pH$	extracellular pH
$_{i}pH$	intracellular pH

References

Awadia, A. K., Haugejorden, O., Bjorvatn, K. and Birkeland, J. M., 1999. Vegetarianism and dental fluorosis among children in a high fluoride area of northern Tanzania. *International Journal of Paediatric Dentistry.* **9**: 3–11.

Ba, Y., Zhang, H., Wang, G., Wen, S., Yang, Y., Zhu, J., Ren, L., Yang, R., Zhu, C., Li, H., Cheng, X. and Cui, L., 2011. Association of dental fluorosis with polymorphisms of estrogen receptor gene in Chinese children. *Biological Trace Element Research.* **143**: 87–96.

Bai, C., Chen, T., Cui, Y., Gong, T., Peng, X. and Cui, H. M., 2010. Effect of high fluorine on the cell cycle and apoptosis of renal cells in chickens. *Biological Trace Element Research.* **138**: 173–180.

Baker, J. L., Sudarsan, N., Weinberg, Z., Roth, A., Stockbridge, R. B. and Breaker, R. R., 2012. Widespread genetic switches and toxicity resistance proteins for fluoride. *Science.* **335**: 233–235.

Blaszczyk, I., Birkner, E., Gutowska, I., Romuk, E. and Chlubek, D., 2012. Influence of methionine and vitamin E on fluoride concentration in bones and teeth of rats exposed to sodium fluoride in drinking water. *Biological Trace Element Research.* **146**: 335–339.

Bratthall, D., Hansel-Petersson, G. and Sundberg, H., 1996. Reasons for the caries decline: what do the experts believe? *European Journal of Oral Sciences.* **104**: 416–422.

Buzalaf, M. A., Caroselli, E. E., de Carvalho, J. G., de Oliveira, R. C., da Silva Cardoso, V. E. and Whitford, G. M., 2005. Bone surface and whole bone as biomarkers for acute fluoride exposure. *Journal of Analytical Toxicology.* **29**: 810–813.

Buzalaf, M. A. R., Leite, A. L., Carvalho, N. T. A., Rodrigues, M. H. C., Takamori, E. R., Niconielo, D. B., Levy, F. M. and Cardoso, V. E. S., 2008. Bioavailability of fluoride administered as sodium fluoride or monofluorophosphate to humans. *Journal of Fluorine Chemistry.* **129**: 691–694.

Buzalaf, M. A. and Whitford, G. M., 2011. Fluoride metabolism. *Monographs in Oral Sciences.* **22**: 20–36.

Cardenas-Gonzalez, M. C., Del Razo, L. M., Barrera-Chimal, J., Jacobo-Estrada, T., Lopez-Bayghen, E., Bobadilla, N. A. and Barbier, O., 2013. Proximal renal tubular injury in rats sub-chronically exposed to low fluoride concentrations. *Toxicology and Applied Pharmacology.* **272**: 888–894.

Cardoso, V. E. S., Whitford, G. M., Aoyama, H. and Buzalaf, M. A. R., 2008. Daily variations in human plasma fluoride concentrations. *Journal of Fluorine Chemistry.* **129**: 1193–1198.

Carvalho, J. G., Leite Ade, L., Peres-Buzalaf, C., Salvato, F., Labate, C. A., Everett, E. T., Whitford, G. M. and Buzalaf, M. A., 2013. Renal proteome in mice with different susceptibilities to fluorosis. *PLoS One.* **8**: e53261.

Carvalho, J. G., Leite, A. L., Yan, D., Everett, E. T., Whitford, G. M. and Buzalaf, M. A., 2009. Influence of genetic background on fluoride metabolism in mice. *Journal of Dental Research.* **88**: 1054–1058.

Chen, Y. X., Lin, M. Q., Xiao, Y. D., Gan, W. M., Min, D. and Chen, C., 1997. Nutrition survey in dental fluorosis-afflicted areas. *Fluoride.* **30**: 77–80.

Correia Sampaio, F., Ramm von der Fehr, F., Arneberg, P., Petrucci Gigante, D. and Hatloy, A., 1999. Dental fluorosis and nutritional status of 6- to 11-year-old children living in rural areas of Paraiba, Brazil. *Caries Research.* **33**: 66–73.

Ekstrand, J. (ed.), 1996. Fluoride Metabolism. Fluoride in Dentistry. Copenhagen, Munksgaard, 13 pp.

Everett, E. T., 2011. Fluoride's effects on the formation of teeth and bones, and the influence of genetics. *Journal of Dental Research.* **90**: 552–560.

Everett, E. T., McHenry, M. A., Reynolds, N., Eggertsson, H., Sullivan, J., Kantmann, C., Martinez-Mier, E. A., Warrick, J. M. and Stookey, G. K., 2002. Dental fluorosis: variability among different inbred mouse strains. *Journal of Dental Research.* **81**: 794–798.

Everett, E. T., Yan, D., Weaver, M., Liu, L., Foroud, T. and Martinez-Mier, E. A., 2009. Detection of dental fluorosis-associated quantitative trait Loci on mouse chromosomes 2 and 11. *Cells Tissues Organs.* **189**: 212–218.

Everett, E. T., Yin, Z., Yan, D. and Zou, F., 2011. Fine mapping of dental fluorosis quantitative trait loci in mice. *European Journal of Oral Sciences.* **119**: 8–12.

Farge, P., Ranchin, B. and Cochat, P., 2006. Four-year follow-up of oral health surveillance in renal transplant children. *Pediatric Nephrology.* **21**: 851–855.

Garcia-Hoyos, F., Barberia, E., Garcia-Camba, P. and Varela, M., 2012. Renal fluoride excretion in children following topical application of fluoride varnish. *European Journal Paediatric Dentistry.* **13**: 280–284.

Guo, X., Sun, G. and Sun, Y., 2003. Oxidative stress from fluoride induced hepatotoxicity in rats. *Fluoride.* **36**: 25–29.

Gutknecht, J. and Walter, A., 1981. Hydrofluoric and nitric acid transport through lipid bilayer membranes. *Biochimica et Biophysica Acta.* **644**: 153–156.

Hallsworth, A. S. and Weatherell, J. A., 1969. The microdistribution, uptake and loss of fluoride in human enamel. *Caries Research.* **3**: 109–118.

Huang, H., Ba, Y., Cui, L., Cheng, X., Zhu, J., Zhang, Y., Yan, P., Zhu, C. and Kilfoy, B., 2008. COL1A2 gene polymorphisms (Pvu II and Rsa I), serum calciotropic hormone levels, and dental fluorosis. *Community Dentistry and Oral Epidemiology.* **36**: 517–522.

Kobayashi, C. A., Leite Ade, L., da Silva, T. L., dos Santos, L. D., Nogueira, F. C., Santos, K. S., de Oliveira, R. C., Palma, M. S., Domont, G. B. and Buzalaf, M. A., 2011. Proteomic analysis of urine in rats chronically exposed to fluoride. *Journal of Biochemical Molecular Toxicology.* **25**: 8–14.

Kobayashi, C. A. N., Leite, A. L., Peres-Buzalaf, C., Carvalho, J. G., Whitford, G. M., Everett, E. T, Siqueira, W. L. and Buzalaf, M. A., 2014. Bone response to fluoride exposure is influenced by genetics. *PlosOne.* **9**: e114343.

Lakshmaiah, N. and Srikantia, S. G., 1977. Fluoride retention in humans on sorghum and rice based diets. *Indian Journal of Medical Research.* **65**: 543–548.

Levy, S. M., Warren, J. J., Phipps, K., Letuchy, E., Broffitt, B., Eichenberger-Gilmore, J., Burns, T. L., Kavand, G., Janz, K. F., Torner, J. C. and Pauley, C. A., 2014. Effects of life-long fluoride intake on bone measures of adolescents: a prospective cohort study. *Journal of Dental Research.* **93**: 353–359.

Li, L., 2003. The biochemistry and physiology of metallic fluoride: action, mechanism, and implications. *Critical Reviews in Oral Biology & Medicine.* **14**: 100–114.

Li, S., Smith, K. D., Davis, J. H., Gordon, P. B., Breaker, R. R. and Strobel, S. A., 2013. Eukaryotic resistance to fluoride toxicity mediated by a widespread family of fluoride export proteins. *Proceedings of the National Academy of Sciences.* **110**: 19018–19023.

Lupo, M., Buzalaf, M. A. and Rigalli, A., 2011. Effect of fluoridated water on plasma insulin levels and glucose homeostasis in rats with renal deficiency. *Biological Trace Element Research.* **140**: 198–207.

Lyaruu, D. M., Bronckers, A. L., Santos, F., Mathias, R. and DenBesten, P., 2008. The effect of fluoride on enamel and dentin formation in the uremic rat incisor. *Pediatric Nephrology.* **23**: 1973–1979.

Messer, H. H., Nopakum, J. and Rudney, J. D., 1989. Influence of pH on intestinal fluoride transport in vitro. *Journal of the Dental Association of Thailand.* **39**: 226–232.

Mousny, M., Banse, X., Wise, L., Everett, E. T., Hancock, R., Vieth, R., Devogelaer, J. P. and Grynpas, M. D., 2006. The genetic influence on bone susceptibility to fluoride. *Bone.* **39**: 1283–1289.

Neuman, W. F. and Neuman, M. W., 1958. *The Chemical Dynamics of Bone Mineral.* University of Chicago Press, Chicago, USA, 209 pp.

Nopakun, J., Messer, H. H. and Voller, V., 1989. Fluoride absorption from the gastrointestinal tract of rats. *Journal of Nutrition.* **119**: 1411–1417.

Ohmi, K., Nakagaki, H., Tsuboi, S., Okumura, A., Sugiyama, T., Thuy, T. T. and Robinson, C., 2005. The effect of fluoridation and its discontinuation on fluoride profiles in the alveolar bone of rat. *Calcified Tissue International.* **77**: 226–232.

Olympio, K. P., Bardal, P. A., Cardoso, V. E., Oliveira, R. C., Bastos, J. R. and Buzalaf, M. A., 2007. Low-fluoride dentifrices with reduced pH: fluoride concentration in whole saliva and bioavailability. *Caries Research.* **41**: 365–370.

Pereira, H. A., Leite, A. D., Charone, S., Lobo, J. G., Cestari, T. M., Peres-Buzalaf, C. and Buzalaf, M. A., 2013. Correction: Proteomic Analysis of Liver in Rats Chronically Exposed to Fluoride. *PLoS One.* **8**: e75343.

Polzik, E. V., Zinger, V. E., Valova, G. A., Kazantsev, V. S. and Yakusheva, M. Y., 1994. A method for estimating individual predisposition to occupational fluorosis. *Fluoride.* **27**: 194–200.

Rao, H. V., Beliles, R. P., Whitford, G. M. and Turner, C. H., 1995. A physiologically based pharmacokinetic model for fluoride uptake by bone. *Regulatory Toxicology and Pharmacology.* **22**: 30–42.

Richards, A., Mosekilde, L. and Sogaard, C. H., 1994. Normal age-related changes in fluoride content of vertebral trabecular bone–relation to bone quality. *Bone.* **15**: 21–26.

Rugg-Gunn, A. J., al-Mohammadi, S. M. and Butler, T. J., 1997. Effects of fluoride level in drinking water, nutritional status, and socio-economic status on the prevalence of developmental defects of dental enamel in permanent teeth in Saudi 14-year-old boys. *Caries Research.* **31**: 259–267.

Schmidt, C. W. and Leuschke, W., 1986. Metabolism of fluorine in terminal renal failure. *Zeitschrift fuer Urologie und Nephrologie.* **79**: 481–484.

Stockbridge, R. B., Lim, H. H., Otten, R., Williams, C., Shane, T., Weinberg, Z. and Miller, C., 2012. Fluoride resistance and transport by riboswitch-controlled CLC antiporters. *Proceedings of the National Academy of Sciences.* **109**: 15289–15294.

Stockbridge, R. B., Robertson, J. L., Kolmakova-Partensky, L. and Miller, C., 2013. A family of fluoride-specific ion channels with dual-topology architecture. *Elife.* **2**: e01084.

Trautner, K. and Siebert, G., 1986. An experimental study of bio-availability of fluoride from dietary sources in man. *Archives of Oral Biology.* **31**: 223–228.

Vasant, R. A. and Amaravadi, V. R. L. N., 2013. A multigrain protein enriched diet mitigates fluoride toxicity. *Journal of Food Sciences and Technology.* **50**: 528–534.

Vieira, A. P., Hancock, R., Dumitriu, M., Limeback, H. and Grynpas, M. D., 2006. Fluoride's effect on human dentin ultrasound velocity (elastic modulus) and tubule size. *European Journal of Oral Sciences.* **114**: 83–88.

Villa, A., Anabalon, M., Cabezas, L. and Rugg-Gunn, A., 2008. Fractional urinary fluoride excretion of young female adults during the diurnal and nocturnal periods. *Caries Research.* **42**: 275–281.

Villa, A., Anabalon, M., Zohouri, V., Maguire, A., Franco, A. M. and Rugg-Gunn, A., 2010. Relationships between fluoride intake, urinary fluoride excretion and fluoride retention in children and adults: an analysis of available data. *Caries Research.* **44**: 60–68.

Wen, S., Li, A., Cui, L., Huang, Q., Chen, H., Guo, X., Luo, Y., Hao, Q., Hou, J. and Ba, Y., 2012. The relationship of PTH Bst BI polymorphism, calciotropic hormone levels, and dental fluorosis of children in China. *Biological Trace Element Research.* **147**: 84–90.

Whitford, G. M., 1994. Intake and metabolism of fluoride. *Advances in Dental Research.* **8**: 5–14.

Whitford, G. M., 1996. The metabolism and toxicity of fluoride. *Monographs in Oral Science.* **16**: 1–153.

Whitford, G. M., 1997. Determinants and mechanisms of enamel fluorosis. *Ciba Foundation Symposium.* **205**: 226–241.

Whitford, G. M., Birdsong-Whitford, N. L. and Finidori, C., 1990. Acute oral toxicity of sodium fluoride and monofluorophosphate separately or in combination in rats. *Caries Research.* **24**: 121–126.

Whitford, G. M. and Pashley, D. H., 1984. Fluoride absorption: the influence of gastric acidity. *Calcified Tissue International.* **36**: 302–307.

Whitford, G. M., Pashley, D. H. and Reynolds, K. E., 1979. Fluoride tissue distribution: short-term kinetics. *American Journal of Physiology.* **236**: F141–F148.

Whitford, G. M., Pashley, D. H. and Stringer, G. I., 1976. Fluoride renal clearance: a pH-dependent event. *American Journal of Physiology.* **230**: 527–532.

Whitford, G. M., Thomas, J. E. and Adair, S. M., 1999. Fluoride in whole saliva, parotid ductal saliva and plasma in children. *Archives of Oral Biology.* **44**: 785–788.

Yoder, K. M., Mabelya, L., Robison, V. A., Dunipace, A. J., Brizendine, E. J. and Stookey, G. K., 1998. Severe dental fluorosis in a Tanzanian population consuming water with negligible fluoride concentration. *Community Dentistry and Oral Epidemiology.* **26**: 382–393.

Zhang, T., Shan, K. R., Tu, X., He, Y., Pei, J. J. and Guan, Z. Z., 2013. Myeloperoxidase activity and its corresponding mRNA expression as well as gene polymorphism in the population living in the coal-burning endemic fluorosis area in Guizhou of China. *Biological Trace Element Research.* **152**: 379–386.

Section III
Analysis

CHAPTER 5

Fluorine Determination in Milk, Tea and Water by High-Resolution, High-Temperature Molecular Absorption Spectrometry

SULEYMAN AKMAN*[a], BERNHARD WELZ[b], NIL OZBEK[a], AND ÉDERSON R. PEREIRA[b]

[a]Department of Chemistry, Faculty of Arts and Sciences, Istanbul Technical University, 34469 Maslak-Istanbul, Turkey; [b]Departamento de Química, Universidade Federal de Santa Catarina, 88040-900 Florianópolis, Santa Catarina, Brazil
*E-mail: akmans@itu.edu.tr

5.1 Introduction

The determination of nonmetals by atomic absorption spectrometry (AAS) is usually not possible because their main resonance lines are in the vacuum-UV region. However, in the 1970s, molecular absorption spectrometry (MAS) was proposed as a technique for the determination of nonmetals using equipment for line source atomic absorption spectrometry (LS AAS) (Dittrich and Townshend, 1986; Welz *et al.*, 2009, 2010). Some diatomic molecules of nonmetals, such as PO or CS, are formed spontaneously in flames or furnaces; alternately they might be "synthesized". For this purpose, a reagent (a

Food and Nutritional Components in Focus No. 6
Fluorine: Chemistry, Analysis, Function and Effects
Edited by Victor R Preedy

Published by the Royal Society of Chemistry, www.rsc.org

molecule-forming element, MFE) has to be added to form a diatomic molecule in the gas phase. Either a continuum source, such as a deuterium lamp, or a hollow cathode lamp (HCL), which emits a line that coincided with the molecular absorption band of the target molecule, were used to measure the absorbance. Since a suitable HCL with an appropriate emission line was not always available, and mostly because of serious background correction problems associated with this approach, LS MAS studies were limited.

After the introduction of commercial instrumentation for high-resolution continuum source atomic absorption spectrometry (HR-CS AAS), MAS studies became much easier and gained more and more importance. These instruments have a high-intensity continuum source lamp, a high-resolution double monochromator consisting of a quartz prism and an echelle grating monochromator, and a CCD array detector that provides a resolution of about 1.5 pm per pixel in the far UV at 200 nm and 3 pm at 400 nm. Using HR-CS AAS equipment, atomic or molecular absorption can be detected and measured at any wavelength between 190 and 900 nm.

In this chapter, after a brief discussion of different methods for the determination of fluorine, the principles of line source atomic absorption spectrometers and high-resolution continuum source atomic absorption spectrometers will be described. Determination of F by LS MAS and especially HR-CS MAS will be discussed. Finally, a detailed literature survey will be presented about F determination in water, milk and tea samples using HR-CS MAS.

5.2 Methods for the Determination of Fluorine/ Fluoride

Fluorine, strictly speaking fluoride, is very easily and commonly determined by fluoride ion selective electrodes (ISE). The instrument is cheap and works in a very similar way to a pH meter. It consists of a potentiometer and a fluoride selective electrode. For the analysis of fluoride by ISE, at first, the instrument is calibrated in the linear range simply upon immersing the electrode into the standard solution, and then the fluoride concentration in the sample is measured directly. However, using this technique, only free F ions (fluoride) rather than total fluorine in the aqueous phase can be measured. This means that covalently bonded F, and hence total F, cannot be detected by ISE. Moreover, the ionic strength of the solution has to be controlled (Harwood, 1969, Kissa, 1987; Hara and Huang, 1997; Shi *et al.*, 2003; Borjigin *et al.*, 2009).

Another technique for the determination of fluoride is ion chromatography (IC). Like ISE, it is specific only to free F ions in the aqueous phase, and requires special columns, ultrapure reagents, troublesome pretreatment of columns and careful ionic strength control. F in covalently bonded molecules cannot be determined. Finally, the instrument is quite expensive (Potter *et al.*, 1986; Michigami *et al.*, 1993; Yiping and Caiyun, 2010).

Other chromatographic techniques, such as high-performance liquid chromatography (HPLC), liquid chromatography–mass spectrometry (LC-MS), gas chromatography (GC) and gas chromatography–mass spectrometry (GC-MS) may be used to determine fluorine; however, only fluorine-containing molecules rather than free fluoride or total fluorine are detected, unless an appropriate sample preparation is applied. Nevertheless, it might be of interest to also determine the F-containing molecules. Finally, determination of F by chromatographic techniques requires appropriate standards; information on the analyte molecule in the library of the instrument, special columns, and the physical properties of the sample has to be suitable to the column material and the detector.

Since the ionization potential of fluorine is very high (17.42 eV) and emission lines for excited F atoms are located in the vacuum-UV range below 100 nm, inductively coupled plasma optical emission spectrometry (ICP-OES) is not suitable for F determination (either from its ionic lines or from emission lines of neutral atoms in the plasma) (Gleisner *et al.*, 2010). Inductively coupled plasma mass spectrometry (ICP-MS) is also not suitable for the determination of fluorine due to the high ionization potential.

5.3 Atomic Absorption Spectrometry

Atomic absorption spectrometry (AAS) is a technique based on the measurement of the absorption of radiation by atoms in the gas phase. AAS can be used for the quantitative determination of around 70 metals and metalloids. Basically, AAS is not suitable for the determination of nonmetals, such as the halogens, because their absorption lines are located in the vacuum-UV range, which is not accessible with conventional atomic absorption spectrometers.

5.3.1 Line Source Atomic Absorption Spectrometers

An atomic absorption spectrometer consists of four basic components, namely (i) an emission source; (ii) an atomizer (iii) a wavelength selector and (iv) a detector with the associated electronics. In LS AAS, as proposed by Walsh in 1955, the high resolution that is necessary for atomic absorption measurements is provided by a line radiation source that usually emits only the spectrum of the target element in the form of very narrow lines with half-widths of a few picometers (Walsh, 1955). This makes it possible to use a relatively low-resolution monochromator that only separates the analytical line from other lines emitted by the radiation source and an integrating detector that measures all the radiation passing the exit slit of the monochromator. Figure 5.1 shows schematically the setup for a line source atomic absorption spectrometer.

i. Emission Source: Hollow cathode lamps (HCL) and electrodeless discharge lamps (EDL) are the most common line sources. Unfortunately, for every element, a special lamp has to be used (although a few multielement HCL are manufactured for a limited number of elements in the same lamp), which makes any kind of simultaneous multielement determination impossible.

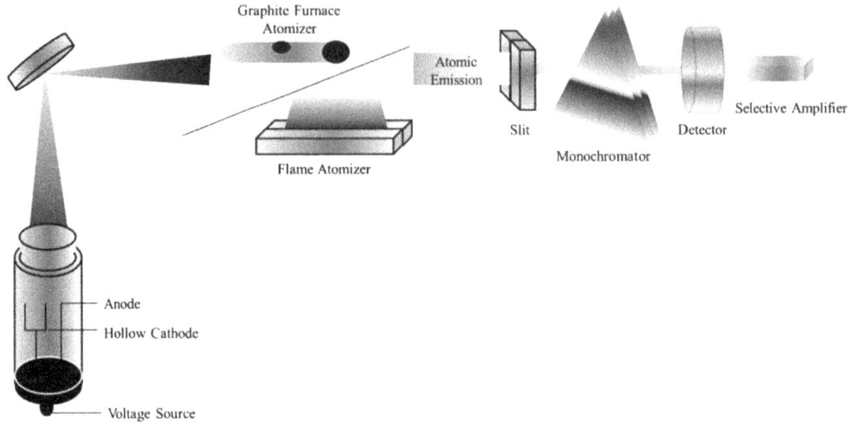

Figure 5.1 Schematic layout of a line source atomic absorption spectrometer.

ii. Atomizers: Two types of atomizers are typically used in AAS, a flame (FAAS) or a graphite tube atomizer (GFAAS).
For flame atomization, two types of flames, namely the air–acetylene and the nitrous oxide–acetylene flame are used, depending on the analyte. The average temperatures of the air–acetylene and nitrous oxide–acetylene flame are around 2300 °C and 2700 °C, respectively, depending on the fuel-to-oxidant ratio. The choice of the flame, its fuel-to-oxidant ratio, and the height of the radiation beam passing through the flame depend on the analyte and the sample matrix.
For a graphite furnace (or electrothermal) atomization, a measured volume (typically 10–50 μL) of a liquid or dissolved sample, or an accurately weighed mass (typically 0.1–1 mg of a solid sample is introduced into a graphite tube (typically 2 cm of length and 0.5 cm diameter). The tubes are heated by passing a low-voltage high current through the tube.
iii. Wavelength selector: In LS AAS, low-to-medium resolution monochromators with gratings having 800 to 2400 lines mm^{-1} are used to separate the analytical wavelength from other lines emitted by the radiation source.
iv. Detector: In LS AAS, photomultiplier tubes are used almost exclusively as detectors.

5.3.2 High-Resolution Continuum-Source Atomic Absorption Spectrometers

Research about AAS using continuum radiation sources (CS AAS) started in the USA in the 1960s, soon after the introduction of LS AAS by Walsh; however, the development of optical and electronic equipment had not yet reached a status to support this technique (Walsh, 1955; Fassel *et al.*, 1966). The breakthrough only occurred in the late 1990s, when Becker-Ross and colleagues described the first high-resolution continuum source atomic

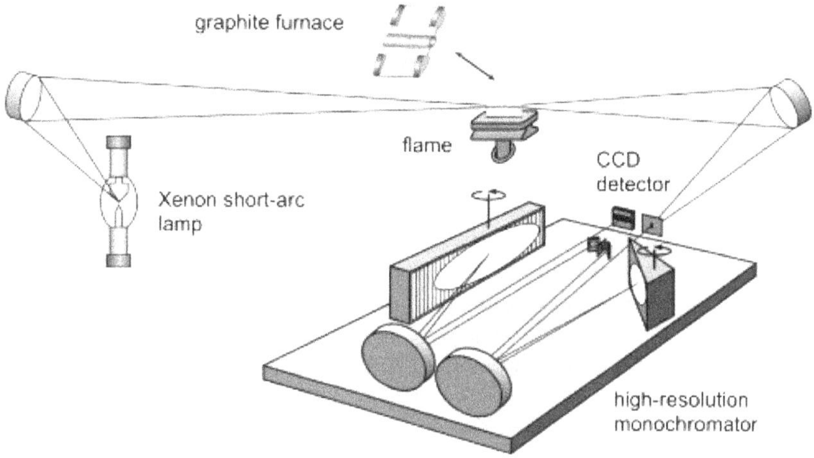

Figure 5.2 Schematic design of a high-resolution continuum source atomic absorption spectrometer. Reproduced from Welz *et al.* (2007), with permission of publisher Elsevier Ltd.

absorption spectrometer that was entirely designed and optimized for the purpose (Becker-Ross *et al.*, 1996; Heitmann *et al.*, 1996). After that it took less than a decade until the first commercial instruments for this technique were introduced; the optical layout of these instruments is shown schematically in Figure 5.2 (Welz *et al.*, 2005). In these instruments, a high-pressure xenon short-arc lamp, operating in a hot-spot mode, is used as a continuum radiation source, emitting an intense continuous spectrum between 190 and 900 nm. A high-resolution double monochromator with a prism pre-monochromator and a high-resolution echelle grating monochromator provide a resolution of $\lambda/\Delta\lambda \approx 175\,000$. A charge-coupled device (CCD) array detector with high quantum efficiency and increased UV sensitivity is used. The detector has 588 pixels, 200 of which are used for analytical purposes and display the entire spectral range that passes the intermediate slit between the prism and the echelle grating monochromator at high resolution. The resolution per pixel is between 1.5 pm per pixel in the far UV, 3 pm at 400 nm and 6 pm at 800 nm. (Frentiu *et al.*, 2013). As the spectral environment of the analytical line is visible in these instruments, and the absorbance spectrum over time is displayed at high resolution, method development and the diagnosis and prevention of interferences become much easier.

The atomizers used for HR-CS AAS are essentially identical with those used in LS AAS, so that it is unnecessary to discuss them further.

5.3.3 Interferences

According to the IUPAC definition, the presence of concomitants in the test sample can cause interferences in the determination of the analyte (IUPAC, 1978). The interference can influence both the attainable precision and the trueness.

Interferences in spectrochemical analysis are generally divided into spectral and nonspectral interferences. Spectral interferences are due to the incomplete isolation of the radiation absorbed by the analyte from other radiation or radiation absorption detected and processed by the electrical measuring system. With nonspectral interferences, the number of free analyte atoms, and thus the measurement signal, is affected directly. Nonspectral interferences are most conveniently classified according to the place or time of their occurrence (Welz *et al.*, 2010).

Both the type and the extent of nonspectral interferences depend to a very high degree on the technique applied and the atomizer employed. All techniques in which the measurement solution or the analyte are transported to the atomizer during the measurement can exhibit *transport interferences*. This interference is typical for FAAS when the viscosity or surface tension of the sample solution influences the aspiration rate of the nebulizer or the effectiveness of nebulization *etc*. This interference is best corrected using the analyte addition or the addition calibration technique. Techniques in which the sample is introduced into the atomizer prior to atomization, such as GFAAS, do not exhibit transport interferences.

Solute volatilization interferences can occur both in FAAS and GF AAS. In the flame technique these interferences are based on a changed rate of volatilization of the aerosol particles if the volatilization of the analyte in the presence, or absence, of a concomitant is incomplete. In GFAAS this interference is based on a change of the volatilization temperature or time of the analyte in the presence of a concomitant. The most efficient way to avoid this interference is the addition of appropriate reagents, such as buffers, releasers or chemical modifiers.

Vapor-phase interferences can occur in all AAS techniques. They are caused by a change in the fraction of the analyte that is dissociated or ionized in the vapor phase. Dissociation interferences are found fairly commonly both in FAAS and GFAAS, caused by a change in the dissociation equilibrium in the absorption volume by concomitants.

Due to the relatively low temperatures prevailing in graphite tube atomizers, no noticeable ionization is observed, so that *ionization interferences* cannot occur. In principle this should also be true for FAAS; however, in the primary reaction zones of the air–acetylene and especially the nitrous oxide–acetylene flame the analyte atoms are mainly atomized by charge transfer from molecular ions of the flame gases, which includes the risk of considerable ionization of easily ionized atoms. The ionization interference can only be removed by adding another easily ionized element, such as K or Cs in great excess to all solutions.

In general, as the same atomizers are used in LS AAS and HR-CS AAS, and all nonspectral interferences have their origin in the atomizer units, these interferences are the same for both types of equipment, and the measures to avoid or correct for these interferences are also the same.

Spectral interferences are additive interferences that cannot be corrected by any calibration technique, such as the analyte addition technique, but only by instrumental methods, which are completely different for LS AAS and HR-CS AAS. For LS AAS there are in essence three distinctly different background correction systems available; deuterium background correction, Smith–Hieftje background correction and using the Zeeman effect.

All the background correction systems used in LS AAS work sequentially, *i.e.*, total absorption and background absorption are measured in rapid sequence and then subtracted from each other to give the atomic absorption. In contrast to this, in HR-CS AAS all measurements of atomic and background absorption are strictly simultaneous as all the 200 pixels used for analytical purposes are read out and evaluated simultaneously. Any change in the absorbance that is identical for all the selected correction pixels, such as fluctuations in the emission of the radiation source or any continuous background absorption is corrected automatically. After this correction there remains any spectrally discontinuous absorption, such as atomic absorption of the analyte or concomitants and molecular absorption of diatomic molecules with rotational fine structure. When this absorption overlaps with the analyte absorption, it can be removed by recording a reference spectrum of the interfering species and using a least-squares algorithm to subtract this spectrum, a process called least squares background correction (LSBC).

The most notorious spectral interference is due to direct line overlap of the analyte line with that of a concomitant element. This interference is quite common in optical emission spectrometry (OES), but there are only a few examples for this interference in AAS, mostly because of the lower temperature of the atomizers and the significantly smaller number of absorption lines, compared to emission lines. Nevertheless, there are examples of this kind of interference in AAS, such as the direct overlap of a secondary iron line with the main absorption line for zinc. This interference cannot be corrected with the systems used in LS AAS, but it can be corrected in HR-CS AAS using LSBC if the interfering element has one more absorption line within the spectral window of the detector, as is the case in the above-mentioned case.

A much more common spectral interference in AAS is that due to the absorption of radiation by diatomic molecules with rotational fine structure, such as PO, NO, SiO, *etc.* More details about this kind of molecular absorption will be given in Section 5.4. If one of these molecular absorption "lines" coincides with the absorption line of the analyte, the measured total absorption will be higher than the atomic absorption. In LS AAS, both, deuterium and Smith–Hieftje background correction cannot correct for this kind of interference; the first one makes an average of the molecular absorption over the spectral window that reaches the detector and the second one measures the background on both sides of the analytical line. Both measurements will almost inevitably differ from the background measured at the analytical line, which depends on the degree of overlap of the two lines. Hence, both, over- and undercorrection might be observed with these systems, as the measured background could be greater or smaller than at the analytical line. Zeeman-effect background correction can in principle correct for this kind of background absorption, as total and background absorption are measured with the same line profile emitted by the radiation source; however, there have been examples reported where these molecules also exhibit a Zeeman effect, which means that the background without and with a magnetic field is not the same. In this case, obviously, the background system cannot work properly, and both, over- or undercorrection might be observed.

In HR-CS AAS, in contrast, this interference can be removed by recording a reference spectrum of the interfering molecule and subtracting it from the sample spectrum using LSBC. This is not a difficult task, as the sample matrix is usually known, and there are not too many diatomic molecules that are thermally stable at the atomization temperatures used in AAS. In addition, the visibility of the spectral environment of the analytical line at high resolution greatly facilitates the discovery and identification of the spectral background.

Further supplementary information on AAS can be found in AAS books (Welz, 1999).

5.4 Molecular Absorption Spectrometry

Up to now, molecular absorption was only treated as a potential interference for AAS measurements; however, it was recognized already in the 1970s that molecular absorption spectra might be used for the indirect determination of nonmetals using AAS equipment (Fuwa *et al.*, 1972; Haraguchi and Fuwa, 1976). At the temperatures of spectroscopic flames, or those that are typically used in graphite furnaces, *i.e.*, at 2000–3000 K, only diatomic and a few triatomic molecules can exist, whereas larger molecules are dissociated readily. Only diatomic molecules will be considered in the following, as they are the only ones that could be successfully used for the determination of nonmetals using molecular absorption spectrometry (MAS).

Molecular spectra consist of transitions between different molecular states, each of which has its own specific energy. Due to the complexity of a molecular system and its additional internal degrees of freedom (vibration and rotation), the number of possible states is far larger than that found in atoms. Consequently, molecular spectra are comprised of many more lines than atomic spectra or of nonstructured bands that extend over relatively large wavelength ranges. Basically there are three types of energy that might be "stored" in a molecule, electronic (E_{el}), vibrational (E_{vib}) and rotational (E_{rot}) energy. The former is based upon Coulomb interaction between nuclei and electrons, in accordance with atomic energies. Vibrational and rotational energy are due to internal motion of the nuclei (Welz *et al.*, 2009).

Electron excitation spectra: In a stable molecular configuration, the outer atomic orbitals overlap and create molecular orbitals that are, at least in part, occupied by electrons. The energy differences between such orbitals are comparable to atomic energies. Therefore, molecular electronic transitions and the resulting electron excitation spectra occur at energies and wavelengths similar to those found in atomic transitions, *i.e.*, in the UV and visible range of the spectrum.

Vibrational spectra: Any molecule has internal energy due to the vibrations of the nuclei with respect to their equilibrium configuration. The amount of vibrational energy E_{vib} is typically one to two orders of magnitude less than the electronic energy E_{el}. Hence, pure vibrational transitions are observed in the infrared (IR) region. For diatomic molecules the only possible vibration is a stretching of the intranuclear axis. For purely harmonic motion of the nuclei, E_{vib} can be described as:

$$E_{vib} = h\nu(v + \tfrac{1}{2}) \text{ with } v = 0, 1, 2 \ldots \quad (5.1)$$

Eqn (5.1) shows that the allowed vibrational energies are equally spaced and separated by hν. The amount of the latter is specific to the nuclei and the strength of their coupling. The number v is called the vibrational quantum number. Within an electronic transition the initial and the final state may possess additional vibrational energy; the resulting absorbance spectrum then exhibits a structure caused by allowed vibrational transitions. For electronic transitions from X to A they are given by:

$$\Delta v = v(\mathrm{A}) - v(\mathrm{X}) = 0, 1, 2 \ldots \quad (5.2)$$

A single vibrational transition $v(\mathrm{X}) \rightarrow v(\mathrm{A})$ is called a band, several bands with a constant Δv value are called a sequence.

Rotational spectra: Rotational energies E_{rot} are usually two to three orders of magnitude lower than the vibrational energies E_{vib}, and the pure rotational spectra are observable in the far-IR to microwave region. For diatomic molecules rotational motion is restricted to an axis perpendicular to the molecular longitudinal axis. Every vibrational state possesses a substructure, which is composed of a large number of rotational states, and the thermal energy of the absorber is high enough to populate many rotational states. Any vibrational band therefore consists of a large number of rotational lines. The spectral distance for two consecutive rotational lines shows that rotational structure of diatomic molecules may exhibit a strongly modulated absorbance signal within a small spectral interval in the order of 10 pm, and the line width is comparable to that of an atomic line (Welz *et al.*, 2009).

It is important to repeat one more time that all the electronic transitions of molecules that occur in the UV and visible range of the spectrum are accompanied by a sequence of vibrational transitions and a multitude of rotational transitions, all of which can in principle be used to measure the absorption of a target molecule, and finally determine the nonmetal that forms this molecule. Some diatomic molecules form spontaneously in flames and/or furnaces, such as PO, CS or NO, others have to be "synthesized" by the addition of an appropriate molecule-forming element (MFE) in large excess. The thermal stability of the target molecule is obviously of importance in order to avoid significant dissociation into the atoms at the temperature of the flame or furnace, and a dissociation energy higher than 500 kJ mol^{-1} is usually considered ideal.

5.4.1 Determination of Fluorine by MAS Using LS AAS Equipment

As mentioned earlier, AAS instruments do not work in the vacuum UV range below 190 nm. Since the resonance line of fluorine is around 95 nm, it is impossible to determine this element with AAS. However, there are some studies in the literature to determine fluorine *via* molecular absorption of diatomic molecules using LS AAS instruments. When fluorine standard solutions and samples are introduced into a flame or a graphite furnace together with a suitable MFE, diatomic molecules of F will form in the gas phase.

Dittrich and colleagues used a dual-channel atomic absorption spectrometer, equipped with a graphite tube atomizer, and hydrogen HCL, which emits a spectral continuum within the range of interest to measure the MAS. The two-line method, using the other channel of the spectrometer, was used to correct for background absorption (Dittrich, 1978, 1979; Dittrich and Vorberg, 1982; Dittrich *et al.*, 1985). These authors investigated several MFE, including Al, Ga and Mg, as shown in Table 5.1. The problem of this approach was that using a continuum source (H_2 HCL), the entire absorption within the spectral band passes of the equipment, which includes all the absorption lines of the diatomic F molecule, but also any other atomic or molecular absorption within this range, was recorded as absorption of the target molecule. As background measurement and correction was at a different wavelength, all these potential interferences could not be detected, opening the door to all kinds of interferences. Obviously, the technique worked well for aqueous standards, but not for real samples with a matrix.

The other authors listed in Table 5.1 used a different approach, working with conventional AAS equipment, using a Pt-HCL, one emission line of which accidentally coincided with an absorption line of the AlF molecule, and conventional deuterium background correction (Tsunoda *et al.*, 1979; Butcher, 1993; Flores *et al.*, 2007). The confusion about the wavelength at which the measurement was made, shown in Table 5.1, is already an indication of the uncertainty of the approach. Obviously, the Pt-HCL only emits one line, which might not overlap perfectly with the absorption line of the AlF molecule, explaining the different wavelengths given by these authors. The second problem is due to the deuterium background correction, which, by definition, can only correct for continuous background. Obviously, besides correcting for continuous background, the system will make an average of any other atomic or molecular absorption that might occur within the spectral window, including all the rotational lines of the AlF molecule and subtract it from the total absorption measured by the Pt-HCL. This clearly resulted in a significant loss of sensitivity even for pure matrix-free solutions.

Table 5.1 Literature Survey for F Determination by LS MAS.

MFE	Wavelength/ nm[a]	Radiation Source	Atomizer	LOD	Ref.
Ga	211.4	H_2	Graphite furnace	1.6 ng	Dittrich (1978)
Al	227.5	H_2	Graphite furnace	0.4 ng	Dittrich (1978)
Mg	268.3	H_2	Graphite furnace	1.1 ng	Dittrich and Vorberg (1982)
Mg	358.2	H_2	Graphite furnace	3.3 ng	Dittrich and Vorberg (1982)
Al	227.5	H_2	Graphite furnace	0.3 ng	Dittrich *et al.* (1985)
Al	227.45	Pt	Graphite furnace	0.04 ng	Tsunoda *et al.* (1979)
Al	227.438	Pt	Air–C_2H_2 flame	13 mg L^{-1}	Butcher (1993)
Al	227.5	Pt	Graphite furnace	0.21 ng	Flores *et al.* (2007)

[a]Wavelength reported by the authors.

5.4.2 Determination of Fluorine by MAS using HR-CS AAS Equipment

Compared to LS AAS equipment, HR-CS AAS instruments have the advantages of high resolution, which easily separates the individual rotational lines of an MAS spectrum, the flexibility to choose any wavelength to measure MAS, and a much more sophisticated background-correction system that makes it possible to correct for any kind of background in between the absorption lines of the molecular spectrum. The first publications about the use of HR-CS FAAS and HR-CS GFAAS equipment for the determination of fluorine *via* MAS were from Huang *et al.* and Heitmann *et al.*, respectively, both using the absorption of the GaF molecule for the determination of fluorine (Heitmann *et al.*, 2006; Huang *et al.*, 2006). Gleisner *et al.* improved the sensitivity of this determination using a greater number of modifiers and a sophisticated temperature program, increasing at the same time the complexity of the procedure (Gleisner *et al.*, 2010). In a later work, Gleisner *et al.* simplified the procedure for the determination of F in toothpaste (Gleisner *et al.*, 2011).

Although the interferences in HR-CS AAS and the measures to avoid or reduce them are essentially the same also in HR-CS MAS, there is one more source of interference with the latter one. In AAS the only problem is to dissociate all compounds of the analyte into atoms, which is usually accomplished using a high enough atomization temperature and maybe some additional modifier in order to facilitate atomization. In MAS, however, we do have the risk of formation of "competitive molecules", *i.e.*, molecules other than the target molecule selected for the determination of the analyte. Obviously, there is the possibility for more than one thermally stable molecule to form for the same analyte, as can be seen from Table 5.1, where several MFE were investigated. Any competitive molecule that forms because of the presence of another MFE in the matrix of the sample will obviously reduce the amount of analyte that will form the target molecule, and hence cause interference.

This kind of interference can be reduced or eliminated by increasing the amount of MFE in order to shift the equilibrium, or even by using a different MFE and changing the target molecule. This has for example been done by going from Ga to Ca as the MFA for samples rich in Ca. An alternative is using the addition calibration technique in case a large number of samples with a similar concentration of the competitive MFE have to be analyzed. In a worst case, the analyte addition technique could be used to solve the problem. Molecular absorption spectrometry was applied for water, milk and tea, which are among the most important sources of F. In the following, the determination of F *via* diatomic molecules formed using an MFE in a flame and a graphite furnace with HR-CS AAS is explained together with a literature survey (Table 5.2).

Gleisner *et al.* (2010) determined fluorine in water samples using the molecular absorption of GaF at 211.248 nm and HR-CS GFAAS; 500 μg Ga was injected as MFE and Zr as a permanent modifier. Figure 5.3 shows the wavelength-resolved molecular absorption spectrum of gallium monofluoride in the vicinity of the band head 211.248 nm. The F concentrations of water samples (drinking water, mineral water and 2 CRM) were determined using the standard calibration technique. The authors stated that in that kind of sample, containing relatively high concentrations of Ca, Mg and chloride, there was no matrix influence.

Ozbek and Akman determined fluorine in water samples by HR-CS GF MAS *via* molecular absorption of the SrF molecule at 651.187 nm without using any modifier (Ozbek and Akman, 2012a). The fluorine concentration of waste water CRM was determined within the uncertainty limits of the certified value for fluorine. In addition, the fluorine concentrations of some water samples reported by the producers on the bottles were found to be within the standard deviation limits of the results. Although the results obtained by the standard calibration and the analyte addition technique were not significantly different, the latter was advised due to the potential risk of interference. Figure 5.4 shows the absorption spectrum for SrF in the vicinity of the 651.187 nm line and the time-resolved absorption spectrum obtained for a water sample (SPS-NUTR WW2) within the same spectral environment.

In another paper by the same authors (Ozbek and Akman, 2013), fluorine in milk samples was determined by HR-CS GF MAS *via* the molecular absorption of calcium monofluoride (CaF) at 606.440 nm, applying a pyrolysis temperature of 700 °C and a vaporization temperature of 2250 °C. The calcium in milk was not sufficient to convert all the fluorine to CaF. Therefore, 40 μg of Ca was added to the samples. In all determinations, an analyte addition technique was applied to prevent the risk of nonspectral interferences due to the matrix.

Mores *et al.* determined fluorine in tea samples *via* molecular absorption of CaF with HR-CS GF MAS at 606.440 nm (Mores *et al.*, 2011). The tea samples were subjected to an acid digestion, an alkaline solubilization and also prepared as conventional infusions with hot water. The authors stated two main advantages of Ca as a MFE; (i) spectral interferences at the working wavelength

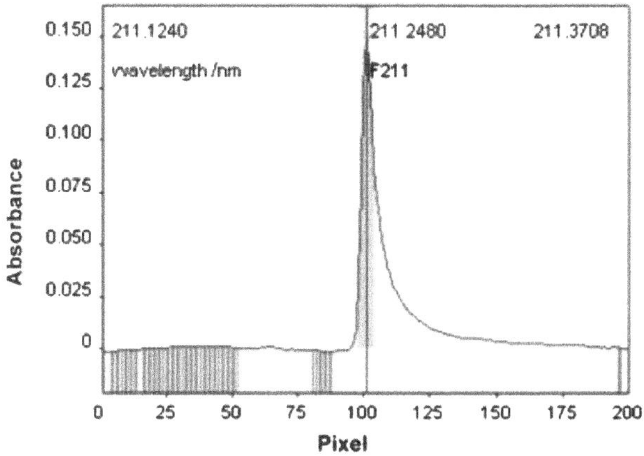

Figure 5.3 Wavelength-resolved molecular absorption spectrum of gallium monofluoride in the vicinity of the band head at 211.248 nm; injection of 20 μL of 20 μg L^{-1} F (0.4 ng F). Reproduced from Gleisner *et al.* (2010) with permission of publisher Elsevier Ltd.

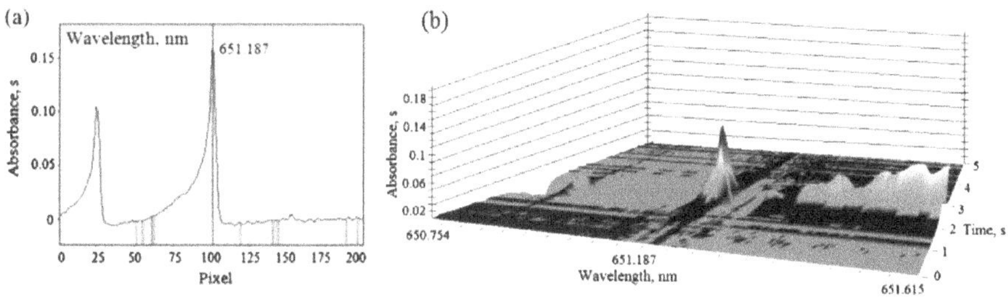

Figure 5.4 (a) Absorption spectra for SrF in the vicinity of 651.187 nm (Sr: 2 mg mL^{-1}; F: 0.1 μg mL^{-1}) (b) Time-resolved absorbance spectrum obtained for a water sample (SPS-NUTR WW2) in the vicinity of the 651.187 nm molecular absorption line of SrF. Sr: 2 mg mL^{-1}. Reproduced from Ozbek and Akman (2012) with permission of publisher Elsevier Ltd.

were extremely unlikely (ii) Ca acted as a molecule-forming agent and also as a chemical modifier. However, Ca reduces the graphite tube lifetime. A linear calibration technique was applied for all determinations. The F concentrations obtained after acid digestion and alkaline solubilization of tea samples were not significantly different. In Figure 5.5, the time-resolved absorption spectrum in the vicinity of the CaF band head at 606.440 nm is shown.

In addition to food and drinks, F has been determined in several other matrices, mostly using gallium as the MFE. Heitmann *et al.* determined F in zinc ore and CRMs, Krüger *et al.* determined F in 5-Fluorouracil in cancer cells, Gleisner *et al.* reported speciation analysis of fluorine in toothpaste samples (Heitmann *et al.*, 2006; Gleisner *et al.*, 2011; Krüger *et al.*, 2012). Bücker and Acker determined F by molecular absorption band of AlF at 227.46 by GF

Table 5.2 Literature Survey for Determination of F in Water, Milk and Tea Samples.

						Graphite Furnace Program			
MFE	Wavelength	Sample	LOD	Linear Range	Mass of MFE	Pyrolysis Temperature	Vaporization Temperature	Notes	Ref.
GaF	211.248 nm	Drinking water, mineral water	5.2 pg	0.04–0.2 ng	500 µg Ga	550	1550	Zr as a permanent modifier,	Gleisner *et al.* (2010)
SrF	651.187 nm	Water	550 pg	0–100 ng	20 µg Sr	800	2200	—	Ozbek and Akman (2012)
CaF	606.440 nm	Tea	1600 pg	5–250 ng	12.6 µg Ca	725	2250	—	Mores *et al.* (2011)
CaF	606.440 nm	Milk, waste water	260 pg	0–100 ng	40 µg Ca	700	2250	Analyte addition calibration technique advised	Ozbek and Akman (2013)

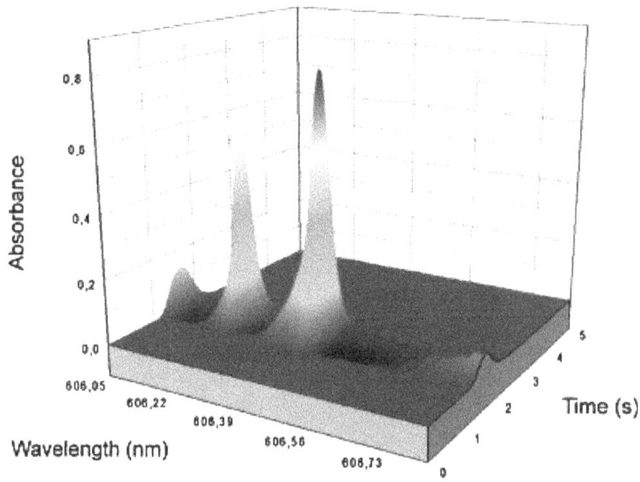

Figure 5.5 Time-resolved absorption spectrum in the vicinity of the CaF band head at 606.440 nm. Reproduced from Mores *et al.* (2011) with permission of publisher Elsevier Ltd.

MAS in synthetic and real process etching solution (Bücker and Acker, 2012). Ozbek and Akman determined F in toothpaste samples by HR CS FAAS using the absorption of AlF at 227.461 nm (Ozbek and Akman, 2012b).

5.5 Conclusion

If only fluoride (F^-) has to be determined in aqueous solution, ISE is the most appropriate method because it is fast, easy, practical and relatively inexpensive, although not free of interferences. IC is suitable for ionic forms of fluorine, including F^-. Other chromatographic techniques as well as molecular absorption methods are not used for the determination of total fluorine; however, fluorine may be present in the molecules determined by those techniques. Every method has its own advantages and disadvantages. HR-CS AAS equipment is principally used for the determination of metals and metalloids; however, total fluorine, *i.e.* covalently bound and its free ionic forms can be determined at trace levels *via* molecular absorption of their diatomic molecules formed in the gas phase of a flame or a graphite furnace. This unique property of HR-CS MAS cannot be achieved by ISE, IC and other techniques.

Summary Points

- Molecular absorption spectrometry is an effective method for halogen determination.
- Line source atomic absorption spectrometry equipment is not adequate for molecular absorption spectrometry.

- High-resolution continuum source atomic absorption spectrometry equipment shows satisfactory results for molecular absorption spectrometry.
- Total F determination can be accomplished by a high-resolution continuum source molecular absorption spectrometer.
- F can be determined using molecular absorption spectrometry of diatomic molecules, such as GaF, CaF, SrF and AlF.

Key Facts

Key Facts of Atomic Absorption Spectrometry

1. Atomic absorption spectrometry is an analytical technique for the determination of more than 70 elements in any matrix.
2. Generally it consists of a radiation source, atomizer, monochromator and detector.
3. In traditional atomic absorption spectrometers (or line source), the radiation source is specific for each element and emits the characteristic emission lines of the analyte.
4. The analyte is atomized in flame or graphite furnace atomizers.
5. After atomization, free atoms absorb ultraviolet or visible radiation and are excited to higher energy levels.
6. The analyte concentration is determined by the measurement of the absorption.
7. Since radiation sources are specific for each element, grating monochromators with moderate resolution can be used.
8. Photomultipliers are used as detectors.
9. AAS is not suitable for the determination of nonmetals, such as the halogens, because their absorption lines are located in the vacuum-UV range, which is not accessible.

Key Facts of High-Resolution Continuum-Source Atomic Absorption Spectrometry

1. In these instruments, a high-pressure xenon short-arc lamp, operating in a hot-spot mode, is used as a continuum radiation source, emitting an intense continuous spectrum between 190 and 900 nm.
2. A charge-coupled device array detector with high quantum efficiency is used as a detector.
3. These instruments have high-resolution double monochromators consisting of a prism and an echelle grating monochromator.
4. As the spectral environment of the analytical line is visible in these instruments, and the absorbance spectrum over time is displayed at high resolution, method development and the diagnosis and prevention of interferences becomes much easier.

5. The atomizers used for high-resolution continuum-source atomic absorption spectrometers are essentially identical with those used in line source atomic absorption spectrometers.

Key Facts of Molecular Absorption Spectrometry

1. In general, the determination of nonmetals such as P, S, F, Cl, Br, I, *etc.* using high-resolution continuum-source atomic absorption spectrometer is based on the measurement of the molecular absorption of a diatomic molecule generated between a metal and the analyte in a flame or graphite furnace.
2. When the solutions of fluorine standards and samples (or solid sample) are introduced into the flame or graphite furnace together with a suitable molecule-forming element, diatomic molecules of F with molecule-forming element are formed in the gas phase.
3. The molecular absorption of the diatomic molecule at one of its rotational lines is measured.

Definitions of Words and Terms

AAS: Atomic absorption spectrometry is a technique based on the measurement of the absorption of radiation by atoms in the gas phase.

FAAS: Atomic absorption spectrometry with a flame atomizer is called flame atomic absorption spectrometry.

GFAAS: Atomic absorption spectrometry with a graphite furnace atomizer is called graphite furnace atomic absorption spectrometry.

LS AAS: Atomic absorption spectrometry that uses line sources, such as a hollow cathode lamp or an electrodeless discharge lamp is called line source atomic absorption spectrometry.

HR-CS AAS: High-resolution continuum-source atomic absorption spectrometry is an absorption spectrometry that uses a continuum source and a high-resolution double monochromator consisting of a prism and an echelle monochromator, and a charge-coupled device (CCD) array detector.

LOD: Limit of detection is the lowest analyte concentration or mass to be separated from background drift with a given confidence level, is calculated as 3 times the standard deviation (σ) of 10 repetitive introduction of a blank to the furnace or flame (3σ) divided by the slope of the calibration graph.

LOQ: Limit of quantification is the analyte concentration or mass based on the same set of measurements, but using 10 times the standard deviation (σ) of 10 repetitive injections of the blank to the furnace or flame, divided by the slope of the calibration graph.

m_0: The characteristic mass, which is used almost exclusively in GF AAS as a measure of sensitivity, is calculated as the mass of analyte that corresponds to an integrated absorbance of 0.0044 s, divided by the slope of the calibration graph.

c_0: The characteristic concentration (c_0) is a measure of sensitivity, mostly used in FAAS, which is calculated as the concentration of the analyte corresponding to an absorbance of 0.0044, divided by the slope of the calibration curve.

MFE: The element added to the sample solutions to form diatomic molecules of a nonmetal in a graphite furnace or a flame to determine nonmetal concentration is called a molecule-forming element.

List of Abbreviations

AAS	Atomic Absorption Spectrometry
CRM	Certificated Reference Material
EDL	Electrodeless Discharge Lamp
GC	Gas Chromatography
GC-MS	Gas Chromatography–Mass Spectrometry
GFAAS	Graphite Furnace Atomic Absorption Spectrometry
HCL	Hollow Cathode Lamp
HPLC	High-Performance Liquid Chromatography
HR-CS AAS	High-Resolution Continuum-Source Atomic Absorption Spectrometry
HR-CS MAS	High-Resolution Continuum-Source Molecular Absorption Spectrometry
IC	Ion Chromatography
ICP-MS	Inductively Coupled Plasma Mass Spectrometry
ICP-OES	Inductively Coupled Plasma Optical Emission Spectrometry
ISE	Ion-Selective Electrode
LC-MS	Liquid Chromatography–Mass Spectrometry
LOD	Limit of Detection
LOQ	Limit of Quantitation
LS AAS	Line Source Atomic Absorption Spectrometry
LS MAS	Line Source Molecular Absorption Spectrometry
m_0	Characteristic Mass
MAS	Molecular Absorption Spectrometry
MFE	Molecule-Forming Element

References

Becker-Ross, H., Florek, S., Heitmann, U. and Weisse, R. 1996. Influence of the spectral bandwidth of the spectrometer on the sensitivity using continuum source AAS. *Fresenius' Journal of Analytical Chemistry.* **355**: 300–303.

Borjigin, S., Ashimura, Y., Yoshioka, T. and Mizoguchi, T. 2009. Determination of fluoride using ion-selective electrodes in the presence of aluminum. *Analytical Science.* **25**: 1437–1443.

Butcher, D. J., 1993. Determination of Fluorine, Chlorine, and Bromine by Molecular Absorption Spectrometry. *Microchemical Journal.* **48**: 303–317.

Bücker, S. and Acker, J., 2012. Spectrometric analysis of process etching solutions of the photovoltaic industry—Determination of HNO3, HF, and H2SiF6 using high-resolution continuum source absorption spectrometry of diatomic molecules and atoms. *Talanta.* **94**: 335–341.

Dittrich, K., 1978. Molecular absorption spectrometry by electrothermal volatilization in a graphite furnace. Part 1. Basis of the method and studies of the molecular absorption of gallium and indium halides. *Analytica Chimica Acta.* **97**: 59–68.

Dittrich, K., 1979. Molekülabsorptionsspektrometrie bei elektro-thermischer verdampfung in einer graphitrohrküvette: Teil 3. Möglichkeiten der Bestimmung von Fluoridspuren durch die Molekülabsorption von AlF-,GaF-InF- und TlF-Molekülen. *Analytica Chimica Acta.* **111**: 123–135.

Dittrich, K., Shkinev, V. M. and Spivakov, B. V., 1985. Molecular absorption spectrometry (MAS) by electrothermal evaporation in a graphite furnace—XIII: Determination of traces of fluoride by mas of AlF after liquid-liquid extraction of fluoride with triphenylantimony(V) dihydroxide. *Talanta.* **32**: 1019–1022.

Dittrich, K. and Townshend, A., 1986. Analysis by Emission, Absorption, and Fluorescence of Small Molecules in the Visible and Ultraviolet Range in Gaseous Phase. *CRC Critical Reviews in Analytical Chemistry.* **16**: 223–279.

Dittrich, K. and Vorberg, B., 1982. Molekülabsorptionsspektrometrie bei elektrothermischer verdampfung in einer graphitrohrküvette: Teil 7. Untersuchung der erdalkali-halogen-molekülabsorptionen, bestimmung von Fluorid- und Chloridspuren durch die molekülabsorption von MgF- und MgCl-molekülen. *Analitica Chimica Acta.* **140**: 237–248.

Fassel, V. A., Mossotti, V. G., Grossman, W. E. L. and Kniseley, R. N., 1966. Evaluation of spectral continua as primary sources in atomic absorption spectroscopy. *Spectrochimica Acta.* **22**: 347–357.

Flores, É. L. M., Barin, J. S., Flores, É. M. M. and Dressler, V. L., 2007. A new approach for fluorine determination by solid sampling graphite furnace molecular absorption spectrometry. *Spectrochimica Acta Part B.* **62**: 918–923.

Frentiu, T., Ponta, M. and Hategan, R. Validation of an analytical method based on the high-resolution continuum source flame atomic absorption spectrometry for the fast-sequential determination of several hazardous/priority hazardous metals in soil. *Chemistry Central Journal.* 2013, 7, 43–53.

Fuwa, K., Haraguchi, H., Okamoto, K. and Nadata, T., 1972. Molecular absorption spectrometry of PO in flame. *Bunseki Kagaku.* **21**: 945–946.

Gleisner, H., Einax, J. W., Morés, S., Welz, B. and Carasek, E., 2011. A fast and accurate method for the determination of total and soluble fluorine in toothpaste using high-resolution graphite furnace molecular absorption spectrometry and its comparison with established techniques. *Journal of Pharmaceutical and Biomedical Analysis.* **54**: 1040–1046.

Gleisner, H., Welz, B. and Einax, J. W., 2010. Optimization of fluorine determination *via.* the molecular absorption of gallium mono-fluoride in a graphite furnace using a high-resolution continuum source spectrometer. *Spectrochimica Acta Part B.* **65**: 864–869.

Hara, H. and Huang, C., 1997. Buffer composition suitable for determining very low fluoride concentrations using a fluoride ion-selective electrode and its application to the continuous analysis of rain water. *Analitica Chimica Acta.* **338**: 141–147.

Haraguchi, H. and Fuwa, K. 1976. Determination of phosphorus by molecular absorption flame spectrometry using the phosphorus monoxide band. *Analitical Chemistry.* **48**: 784–786.

Harwood, J. E., 1969. The use of an ion-selective electrode for routine fluoride analyses on water samples. *Water Research.* **3**: 273–280.

Heitmann U., Becker-Ross, H., Florek, S., Huang, M. D. and Okruss, M., 2006. Determination of fluorine, chlorine, and bromine by molecular absorption spectrometry. *Journal of Analytical Atomic Spectroscopy.* **21**: 1314.

Heitmann, U., Schütz, M., Becker-Ross, H. and Florek S., 1996. Measurements on the Zeeman-splitting of analytical lines by means of a continuum source graphite furnace atomic absorption spectrometer with a linear charge coupled device array. *Spectrochimica Acta Part B.* **51**: 1095–1105.

Huang, M. D., Becker-Ross, H., Florek, S., Heitmann, U. and Okruss, M. 2006. Determination of halogens *via.* molecules in the air–acetylene flame using high-resolution continuum source absorption spectrometry, Part II: Chlorine. *Spectrochimica Acta Part B.* **61**: 959–964.

IUPAC 1978. Nomenclature, symbols, units and their usage in spectrochemical analysis—III. Analytical flame spectroscopy and associated non-flame procedures. *Spectrochimica Acta Part B.* **33**: 247–269.

Kissa, E., 1987. Determination of inorganic fluoride in blood with a fluoride ion-selective electrode. *Clinical Chemistry.* **33**: 253–255.

Krüger, M., Huang, M.-D., Becker-Ros, H., Florek, S., Ott, I. and Gust, R. 2012. Quantification of the fluorine containing drug 5-fluorouracil in cancer cells by GaF molecular absorption *via.* high-resolution continuum source molecular absorption spectrometry. *Spectrochimica Acta Part B.* **69**: 50–55.

Michigami, Y., Kuroda, Y., Ueda, K. and Yamamoto, Y. 1993. Determination of urinary fluoride by ion chromatography. *Analitica Chimica Acta.* **274**: 299–302.

Morés, S., Monteiro, G. C., Silva Santos, F., Carasek, E. and Welz, B. 2011. Determination of fluorine in tea using high-resolution molecular absorption spectrometry with electrothermal vaporization of the calcium mono-fluoride CaF. *Talanta.* **85**: 2681–2685.

Ozbek, N. and Akman, S., 2012a. Method development for the determination of fluorine in water samples *via* the molecular absorption of strontium monofluoride formed in an electrothermal atomizer. *Spectrochimica Acta Part B.* **69**: 32–37.

Ozbek, N. and Akman, S. 2012b. Method development for the determination of fluorine in toothpaste *via* molecular absorption of aluminum mono fluoride using a high-resolution continuum source nitrous oxide/acetylene flame atomic absorption spectrophotometer. *Talanta.* **94**: 246–250.

Ozbek, N. and Akman, S., 2013. Determination of fluorine in milk samples *via* calcium-monofluoride by electrothermal molecular absorption spectrometry. *Food Chemistry.* **138**: 650–654.

Potter, J. J., Hilliker, A. E. and Breen, G. J., 1986. Determination of fluoride and monofluorophosphate in toothpastes by ion chromatography. *Journal of Chromatography A.* **367**: 423–427.

Shi, S., Zhu, Z., Suo, R. and Wang, C. 2003. Rapid method for the determination of trace fluoride and activation of ion-selective electrode. *Analytical Science.* **19**: 671–673.

Tsunoda, K. Chiba, K. Haraguchi, H. Fuwa, K., 1979. Platinum atomic lines for determination of ultratrace fluoride by aluminum monofluoride molecular absorption spectrometry. *Analytical Chemistry.* **51**: 2059–2061.

Walsh, A., 1955. The application of atomic absorption spectra to chemical analysis, *Spectrochimica Acta.* **7**: 108–117.

Welz, B., Becker-Ross, H., Florek, S. and Heitmann, U., 2005. High Resolution Continuum Source AAS, Wiley-VCH, Weinheim, Germany.

Welz, B., Borges, D. L. G., Lepri, F. G., Vale, M. G. R. and Heitmann, U., 2007. High-resolution continuum source electrothermal atomic absorption spectrometry — An analytical and diagnostic tool for trace analysis. *Spectrochimica Acta Part B.* **62**: 873–883.

Welz, B., Lepri, F. G., Araujo, R. G. O., Ferreira, S. L. C., Huang, M.-D., Okruss, M. Becker-Ross, H., 2009. Determination of phosphorus, sulfur and the halogens using high-temperature molecular absorption spectrometry in flames and furnaces–A review. *Analitica Chimica Acta.* **647**: 137–148.

Welz, B., Morés, S., Carasek, E., Vale, M. G. R., Okruss, M. and Becker-Ross, H., 2010. High-resolution continuum source atomic and molecular absorption spectrometry — a review. *Applied Spectroscopy Reviews.* **45**: 327–354.

Welz, B. and Sperling, M., 1999. Atomic Absorption Spectrometry, Wiley-VCH, Weinheim, Germany.

Yiping H., Caiyun W., 2010. Ion chromatography for rapid and sensitive determination of fluoride in milk after headspace single-drop microextraction with in situ generation of volatile hydrogen fluoride. *Analitica Chimica Acta.* **661**: 161–166.

CHAPTER 6

Blood, Plasma and Bone Fluoride Measurement

MAELA LUPO[a] AND ALFREDO RIGALLI*[a]

[a]Bone Biology Laboratory, School of Medicine, Rosario National University, Santa Fe 3100, 2000 Rosario, Argentina
*E-mail: arigalli@conicet.gov.ar

6.1 Methodologies for Fluoride Measurement

Fluorine is an element with no specific function in the human body, but has been assigned beneficial and adverse effects on different systems. It has been shown that an adequate intake may enhance mineralization and resistance of tooth enamel (Magalhães *et al.*, 2011) and bone tissue (Brun *et al.*, 2010). Adverse effects of fluoride are innumerable and the origin of them would be an increase in oxidative stress induced by an imbalance between pro and antioxidant substances originated by inhibition of the respiratory chain (Fina *et al.*, 2012). The pharmacokinetic of fluoride has been studied and is well characterized (Rigalli *et al.*, 1994). The oral intake is the main form of fluoride incorporation to the human body, either in food or dissolved in water. The fluoride concentration in plasma is in equilibrium with the soft tissues and depends primarily on the intake, the renal function and the bone tissue uptake (Figure 6.1).

Any factor that affects any of these functions will influence the plasma concentration of fluoride. The effects of fluoride are not always dependent on

Food and Nutritional Components in Focus No. 6
Fluorine: Chemistry, Analysis, Function and Effects
Edited by Victor R Preedy

Published by the Royal Society of Chemistry, www.rsc.org

the plasma concentration, however, the knowledge of the value of fluoride concentration in plasma and urine are very important in interpreting their effects on the biological level.

The potentiometric determination using an ion-selective electrode is the fastest, most reliable and economical method to measure the concentration of fluoride, but there are also colorimetric techniques. The potentiometric method is based on the use of a fluoride-selective electrode. Although the electrode is only sensitive to fluoride ions in aqueous solution, the potentiometric technique can be applied to any type of sample, directly or combined with the isothermal distillation, described later. The great advantage of the potentiometric method over the other is that it can be optimized for a large number of samples, it uses a small amount of sample, measures a wide range of concentrations with high precision and it is less expensive, taking into account the investment in the electrode.

Measuring the concentration of fluoride in water or urine is a simple and fast method, whenever the concentration is above the detection limit of the electrode. However, fluoride measurement is more complicated when the complexity of the sample increases due to the presence of other ions, pH outside the range 4–6, the presence of protein or fluoride concentration lower than the detection limit. For instance, fluoride measurements in blood, plasma or serum are complicated determinations. Moreover, the measurement of fluoride content in solid samples such as bone, tooth, food, feces, soil samples, *etc*. adds further difficulty to the measurement and interpretation of the results.

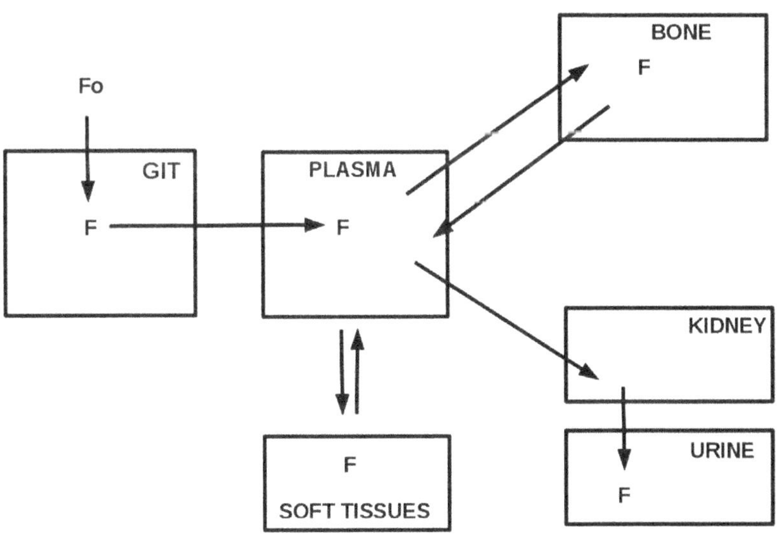

Figure 6.1 Pharmacokinetic of fluoride. Pharmacokinetic of fluoride. Fo: oral dose, F: fluoride concentration in the gastrointestinal tract (GIT), blood, soft tissues, bone and urine.

6.2 Electrode Description

The fluoride-selective electrode consists of a lanthanum fluoride crystal with a very low percentage of other elements, such as europium. The crystal serves as an ion-conductive membrane in which only the fluoride ion can act as mobile (Vogel *et al.*, 1983). The glass wall separates the sample from a reference fluoride solution inside the electrode (Figure 6.2).

There are two types of electrodes: the fluoride combination electrode, combining the fluoride electrode and a reference electrode, and the half-cell fluoride electrode that needs to be combined with a reference electrode. In any of the above mentioned, when the fluoride-selective electrode is placed in contact with the solution, the electric potential difference developed is a linear function of the logarithm of the concentration of fluoride in the sample. The electric potential measured depends on the concentration of fluoride (C) in the solution as described by the Nernst equation:

$$E = E^0 - S \log C$$

E = electric potential measured
E^0 = reference electric potential
S = slope

The fluoride electrode is sensitive to fluoride and its concentration depends on the pH of the medium. When the pH is lower than 3.5 the fluoride concentration decreases as fluoride becomes hydrofluoric acid (HF). When the pH is greater than 6, the fluoride concentration is unchanged but the electrode

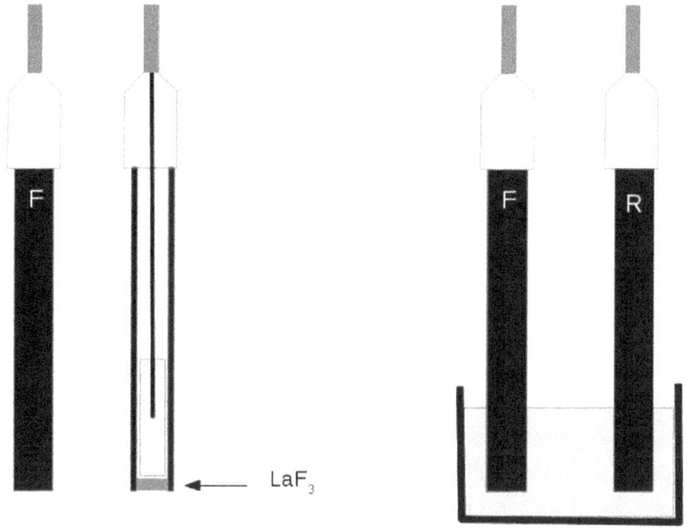

Figure 6.2 Fluoride selective electrode. Left: Exterior view and longitudinal section of a half-cell fluoride selective electrode (F). Right: half-cell selective fluoride electrode with a reference electrode (R).

gives erroneous measurements because it is sensitive to the hydroxyl ion. Furthermore, the fluoride concentration depends on the presence of other ions in the solution with which it may form complex ions, such as Al^{3+}, Fe^{3+}, Cr^{3+}, among others. The ionic strength of the samples must be adjusted by the addition of a total ionic strength adjustor buffer (TISAB). This buffer contains complexing agents that bind with the ions that react with fluoride, so that these ions remain masked. Therefore, it is needed to avoid the effects of pH, ionic strength and fluoride ion complexing. The slope of the Nernst equation is a direct function of the temperature, for this reason the measurement of standard and sample solutions must be performed at the same temperature.

Importantly, the electrode has a linear relationship between the measured electric potential difference and the logarithm of the concentration, within a range from 1 mole L^{-1} to 10^{-5} mole L^{-1}. For concentrations lower than 10^{-5} mole L^{-1} the function is no longer linear. Below a certain concentration value, known as the electrode detection limit, the electrode is insensitive to fluoride. Typically, the electrode detection limit is between 10^{-7} and 10^{-6} mole L^{-1}.

The electrode response is not instantaneous, and the measurement is carried out by immersing the electrode in a solution having a voltage-settling time, which is generally lower than 1 min.

Importantly, the electrode even in the presence of a TISAB may not respond predictably in the presence of proteins and other substances present in biological samples. Processing solutions of known concentration as quality control is advisable to determine the accuracy of the measurement.

6.3 Isothermal Distillation

The isothermal distillation is a technique that is used in combination with the potentiometric determination of fluoride for separating fluoride from the sample and recovering it in a solution of known ionic strength. Isothermal distillation is based on the fact that in an acidic medium fluoride produces HF, a volatile compound with a boiling point of 19 °C. Hydrofluoric acid reacts with an alkaline solution to form a salt. It is common to make HF react with a sodium hydroxide solution (NaOH) to form sodium fluoride (NaF), which is a water-soluble salt and does not interfere with the measurement of the electrode.

Isothermal distillation not only isolates fluoride from complex samples where the electrode could not be used, but also concentrates fluoride from the samples. By using a smaller volume of alkaline solution than that of the sample, fluoride is concentrated by a factor equal to the ratio of the volumes, as described by the following equation:

$$\text{concentration factor} = \frac{\text{sample volume}}{\text{alkaline solution volume}}$$

In order to ensure that fluoride sample is converted into HF, the sample should be treated with an acid, and there are different possibilities: sulfuric acid, hydrochloric acid or phosphoric acid. Hexamethyldisiloxane can be added, which allows the complete distillation of fluoride in 24 h at room temperature, by a compound formation between fluoride and silicon, which are volatile at room temperature. When operating at 60 °C the use of hexamethyldisiloxane can be replaced by using concentrated phosphoric acid. Sulfuric acid and hydrochloric acid cannot be used at 60 °C as sulfuric acid has an oxidizing action on other substances in the samples and hydrochloric acid transforms into gas due to its low boiling point that causes neutralization of the alkaline solution. When performing an isothermal distillation two chemical reactions occur: one in the sample and one in the alkaline solution.

In the sample reaction, HF is generated by reaction with the hydronium ion produced by the acid (hydrochloric, sulfuric or phosphoric acid)

$$H_3O^+ + F^- \longrightarrow HF + H_2O$$

In the alkaline solution HF is converted into NaF as described in the following equation:

$$HF + NaOH \longrightarrow NaF + H_2O$$

The alkaline solution after having reacted with HF remains alkaline because there is an excess of NaOH with respect to the fluoride of the sample. Thus after the reaction is completed, the alkaline solution should be neutralized to pH = 5, for later potentiometric measurement.

Isothermal distillation is carried out in a distillation chamber that is a sealed device that allows the two aforementioned reactions to occur consecutively. The device must be plastic, since the HF reacts with glass, sealed and also to allow the mixing of the reagents once closed. The 3 mL cryovials can act as a distillation chamber because they meet the above conditions and allow the processing of large numbers of samples. One drop of NaOH is placed inside the cap of the cryovial that will react with the HF released from the sample after reacting with the phosphoric acid (Figure 6.3).

The required volumes of plasma or serum can be variable but volumes between 20 and 100 µL are recommended. In the case of bone samples of 2 to 10 mg of bone or ash obtained after incineration process are used. Inside the cap 5 µL of NaOH 1.5 mole L^{-1} are added and with the sample, 100 µL of concentrated phosphoric acid is added to the distillation chamber tube (Figure 6.3). After the sample and the phosphoric acid addition the tube is capped and placed at 60 °C for 24 h. After this period of time, the chamber is uncovered and 60 µL of glacial acetic acid diluted 1/120 is added inside the cap in order to adjust the pH to 5. In the above conditions the recovery is of 100%. Standards solutions containing NaF 0.1–1000 nmole dissolved in 100 µL of distilled water are used.

The combination of potentiometry and isothermal distillation depends on the complexity and origin of the samples. If fluorine is bound to carbon, potentiometry does not detect it, even with prior application of isothermal distillation. In this case the sample must first be incinerated at 500 °C and

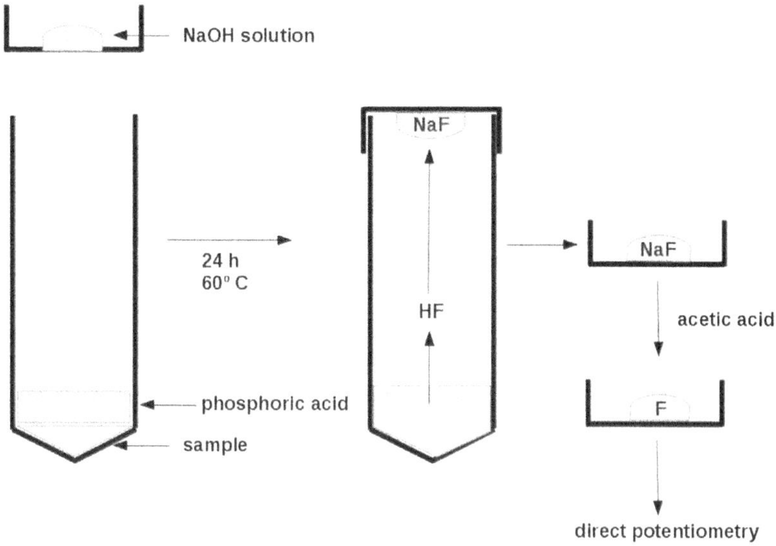

Figure 6.3 Isothermal distillation of fluoride. Scheme of the distillation chamber and the procedure for the isolation of fluoride by isothermal distillation.

then proceed to measure fluoride in the ashes. Different chemical forms in which fluoride can be found in biological samples and their measurement are described in the following paragraph.

6.4 Different Chemical Forms in Which Fluoride Can Be Found in Biological Samples

We must remember that the fluoride selective electrode is only sensitive to fluoride in aqueous solution at a specific concentration range. The main problem faced in determining the concentration of fluoride is that fluoride can be found in different chemical forms, many of which are undetectable by the electrode. A further complication is that many samples are solid, fluoride is in the presence of other compounds or at concentrations below the detection limit of the electrode.

Fluoride can be found in samples basically in three chemical forms (Pera *et al.*, 2007): fluoride (F^-), nonfluoride compounds (NFC) and covalently bound fluorine (CBF). The sum of the three forms will be call the total fluorine content (TFC).

The F^- ion is a monatomic chemical form. The ion can be found in combination with cations that in some cases generate soluble salts, as the case of NaF or may be combined with other cations forming insoluble salts such as calcium fluoride (CaF_2). In the case of soluble salts the measurement is performed by direct potentiometry (the solution is only added with a TISAB) with the ion-specific electrode, given the precautions with regard to pH, ionic strength, presence of complexing agents and detection limit of the electrode. The fluoride that is bound to cations forming insoluble or poorly differentiated compounds is not detected by direct potentiometry.

The NFC are those compounds in which fluoride is not detected by the electrode and should be transformed into F^- by the isothermal distillation technique. Included in NFC are poorly soluble compounds (such as calcium fluoride and fluorapatite), soluble but slightly dissociated fluoride compounds (including complex ions of aluminum and other metals) and fluorine bonded by chemical bonds that are cleaved by the treatment with acids, such as monofluorophosphate (MFP) and the MFP bound to proteins.

By treating a sample with isothermal distillation, both F^- and NFC are converted into HF and then measured together. Acid labile fluoride (ALF) is called the sum of the F^- and NFC. In a sample containing both F^- and NFC, chemical reactions that occur during distillation are:

In the sample:

$$F^- + \text{acid} \longrightarrow HF$$

$$\text{NFC} + \text{acid} \longrightarrow HF$$

In alkaline solution, HF is converted into NaF. The aforementioned NaF is originated from the NFC and the F^-, and is the one called ALF.

$$HF + NaOH \longrightarrow NaF + H_2O$$

Within CBF are all those chemical forms of fluorine that are attached to other molecules by covalent bonds that are not cleaved by the treatment with acids. As an example of CBF, organic fluorine compounds in the carbon skeleton such as betamethasone and flurbiprofen can be mentioned. In order to measure CBF, prior to the measurement it is necessary to separate fluorine atom from the molecule. In the case of a sample with organic molecules, incineration is appropriate. Once the material is incinerated, fluorine is transformed into F^- and remains in a solid phase bound to other ions. When these compounds are soluble, they are dissolved in water and direct potentiometry is applied. If after incineration an insoluble or slightly dissociated material is obtained, it should then be treated with isothermal distillation and finally measured by direct potentiometry. In a biological sample F^-, NFC and CBF may be found. The sum of the three forms is the TFC.

The measurement of the various forms of fluorine in a sample requires different treatments. The measurement by direct potentiometry allows determination of the value of F^- in a sample. If the sample is subjected to isothermal distillation and then direct potentiometry, the value of ALF would be known. Finally, if the same sample is subjected to incineration, isothermal distillation and then to direct potentiometry, the value of TFC is obtained. The difference between the value of ALF and F^- allows the value of NFC to be determined. While the difference between TCF and ALF allows the determination of the value of CBF. Figure 6.4 shows the different forms of fluorine and the methods for measurement.

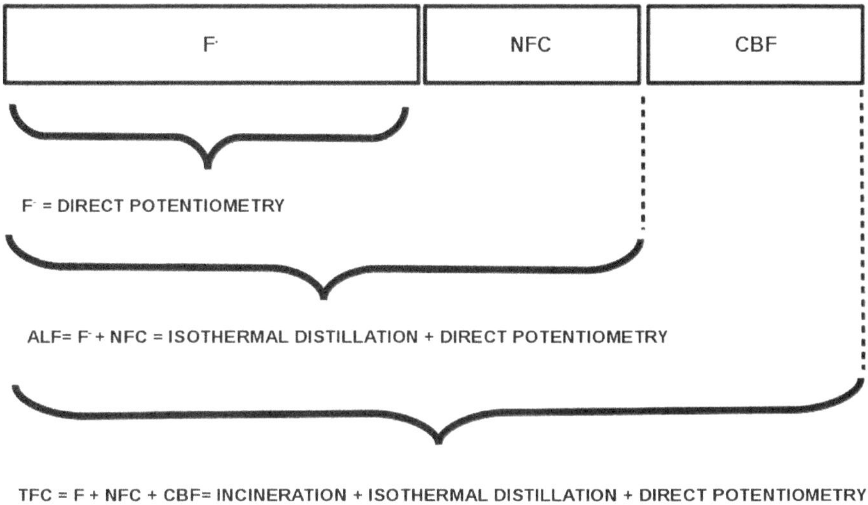

Figure 6.4 Chemical forms of fluorine in biological samples. Different chemical forms of fluorine and the different measurements. F^-: fluoride, ALF: acid labile fluorine, NFC: nonfluoride compounds, CBF: covalently bound fluorine, TFC: Total fluorine content.

6.5 Blood Fluorine Concentration

In blood and plasma fluorine can be found in different chemical forms and the measurement technique applied depends on the presence of those forms, as indicated above. The chemical form in which the fluorine is present in blood will depend on the chemical compound to which it is exposed. The presence of proteins and other substances, as well as a low concentration makes it impossible to perform direct potentiometry. There are studies that have described adaptations of technical and laboratory resources that allow measurement of the concentration of fluoride in blood and plasma directly with the electrode (Fry and Taves, 1970; Jardillier and Desmet, 1973; Melton *et al.*, 1973; Fuchs *et al.*, 1975; Villa, 1988). However, there are not many studies that have used these techniques.

The most widely currently used technique for determining plasma fluorine concentration is the direct potentiometry after isothermal distillation. This combination of techniques measures the concentration of F^- and NFC, therefore it measures the ALF. This technique was described by Taves (1968) and was then slightly modified by different authors (Whitford 1996; Rigalli *et al.*, 1999a). Importantly, when applying these techniques CBF compounds are not measured.

The presence of F^-, NFC or CBF can vary according to the treatment.

6.5.1 Blood Fluoride Concentration after NaF Intake

When the ingested compound is NaF, for example on treatment with the salt or fluoridated water intake, the predominant species in plasma is F^- (Carlson *et al.*, 1960; Rigalli *et al.*, 1994, 1997), the concentration of NFC being negligible. Therefore, the measurement by applying direct potentiometry after isothermal distillation allows determination of the concentration of F^-.

When NaF is administered orally or by using fluoridated water, fluoride is absorbed from the stomach and intestine, and passes to blood (Figure 6.5). The fluoride that is found in blood is removed from this compartment by incorporation into bone as fluorapatite or excreted in urine as fluoride.

6.5.2 Blood Fluorine Concentration after Monofluorophosphate Intake

When MFP is orally administered, a fraction of the drug is hydrolyzed to fluoride in the intestine where it is absorbed (Rigalli *et al.*, 1997). However, a fraction of the MFP is absorbed in the stomach without being hydrolyzed and binds mainly to serum alpha-macroglobulin (Figure 6.5). The MFP linked to alpha-macroglobulin constitutes part of the NFC of plasma; this fraction is taken up by the liver and bone by scavenger receptors (Esteban *et al.*, 1999). CBF is part of the bone fluorine content in treatments with MFP, as it will be

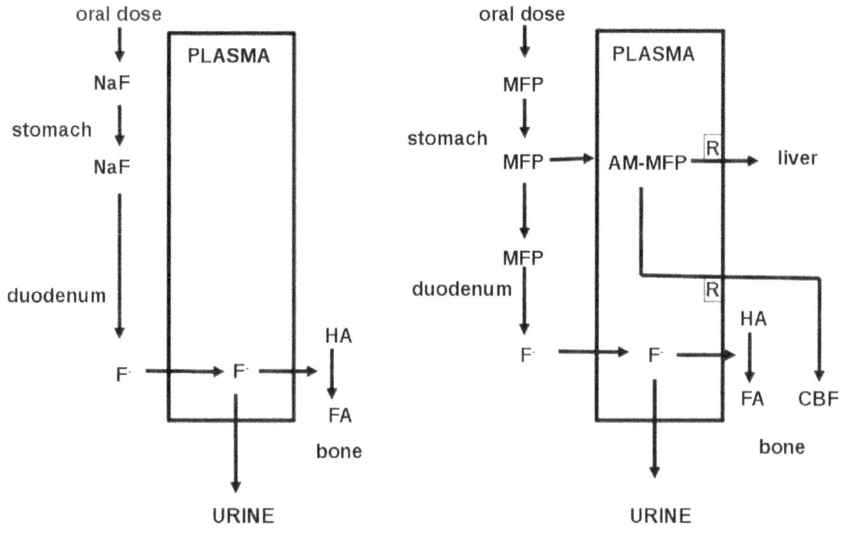

Figure 6.5 Pharmacokinetic of sodium fluoride and sodium monofluorophosphate. Pharmacokinetic of NaF (left) and MFP (right). HA: hydroxyapatite, FA: fluoroapatite, AM-MFP: alpha-macroglobulin-MFP complex, R: receptor of alpha-macroglobulin.

described below. In patients receiving MFP the presence of fluorine bound to peptides in plasma was demonstrated (Rigalli *et al.*, 1999b). This originates from the metabolism of MFP joined to alpha-macroglobulin (Esteban *et al.*, 1999). NFC and F^- concentration after the administration of MFP are modified by the calcium content of the diet. Coadministration of MFP and calcium increases the NFC fraction and decreases F^-. This modification is the consequence of a decrease in the hydrolytic activity of the intestinal alkaline phosphatase induced by calcium. In the presence of calcium, increases in MFP concentration and in the rate of absorption of MFP in the intestine were observed (Beinlich *et al.*, 2003).

When performing the determination using direct potentiometry after isothermal distillation the value of ALF is obtain, which in this case corresponds to the sum of F^- and NFC (MFP bound to alpha-macroglobulin). To determine the concentration of F^- and protein-bound fraction (NFC) a plasma ultrafiltration should be performed, using membranes that remove proteins. Then plasma and plasma ultrafiltrate are subjected to direct potentiometry after isothermal distillation. The concentration of ALF in plasma is the sum of F^- and NFC. The concentration of ALF in the plasma ultrafiltrate indicates the concentration of F^-. The difference between the measured value of ALF in plasma and ALF in the ultrafiltrate is the fluorine bound to proteins (Figure 6.6).

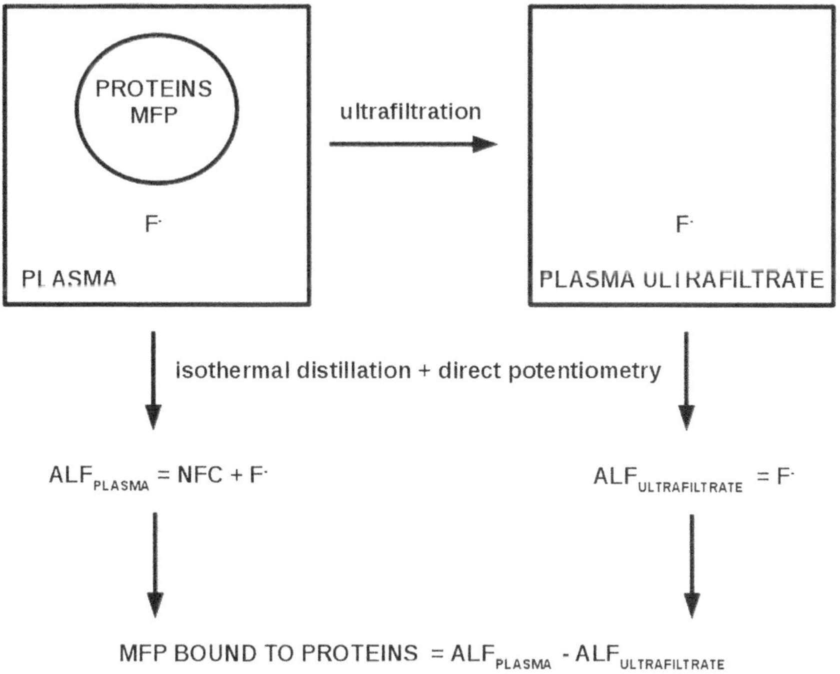

Figure 6.6 Measurement of fluoride and monofluorophosphate bound to proteins. Measurement of fluoride (F^-) and MFP bound to proteins in plasma. ALF: acid labile fluorine, NFC: nonfluoride compounds.

6.5.3 Variables That Influence Plasma Fluoride Levels

The literature generally reports values of fluoride plasma levels and those values generally correspond to individuals consuming fluoridated water, fluoridated toothpaste or are medicated with NaF. When other compounds that can modify the values of CBF and/or NFC are administered, the value of fluorine in plasma should be taken into account.

Plasma fluoride levels depend on several factors: renal function (Lupo *et al.*, 2011), bone turnover and dose (Waterhouse *et al.*, 1980). Plasma fluoride levels were found positively correlated to bone fluorine content. However, in rats treated with NaF inversely related values were found with respect to bone fluorine content (Rigalli *et al.*, 1992; Rigalli *et al.*, 1995). More recent studies indicated that plasma fluoride levels are inversely related to the osteogenic stimulus. A significant decrease in plasma fluoride levels was observed in rats with increased bone formation induced by pharmacological treatments (Brance *et al.*, 2013). This apparent discrepancy might be due to different states of the bone remodeling process. In previous studies where a positive relationship between plasma fluoride levels and bone fluoride content was observed, bone remodeling was not measured. The correct interpretation of the results would require the measurement of markers of bone remodeling or the rate of incorporation of fluoride to bone (Lupo *et al.*, 2011).

In humans, plasma fluoride levels increased with age in both sexes with a marked increase in women after menopause (Husdan *et al.*, 1976). This sudden increase was associated with an increase in bone remodeling (Itai *et al.*, 2010).

The type of tablet used for the administration of NaF should be considered when interpreting the values of plasma fluoride levels. A study with human volunteers receiving a dose of 9 mg of F^- as NaF in enteric-coated tablets or immediate-release tablets was carried out. No difference was found in the time of maximum concentration of plasma fluoride levels (130–140 min), although the plasma concentration peak was significantly lower in the case of enteric-coated tablets (2.3 µmole L^{-1}) with respect to immediate release tablets (4.4 µmole L^{-1}). Moreover, the area under the curve of plasma fluoride levels was three times higher in the case of immediate-release tablets (1.45 mmole L^{-1} min^{-1}). These results were coincident with lower urinary excretion of fluoride in the case of enteric-coated tablet. This difference was attributed to a sharp increase in fluoride concentration on the duodenum, due to the dissolution of the enteric-coated tablet in this area of the digestive tract, resulting in inhibition of the fluoride intestinal absorption process. In contrast, the immediate-release tablet starts dissolving after intake, therefore enables gastric absorption and lower concentrations in the gastrointestinal lumen. In rats, it was shown that high concentrations of fluoride in the duodenal lumen will produce inhibition of the uptake of fluoride, a lower area under the curve of plasma fluoride levels, increased faecal fluoride excretion and decreased bone fluorine content, when the oral dose exceeded 40 µmole per day (Rigalli *et al.*, 2001). Less absorption of fluoride ion at higher concentrations was coincident with a decrease in the metabolic activity of the intestinal mucosa, as evidenced by a lower oxygen consumption of the mucosa, and probably a decrease in the energy-producing processes.

When the concentration of fluorine in blood is tested after the administration of drugs such as MFP, the different forms of fluoride compounds should be taken into account.

After administration of NaF or MFP in human volunteers, no significant differences in the area under the curve (AUC) of plasma fluoride levels were observed. However, the value of AUC of ALF in the case of MFP administration was twice (2930 μmole L^{-1} min^{-1}) than that obtained after the administration of NaF (1725 μmole L^{-1} min^{-1}), due to the presence of MFP linked to alpha-macroglobulin. However, there was no difference in the time at which the maximum was reached in plasma. The treatments in humans with NaF (30 mg F^- per day) or MFP (15 mg F^- per day) generate values of ALF between 2 and 8 μmol L^{-1}. While in the treatment with NaF the value of ALF was due to the presence of F^-, in treatments with MFP the value of ALF was a consequence of an increase in NFC (Rigalli *et al.*, 1999b). Plasma fluoride levels increased after the administration of a single dose of NaF or MFP in rats. However, when MFP was administered chronically the plasma NFC level increased and fluoride levels did not exceed 2 μmole L^{-1} (Rigalli *et al.*, 1995).

6.6 Bone Fluorine Content

Measurement of fluorine in bone is usually done by the incineration process followed by direct potentiometry after isothermal distillation; therefore the TFC is measured. In bone, with rare exceptions, fluoride is in the form of fluorapatite. As a consequence, we are going to use fluorapatite, bone fluoride content or total bone fluoride content as equivalent. In cases where other compounds are present it will be indicated.

In bone, fluoride replaces the hydroxyl group in the hydroxyapatite crystal: $Ca_{10}(PO_4)_6(OH)_2$, forming fluorapatite. $Ca_{10}(PO_4)_6F_2$ (Posner *et al.*, 1963). Bone fluoride content depends on the age, amount of fluoride in the diet and the period during which the intake is performed.

The half-life of fluoride in human bone is 8–9 years (Forbes *et al.*, 1978). In young rats 50–70% of the fluoride released from bone by the resorption process is deposited back into the bone (Guo *et al.*, 1988).

The effect of fluoride on bone tissue is controversial, depending on dose, duration of treatment, age of the animal, *etc.* When ingesting significant amounts of fluoride, the bone is poorly mineralized with poor histological organization, high osteoid content (Vigorita and Suda, 1983), less deformability and increased fragility (Stein and Granik, 1980). Fluoride may affect collagen synthesis (Uslu, 1983), increase osteoblast numbers (Farley *et al.*, 1983), increase osteoclast activity and generate a less porous bone. However, it has been observed that women who consumed drinking water with fluoride concentration greater than 2 ppm showed a reduced frequency of vertebral fracture than women in populations where the fluoride water content was lower (Bernstein *et al.*, 1966). There has also been an increase in mineral density of lumbar spine and femoral neck (de la Sota *et al.*, 1997). In populations where the drinking water contains 1 ppm of fluoride, the frequency of femoral fractures was significantly lower than in populations with water without fluoride (Simonen and Laitinen, 1985). No difference was found in skeletal maturation in young women living in areas with and without fluoride in drinking water (Wenzel *et al.*, 1982).

Many studies have described increased bone tissue, decreased bone resorption, increased density, decreased bone pain and decreased frequency of fractures in chronic treatment with fluoride (Bernstein and Cohen, 1967; Briancon and Meunier, 1981). In these studies improvements were observed after at least six months of treatment (Schulz *et al.*, 1986).

An increase in bone fluoride content was observed with high doses of NaF and long treatments. However, if the dose is too high inhibition of gastrointestinal absorption processes occur. In rats, fluoride intakes higher than 8 mg F^-kg^{-1} bw per day as NaF inhibit absorption in the intestine, showing lower levels of bone fluoride content and higher fluoride values in feces than with a dose of 4 mg F^-kg^{-1} bw per day (Rigalli *et al.*, 2001).

Bone fluoride content depends on the process of remodeling of the bone tissue. In rats with renal failure and increased bone remodeling, increased incorporation of fluoride to the bone, at the same dose was observed (Lupo *et al.*, 2011).

Physical activity increases bone formation. The bone fluoride content of sedentary rats was significantly lower than the value found in rats receiving the same dose of NaF in the drinking water and performed physical activity for 30 min per day on a treadmill (Lombarte *et al.*, 2013).

The bone fluorine content of rats treated with MFP did not differ from the value found in rats treated with NaF for 30 days (Brun *et al.*, 2010). However, a fraction of CBF was found in rats treated with MFP while in rats treated with NaF it was not found. This fraction might originate from some unknown process of the metabolism of MFP bound to alpha-macroglobulins.

Summary Points

- Biological fluoride effects are dependent on fluoride concentration.
- Fluoride concentration can be measured by potentiometric methods.
- After the intake of substances that contain fluoride, *e.g.* sodium fluoride, fluoride is the predominant chemical form of fluorine in blood, tissues and urine.
- Blood, plasma, serum, soft tissues and bone can contain fluorine in different chemical forms and concentrations. All of them can be measured by combining a potentiometric method with an ion (fluoride) selective electrode, isothermal distillation and incineration of the sample.

- Plasma, blood and serum fluoride levels depend on the fluorine-containing substance that has been ingested, the dose, the bone remodeling process and the renal function.
- When monofluorophosphate is administered, part of the fluorine is bound to proteins in plasma. As a consequence, in plasma fluoride and fluorine bound to proteins are found. Isothermal distillation and direct potentiometry must be used to determine the total concentration of fluorine en plasma in these treatments.

Key Facts of Fluoride Measurement

- Fluoride is a common contaminant of drinking water. Fluoride measurement in water allows us to determine whether it is appropriate for drinking and cooking.
- Side effects of fluoride are a function of plasma fluoride concentration.
- Urinary excretion of fluoride is a measurement of fluoride exposure.
- Fluorosis is an illness produced by high intakes of fluoride, especially by drinking water with high concentration of fluoride.
- Fluoride is not only present in water, it can also be found in foods. The concentration of fluoride in foods must be considered when fluorosis is present in a population.
- Dentifrices, varnishes and mouth rinses have high concentrations of fluoride and can contribute to high fluoride intake and fluorosis.

Definitions of Words and Terms

Acid labile fluoride (ALF): the sum of all chemical forms of fluorine that produces hydrofluoric acid when the sample is treated with a strong acid.
Bone remodeling: The process that occurs in bone tissue as the consequence of the activities of osteoclasts and osteoblasts. After a microfracture or the action of parathyroid hormone, osteoclasts are activated and remove bone by a process called bone resorption, then osteoblasts produce new bone tissue, which finally calcifies.
Covalently bound fluorine (CBF): Chemical forms of fluorine that are bound to other atoms by covalent bonds, which are not cleaved by the treatment with strong acids.
Detection limit: The lowest concentration of a substance that can produce a response in a measurement instrument, different from the absence of the substance.
Direct potentiometry: A potentiometric measurement when the electrodes are used without previous treatment of the sample.
Fluoride: Anionic monoatomic chemical form of fluorine.
Fluoroapatite: A mineral with the formula $Ca_{10}(PO_4)_6F_2$ that is present in bone mineral, when the process of mineralization occurs in the presence of fluoride.

Ion-selective electrode: A sensor that converts the concentration of a specific ion into a value of electric potential.

Isothermal distillation: A chemical process that allows separating a substance from a sample based on the boiling point of the substance. In the process, the substance is released from the sample by the treatment with another substance, and then it is trapped in another substance that is inside the isothermal distillation chamber.

Nernst equation: Mathematical equation that states that the electric potential developed by an ion-selective electrode is a linear function of the logarithm of the concentration of a substance.

Nonfluoride compounds (NFC): Atoms of fluorine in a molecule that are not detected by an ion-selective electrode but produce hydrofluoric acid by the treatment of the molecule with a strong acid.

Parts per million (ppm): Concentration of a substance in a sample that express the amount of milligrams of the substance in a kilogram of sample. For example, drinking water with 5 ppm fluoride. 5 ppm indicates that 1 kilogram of water (or 1 liter) contains 5 mg of fluoride.

Potentiometric method: Chemical measurement of a substance that uses an ion-selective electrode and converts the concentration of the substance into an electric potential.

Total ionic strength adjustor buffer (TISAB): Buffer used in potentiometric methods that allow adjusting the ionic strength and pH of the sample. TISAB also has substances that avoid the interference of other substances in the process of measurement.

Total fluoride content (TFC): The sum of all the chemical forms of fluorine found in a sample.

List of Abbreviations

ALF	Acid labile fluoride
AUC	Area under the curve
bw	Body weight
°C	Degree Celsius
CBF	Covalently bound fluorine
F^-	Fluoride
HF	Hydrofluoric acid
L	liter
MFP	Monofluorophosphate
NFC	Nonfluoride compound
NaF	Sodium fluoride
NaOH	Sodium hydroxide
ppm	Parts per million
TFC	Total fluorine content
TISAB	Total ionic strength adjustor buffer

References

Beinlich, A. D., Brun, L. R., Rigalli, A. and Puche, R. C., 2003. Intestinal absorption of disodium monofluorophosphate in the rat as affected by concurrent administration of calcium. *Arzneimittel-Forschung/Drug Research.* **53**: 584–589.

Bernstein, D. S., Sadowsky, N., Hegsted, D. M., Guri, C. D. and Stare, F. J., 1966. Prevalence of osteoporosis in high and low fluoride areas in North Dakota. *JAMA: The Journal of the American Medical Association.* **198**: 499–504.

Bernstein, D. S. and Cohen, P., 1967. Use of sodium fluoride in the treatment of osteoporosis. *The Journal of Clinical Endocrinology and Metabolism.* **27**: 197–210.

Brance, M. L., Brun, L. R., Di Loreto, V. E., Lupo, M. and Rigalli, A., 2013. Climacteric: The journal of the International Menopause Society. (in press).

Briancon, D. and Meunier, P. J., 1981. Treatment of osteoporosis with fluoride, calcium and vitamin D. *The Orthopedic Clinics of North America.* **12**: 629–648.

Brun, L. R., Pera, L. and Rigalli, A., 2010. Bone morphometry and differences in bone fluorine containing compounds in rats treated with NaF and MFP. *Biomedicine and Pharmacotherapy.* **64**: 1–6.

Carlson, C. H., Armstron, W. D. and Singer, L., 1960. Distribution and excretion of radiofluoride in the human. *Proceedings of the Society for Experimental Biology and Medicine.* **104**: 235–239.

Esteban, L., Rigalli, A. and Puche, R. C., 1999. Metabolism of the complex monofluorophosphate-alpha-2-macroglobulin in the rat. *Medicina.* **59**: 151–156.

Farley, J. R., Wergedal, J. E. and Baylink, D. J., 1983. Fluoride directly stimulates proliferation and alkaline phosphatase activity of bone forming cells. *Science.* **222**: 330–332.

Fina, B., Brance, M. L., Brun, L. R. and Rigalli, A., 2012. Fluoride inhibition of oxygen consumption and increased oxidative stress in rats. *Fluoride.* **45**: 343–348.

Forbes, G. B., Taves, D. R., Smith, F. A. and Kilpper, R. W., 1978. Bone mineral turnover in a patient with osteogenesis imperfecta estimated by fluoride excretion. *Calcified Tissue Research.* **25**: 283–288.

Fry, B. W. and Taves, D. R., 1970. Serum fluoride analysis with the fluoride electrode. *The Journal of Laboratory and Clinical Medicine.* **75**: 1020–1025.

Fuchs, C., Dorn, D., Fuchs, C. A., Henning, H. V., Mc Intosh, C., Scheler, F. and Stennart, M., 1975. Fluoride determination in plasma by ion selective electrodes: a simplified method for the clinical laboratory. *Clinica Chimica Acta.* **60**: 157–167.

Guo, M. K., Nopakun, J., Messer, H. H., Ophaug, R. and Singer, L., 1988. Retention of skeletal fluoride during bone turnover in rats. *The Journal of Nutrition.* **118**: 362–366.

Husdan, H., Vogl, R., Oreopoulus, D., Gryfe, C. and Rapoport, A., 1976. Serum ionic fluoride: normal range and relationship to age and sex. *Clinical Chemistry.* **22**: 1884–1888.

Itai, K., Onoda, T., Nohara, M., Ohsawa, M., Tanno, K., Sato, T., Kuribayashi, T. and Okayama, A., 2010. Serum ionic fluoride concentrations are related to renal function and menopause status but not to age in a Japanese general population. *Clinica Chimica Acta.* **411**: 263–266.

Jardillier, J. C. and Desmet, G., 1973. Etude du fluor sérique et de ses combinaisons par une technique utilizant une électrode spécifique. *Clinica Chimica Acta.* **47**: 357–363.

Lombarte, M., Fina, B. L., Lupo, M., Buzalaf, M. A. and Rigalli, A., 2013. Physical exercise ameliorates the toxic effect of fluoride on the insulin-glucose system. *The Journal of Endocrinology.* **218**: 99–103.

Lupo, M., Buzalaf, M. A. and Rigalli, A., 2011. Effect of fluoridated water on plasma insulin levels and glucose homeostasis in rats with renal deficiency. *Biological Trace Element Research.* **140**: 198–207.

Magalhães, A. C., Wiegand, A., Rios, D., Buzalaf, M. A. and Lussi, A., 2011. Fluoride in dental erosion. In: Buzalaf M. A. R (ed.) *Fluoride and the Oral Environment.* Karger, Basel, Switzerland, pp. 158–170.

Melton, J. R., Hoover, W. L. and Ayers, J. L., 1973. Known addition procedure for determining fluoride in feeds with an ion-specific electrode. *Journal – Association of Official Analytical Chemists.* **57**: 508–510.

Pera, L. I., Puche, R. C. and Rigalli, A., 2007. Potentiometric measurement of ionic, acid labile and covalently bound fluorine. *Journal of Fluorine Chemistry.* **128**: 1425–1428.

Posner, A. S., Eanes, E. D., Harper, R. A. and Zipkin, I., 1963. X-Ray diffraction analysis of the effect of fluoride on human bone apatite. *Archives of Oral Biology.* **8**: 549–570.

Rigalli, A., Ballina, J. C. and Puche, R. C., 1992. Bone mass increase and glucose tolerance in rats chronically treated with sodium fluoride. *Bone and Mineral.* **16**: 101–108.

Rigalli, A., Ballina, J. C., Beinlich, A. D., Alloatti, R. and Puche, R. C., 1994. Pharmacokinetics differences between sodium fluoride and sodium monofluorphosphate and comparative bone mass increasing activity of both compounds, in the rat. *Arzneimittel-Forschung/Drug Research.* **44**: 762–766.

Rigalli, A., Alloatti, R., Menoyo, I. and Puche, R. C., 1995. Comparative study of the effect of sodium fluoride and sodium monofluorophosphate on glucose homeostasis in the rat. *Arzneimittel-Forschung/Drug Research.* **45**: 289–292.

Rigalli, A., Esteban, L., Pera, L. and Puche, R. C., 1997. Binding of monofluorophosphate to alpha-2-macroglobulin and C3. *Calcified Tissue International.* **60**: 86–89.

Rigalli, A., Alloatti, R. and Puche, R. C., 1999a. Measurement of total and diffusible serum fluoride. *Journal of Clinical Laboratory Analysis.* **13**: 151–157.

Rigalli, A., Pera, L., Morosano, M., Masoni, A., Bocanera, R., Tozzini, R. and Puche, R. C., 1999b. In postmenopausal osteoporosis the bone increasing effect of monofluorophosphate is not dependent on serum fluoride. *Medicina (B Aires).* **59**: 157–161.

Rigalli, A., Beinlich, A. and Puche, R. C., 2001. Intestinal absorption of fluoride at high luminal concentration of fluoride. *Arzneimittel-Forschung/Drug Research.* **51**: 151–155.

Schulz, E. E., Engstrom, H., Sauser, D. D. and Baylink, D. J., 1986. Osteoporosis: radiographic detection of fluoride-induced extra-axial bone formation. *Radiology.* **159**: 457–462.

Simonen, O. and Laitinen, O., 1985. Does fluoridation of drinking water prevent bone fragility and osteoporosis? *Lancet.* **2**: 432–433.

de la Sota, M., Puche, R., Rigalli, A., Fernandez, L. M., Benassati, S. and Boland, R., 1997. Bone mass and glucose homeostasis of subjects with high spontaneous fluoride intake. *Medicina.* **57**: 417–420.

Stein, I. D. and Granik, G., 1980. Human vertebral bone: relation of strength, porosity and mineralization to fluoride content. *Calcified Tissue International.* **32**: 189–194.

Taves, D., 1968. Separation of fluoride by rapid diffusion using hexamethyldisiloxane. *Talanta.* **15**: 969–974.

Uslu, B., 1983. Effect of fluoride on collagen synthesis in the rat. *Research in Experimental Medicine.* **182**: 7–12.

Vigorita, V. J. and Suda, M. K., 1983. The microscopic morphology of fluoride induced bone. *Clinical Orthopaedics and Related Research.* **177**: 274–282.

Villa, A. E., 1988. Rapid method for determining very low fluoride concentration using an ion selective electrode. *The Analyst.* **113**: 1299–1303.

Vogel, G. L., Chow, L. C. and Brown, W. E., 1983. A microanalytical procedure for the determination of calcium, phosphate and fluoride in enamel biopsy samples. *Caries Research.* **17**: 23–31.

Waterhouse, C., Taves, D. and Munzer, A., 1980. Serum inorganic fluoride: changes related to previous fluoride intake, renal function and bone resorption. *Clinical Science (London).* **58**: 145–152.

Wenzel, A., Thylstrup, A. and Melsen, B., 1982. Skeletal development and dental fluorosis en 12-14 year old Danils girls from a fluoride and nonfluoride comunity. *Scandinavian Journal of Dental Research.* **90**: 83–88.

Whitford, G. M., 1996. The Metabolism and Toxicity of Fluoride, Karger, Basel, Switzerland, 158 pp.

Section IV
Function and Effects

CHAPTER 7

Fluoride Accumulation in Crops and Vegetables: Indian Perspectives

SRIMANTA GUPTA*[a] AND DALI MONDAL[a]

[a]Department of Environmental Science, The University of Burdwan, Burdwan, 713 104, West Bengal, India
*E-mail: srimantagupta@yahoo.co.in

7.1 Introduction

The availability of clean water is essential. It serves as the basic component in different spheres of human life for a large number of habitants (Ayoob and Gupta, 2006). In many regions of the world populations are dependent on groundwater for agriculture and drinking (Amini *et al.*, 2008; Gupta and Banerjee, 2009). Unfortunately, groundwater may be contaminated from natural sources by aquifer rocks. One of the most widespread contaminant is fluoride (F) and it is estimated that many millions of people are affected worldwide (WHO, 2008).

It has been estimated that nearly 25 million people in India are already affected by F while over 66 million more, including 7 million children, in 20 states (Figure 7.1) of the country are susceptible to it. Out of the total of 569 districts of India, 205 (36%) have F contamination in their drinking water. Fluoride contamination in the drinking water of these districts is mostly geogenic in nature though there are a few pockets where brick kiln, alumina, phosphate and fertilizers industries are the culprit. As per the

Food and Nutritional Components in Focus No. 6
Fluorine: Chemistry, Analysis, Function and Effects
Edited by Victor R Preedy

Published by the Royal Society of Chemistry, www.rsc.org

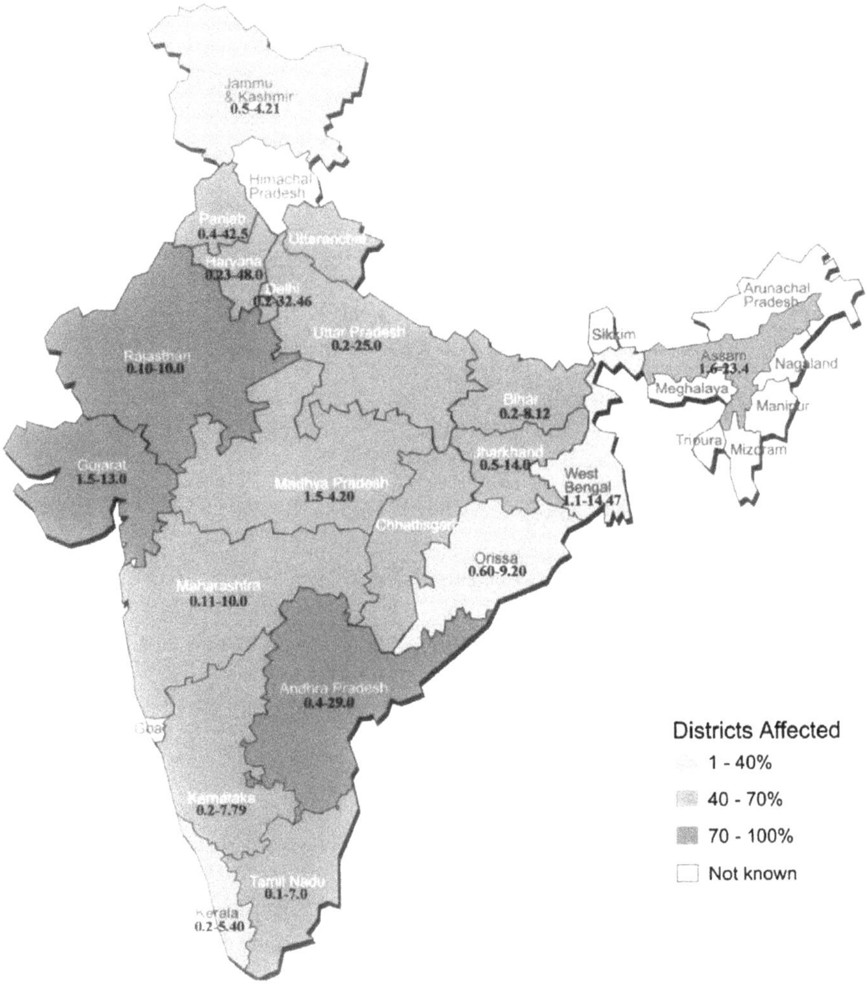

Figure 7.1 Fluorosis affected states in India. The number within each state is the fluoride range in drinking water. Rajasthan, Gujarat and Andhra Pradesh are the most affected states in India. (Source: UNICEF State of Art Report, 1999 and FR and RDF data bank.)

records of the Central Ground Water Board (CGWB), Government of India, all districts of Rajasthan, 15 districts of Andhra Pradesh, 27 districts each of Orissa and Gujarat, 18 districts of Uttar Pradesh, 13 districts of Punjab, 12 districts each of Karnataka and Haryana and 10 districts each of Madhya Pradesh and Maharashtra have high F concentration (above the permissible limits). In a country like India, the majority of the population (85%) lives in villages and depends mainly on groundwater resources both for drinking and agriculture purpose. Approximately 65% (50–80%) of the irrigated land is under groundwater consumption (Raju, 1998).

Fluoride is not an essential element for plants but is essential for animals and humans. The continuous ingestion by animals in excessive amounts of F may lead to fluorosis and the suboptimal levels in the diet can have an equally damaging effect (Kabata-Pendias, 2001). The World Health Organisation (WHO, 1984) and Bureau of Indian Standards (BIS, 2003) have laid down the maximum permissible limits of F in drinking water as 1.5 mg L^{-1} and 1.0 mg L^{-1}, respectively, but there is no stringent threshold limits of F in soil and plants above which the ingestion may be detrimental to human health. The major part of F ingested in areas endemic to fluorosis is water, although some food materials contribute considerable amounts to total intake (Singh *et al.*, 1993). Crops grown in close proximity to certain factories, especially those producing phosphate fertilizer, steel, aluminum, magnesium, glass and bricks contain more F since F and heavy-metal-containing dust emitted into the air settle on plants (Oelschlager, 1974). The gaseous F emitted from these industrial sources is known to have phytotoxic effects on plants. Plants can also incorporate F from contaminated soils (Arnesen, 1997). The fluoride thus absorbed is translocated to the shoots, causing physiological, biochemical and structural damage and even cell death, depending on the concentration in the cell sap (Miller, 1993). Some plants accumulate F even at higher concentration (up to 4000 μg F g^{-1}) do not show signs of toxicity (Sheldrake *et al.*, 1978). Most other plants show sign of toxicity at much lower concentration (Brewer, 1966), while some species are extremely sensitive to concentration <20 μg F g^{-1} (Istas and Alaerts, 1974). Fluoride is more soluble in acid soils due to which its uptake by plants is enhanced. Liming the soil with calcium oxide reduces F accumulation (Bear, 1954). Plants grown in clay loam take up less F than those grown in sand. Plants do not require F, and tissues concentration from uncontaminated soils rarely exceed 30 mg kg^{-1}, dry mass (Kabata-Pendias, 2001). The fluoride content of both leafy and root vegetables usually do not differ appreciably from those of cereals, with the exception of spinach that is usually enriched in F and it is known as a good accumulator of F (Madhavan and Subraminian, 2006). Among items of plant origin, tea leaves contain significant amounts (30–300 mg g^{-1} dry weight) of F. Bioavailability of F to plants could be influenced by several metal ions due to complexation, precipitation, pH changes, *etc.*

From an Indian perspective very limited field-based research work has been done in the aspect of F accumulation in crops and vegetables (Gupta and Banerjee, 2011; Saini *et al.*, 2013). Most of the research studies have been done in the aspect of physiological and biochemical effects and that too also in the laboratory condition or in pot culture (Brahmbhatt and Patel, 2013). It is thought that substantial amounts of F may be ingested by crops and vegetables cultivated with F-contaminated irrigation water. In some states of India particularly Rajasthan, Uttar Pradesh and Andhra Pradesh F-contaminated groundwater is the major source of irrigation water for cultivation (Gupta and Banerjee, 2009). Fluoride uptake by plants depends on the biogeochemical factors, affecting its availability in soils. With this background this book

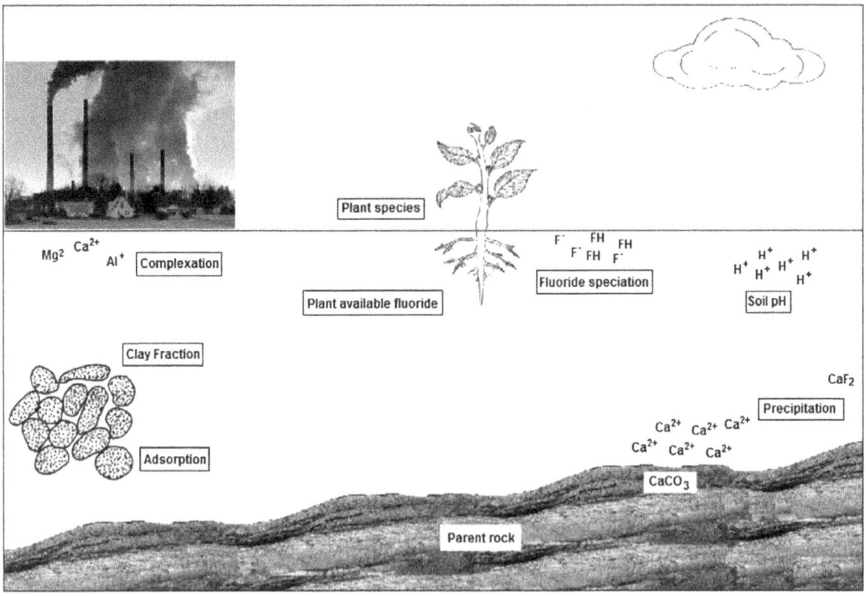

Figure 7.2 Schematic diagram showing factors and processes influencing fluoride availability in plants. Fluoride emitted from industries become complexed with Al^{3+}, Mg^{2+} and Ca^{2+}. In acidic soil F became available to plant. Fluoride adsorbed at the surface of clay particle by adsorption process. Presence of excess Ca^{2+} may induce precipitation of F as CaF_2 by precipitation process. (Source: own artwork.)

chapter focuses on the fate of F in soils, its uptake, and distribution and accumulation in different crops and vegetables of Indian origin. Additionally, a separate case study is also discussed for better understanding of its accumulation scenario in different crops and vegetables as well as its quantitative intake thorough diet. Figure 7.2 illustrates the factors and processes that influence the F availability in soils, irrigated with F contaminated water discussed in this chapter.

7.2 Fluoride in Soil

For all soils, it is the soluble F content that is biologically important to plants and animals. The solubility of F in soil is controlled mainly through F adsorption by inorganic constituents of the soil and soil pH (Loganathan *et al.*, 2003). In normal soil, the F is strongly adsorbed to the soil (Barrow, 1986) and hence plant uptake of F is generally minimal (McLaughlin *et al.*, 1997). The greater solubility of F under acidic conditions was explained by the formation of AlF_x complexes, whereas under alkaline conditions by desorption of free F as a result of repulsion by the negatively charged surfaces (Wenzel and Blum, 1992).

7.2.1 Mobility of Fluoride in Soil Solution

The mobility of F in soils depends on various factors such as initial concentration of F, solubility of F-bearing minerals, types of soil, pH of soil and other dissolved cation/anions in the soil solution. The solubility of F is controlled by equilibrium with F-containing minerals. In calcareous soil (pH ranges 7.0 to 8.5) a continuous supply of soluble calcium will induce the fluorite precipitation when this type of soil is irrigated with high-F water. The presence of large amounts of clay size fraction and aluminum hydroxide also affect the absorption of F in the soil (Arnesen and Krogstad, 1998). Aluminium hydroxide has a pH range of 7.1 to 9.5 and alumina has PZC of 8.0 (Mullar, 2005). So F has a high affinity to adsorb on these minerals in the neutral pH range. However, anion exchange is the major mechanism for the retention of F on aluminum hydroxide. The general reaction of anion exchange of F with minerals (Huang and Jackson, 1965) can be written as:

$$\text{Mineral} - \text{OH} + \text{F} \longrightarrow \text{Mineral} - \text{F} + \text{OH}^-$$

The presence of clay minerals in soil plays an important role in retaining F in soils in the pH range between 5.5 and 6.5.

7.2.2 Pathways of Fluoride Uptake by Plants

Most of the soluble F fraction in the soil is transported to plants through an apoplastic transport mechanism in the root systems (Figure 7.3). This mechanism is more dominant because cell walls also having negative charge promote the transport of negatively charged F ions. Apart from this, the low permeability of F through cell membrane also hinders diffusion in the cells. The apoplastic (nonliving) pathway provides a route toward the vascular stele through free spaces and cell walls of the epidermis and cortex. An additional apoplastic route that allows direct access to the xylem and phloem is along the margins of secondary roots. Secondary roots develop from the pericycle, a cell layer just inside the endodermis. The endodermis is characterized by the casparian strip, a suberized layer that hinders F movement in the symplast in order to enter the vascular system. Since secondary roots grow through the endodermis, a direct pathway to the xylem and phloem is available that bypasses the casparian strips and allows F to enter the vascular system without moving into the symplast (living tissue).

7.2.3 Factors Affecting Plant Uptake of Fluoride

In general, it is expected that F would be more readily taken up by plant at a pH level where F is more soluble. Two factors are responsible for pH-dependent F uptake: (i) Anionic exclusion of F due to the presence of negative charge in the cell wall and lower permeability coefficient of F ion than that of HF and (ii) Least F is dissolved in soil solution at a pH level of 6 and therefore less is available for plants. According to Stevens *et al.* (1995) at lower pH

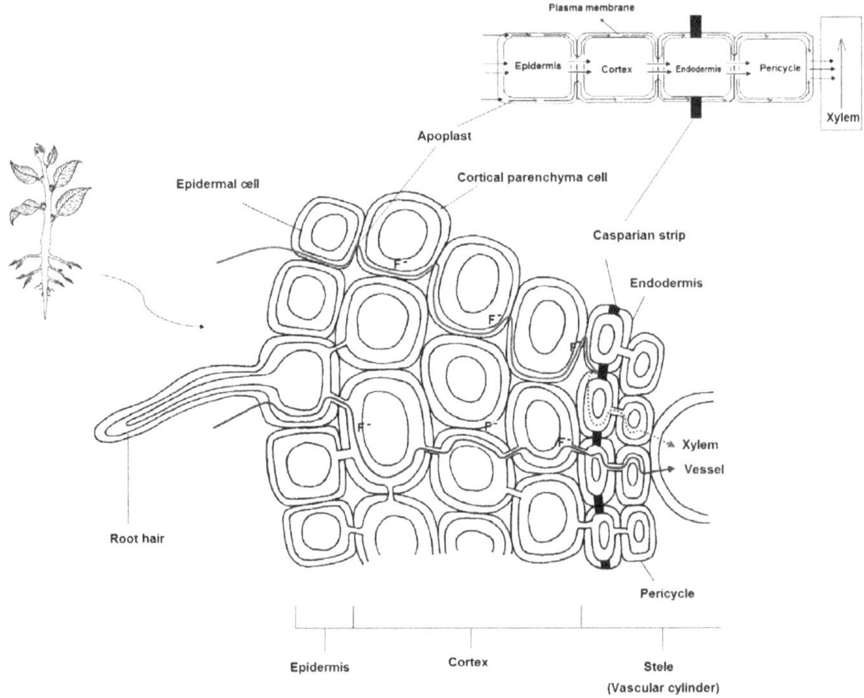

Figure 7.3 Pathways of fluoride in the root. This figure illustrates apoplastic pathways of fluoride through the root cell. (Source: own artwork.)

values the F content in the shoot is predicted to increase. The risk of plant uptake of F is very low in soils with neutral or high pH and increases with decreasing pH. Higher Aluminium content in the tissue will have an affinity to bind greater amount of F as complexes such as Al $(OH)_2F$. The calcium content in soil also has an influence on the concentration of F in plants. In calcium-rich soil a higher amount of F is observed in plant roots.

7.3 Fluoride Accumulations in Crops and Vegetables

Nearly all food contains small quantities of F and the total daily intake through any average human diet is small, except in endemic regions. The contribution of food to the total daily intake of F varies from region to region. Fluoride in soil and water used for irrigation ultimately finds its way into food items. Fluoride exerts residual toxicity due to its nonbiodegradable nature and passes on to each phase of the trophic level through the food chain. The food items carrying high levels of F are paddy (*Oryza sativa*), wheat (*Triticum sativum*), mustard (*Brassica juncea*), methi (*Trigonella foenum-graecum*), mentha (*Mentha arvensis*), maize (*Zea mays*), bathua (*Chenopodium alba*), pea (*Pisum sativum*), moong (*Vigna radiate)*, chickpea

Table 7.1 Fluoride content in agricultural crops and of different parts of India. Fluoride content in different food items has been studied in detail by various investigations. A, B, C and D indicates Kolkata, Mumbai, Podli (Andhra Pradesh) and Andhra Pradesh respectively having F in irrigation water of 0.4, 0.15–0.48, 2.5–5.0 and 1–12 mg L^{-1}, respectively. (Source: Susheela (ed), 1993).

	Fluoride in (mg kg^{-1})			
Crops and Vegetables	**A**	**B**	**C**	**D**
Cereal				
Wheat	4.6	2.59–3.3	—	—
Rice	5.9	3.27–14.03	29.3	—
Pearl millet	—	1.72–2.23	2.82	74.0
Maize	5.6	—	—	—
Pulses and Legumes				
Bengal gram	6.2	3.84–4.84	—	14.8
Green gram dal	2.5	2.34–4.84	—	21.2
Red gram dal	3.7	2.34–4.84	—	52.8
Soyabean	4.0	—	—	—
Leafy Vegetables				
Spinach	2.0	0.77–4.14	—	—
Cabbage	3.3	1.28–2.29	—	—

(*Cicer arietinum*), ridge gourd (*Luffa acutangula*), lady's finger (*Abelmoschus esculentus*), tomato (*Lycopersicon esculentum*), pumpkin (*Cucurbita moschata*), bottle gourd (*Lagenaria siceraria*), cauliflower (*Brassica oleracea var. Botrytis*), cabbage (*Brassica oleracea var. Capitata*), carrot (*Daucus carot*), brinjal (*Solanum melongena*), beet (*Beta vulgaris*), potato (*Solanum tuberosum*), kachri (*Citrullus melo var. Momordica*), and onion (*Allium cepa*). Although it is very well recognized that F enters the body through various food items, the data available in the country is rather scanty. In general, it can be said that F accumulation in the roots is higher than in leaves, stems and seeds (Gupta and Banerjee, 2009). However, in tea plants the data show a completely different behavior, as F is accumulated in the leaves and the concentration in the roots is lower. The fluoride concentration in plant shoots differs from species to species. Due to the apoplastic pathway through the leaks of the casparian strip, the F concentration of highly branched root systems seems to have higher uptake of F in the leaves *via* xylem transport. This is probably due to the relatively low permeability through the endodermis. However, the information is based on agricultural crop analysis of certain regions/ locations. The data thus emerging from a location is unlikely to be valid for other regions or locations. The main reason for this being the F content of water and soil being variable from location to location. There is a need for F analysis of agricultural crops district-wise, which will add invaluable information to educate and create awareness on the importance of selection of agricultural crops which are health-promoting crops and vegetables. The fluoride content in various food items analyzed under the Fluoride Mission Programme of India is presented in Table 7.1.

7.3.1 Indian Scenerio

7.3.1.1 Rajasthan

This is one of the most severely F-affected states in India and is considered to be endemic to fluorosis in all 33 districts. Reported work from Nagaur, Dausa and Tonk districts of Rajasthan shows that among 6 types of leafy vegetables, bathua, mentha and spinach were found to accumulate higher amounts of F – 67.30, 43.29 and 35.27 mg kg^{-1} F, respectively, when irrigated with 3.54 to 11.82 mg L^{-1} F in irrigated water (Table 7.2). Fluoride normally accumulates throughout the lifetime of leaves. Old leaves have higher F content than the young leaves due to their longer time of exposure (Vike and Habjorg, 1995). Sigh *et al.* (1995) also reported that leafy and root vegetables accumulate more F. There was a significant correlation between the total F content and labile F content in leafy vegetables at (r = 0.596) at the 0.05 level of significance. In comparison to leafy vegetables the total F content (Average) in crops observed that pearl millet 20.92 6 mg kg^{-1} > mustard 12.30 6 mg kg^{-1} > wheat 6.32 6 mg kg^{-1}.

7.3.1.2 Uttar Pradesh

The Unnao district of Uttar Pradesh has been reported widely for the occurrence of F in groundwater (Kumar *et al.*, 2004), which is used for drinking as well as irrigating the vegetables and cereal crops. In this area a study was conducted (Jha *et al.*, 2011) to evaluate the F ingestion through vegetables and cereal crops grown in the vicinity of the brick kilns and sodic area having pH > 8.5 and SAR > 15. The fluoride concentrations in vegetables and cereal crops such as paddy and wheat are presented in Table 7.3. It was found that F concentration was highest in the vegetables and cereal crops grown near the vicinity of brick kilns followed by the areas dominated by sodic soil. In both the areas, the F content in spinach, mentha, ridge gourd and lady finger was higher than other vegetables such as tomato, pumpkin, bottle gourd, cauliflower, cabbage and carrot. However, in cereal crops, the F accumulation was found to be more in wheat as compared to paddy. The exposure dose of F was determined using estimated daily intake (EDI) and bioconcentration factor (BCF) of F. The children of age group 3–14 years in the potentially fluoridated area were found to be at the risk of fluorosis. The mean BCF value of F was the highest in mint (36.6 mg kg^{-1} $_{dwt}$ plant per mg kg^{-1} $_{dwt}$ soil), followed by spinach (33.99 mg kg^{-1} $_{dwt}$ plant per mg kg^{-1} $_{dwt}$ soil).

7.3.1.3 Andhra Pradesh

An exhaustive study has been conducted and results published that focuses on the F content of crops and other items grown/available in Anantapur district in Andhra Pradesh. The highlight of the publication is 32 locally grown

Table 7.2 Fluoride accumulation scenario in crops and vegetables of different endemic areas of Rajasthan, India. This table represents accumulation of F by different crops and vegetables collected from different endemic villages of Nagur, Dasua and Tonk district of Rajasthan, India. In this three districts F content in irrigation water varies from 0.92–14.62 mg L^{-1}, 5.1–14.7 mg L^{-1} and 0.3–9.8 mg L^{-1} respectively. Each value represents mean ± SD of F concentration (mg kg^{-1}) in crops and vegetables. Data obtained from Yadav *et al.* (2013) and Saini *et al.* (2013) with permission from the publisher.

District	Name of the Tehsils	Name of the Villages	Crops and Vegetables (mg kg^{-1})																			
Nagaur	Nawa		Pearl millet	Wheat	Mustard	Paddy	Barley	Chickpea	Mentha	Spinach	Radish leaves	Mustard leaves	Methi	Bathua	Moong	Gwarfali	Pea	Kachri	Onion	Radish	Tomato	Potato
		Sirsi	2.76 ± 0.22	—	—	—	—	—	—	—	—	—	12.6 ± 0.35	—	—	—	—	—	—	—	—	—
		Lohrana	3.84 ± 0.20	—	—	14.44 ± 0.19	—	—	—	—	—	—	13.06 ± 0.21	—	—	—	—	—	—	—	—	—
		Chosla	3.70 ± 0.10	—	—	—	3.84 ± 0.15	15.88 ± 0.12	—	—	—	—	—	—	—	—	—	—	—	—	—	—
		Piperali	3.84 ± 0.17	—	—	—	—	—	—	—	—	—	—	—	—	—	—	—	—	—	—	—
		Trisinghiya	2.76 ± 0.25	—	—	15.88 ± 0.12	—	—	—	—	—	—	10.46 ± 0.19	—	—	—	—	—	—	—	—	—
		Loonwa	3.70 ± 0.16	—	—	—	—	—	—	—	—	—	—	—	—	—	—	—	—	—	—	—
		Lalas	2.88 ± 0.20	6.96 ± 0.21	14.44 ± 0.18	—	4.84 ± 0.12	11.74 ± 0.23	—	23.12 ± 0.19	17.36 ± 0.13	24.86 ± 0.20	18.24 ± 0.15	—	—	—	—	—	—	21.42 ± 0.11	—	—
		Gigore	—	2.76 ± 0.21	10.10 ± 0.23	—	1.98 ± 0.21	7.80 ± 0.11	—	19.80 ± 0.19	10.46 ± 0.15	24.0 ± 0.26	—	13.94 ± 0.16	—	—	—	—	8.08 ± 0.17	15.32 ± 0.25	—	—
		Chitawa	1.88 ± 0.14	2.88 ± 0.11	—	—	—	11.34 ± 0.10	—	16.76 ± 0.18	—	—	—	—	—	—	—	13.94 ± 0.13	—	—	—	—
		Govindi (Rulaniyo ki Dhani)	4.66 ± 0.22	—	—	—	5.66 ± 0.26	—	—	—	—	—	—	—	—	—	—	—	—	—	—	—
		Rawa	3.22 ± 0.17	—	—	—	—	—	—	—	—	—	—	—	10.70 ± 0.23	13.24 ± 0.11	—	—	—	—	—	—
		Anandpura	6.18 ± 0.24	3.22 ± 0.13	—	15.98 ± 0.22	—	—	—	15.98 ± 0.18	11.94 ± 0.16	—	—	—	—	—	—	—	23.92 ± 0.20	14.58 ± 0.10	—	—
		Shivdanpura	7.24 ± 0.21	—	—	18.98 ± 0.23	—	—	—	25.70 ± 0.12	10.70 ± 0.15	—	18.98 ± 0.10	—	—	—	—	—	17.44 ± 0.19	17.44 ± 0.19	—	—
		Sujanpura	—	—	—	—	—	—	—	—	20.56 ± 0.16	—		13.24 ± 0.20	—	—	8.34 ± 0.11	—		22.20 ± 0.19	—	—
Dausa		Bairwa mohalla hingotia	—	3.24 ± 0.01	—	—	—	—	—	—	—	—	—	—	—	—	—	—	—	—	1.10 ± 0.23	1.22 ± 0.24
		Bairwa Dausa	—	14.2 ± 0.19	—	—	—	—	—	—	—	—	—	—	—	—	—	—	—	—	4.09 ± 0.12	2.89 ± 1.12
		Jirotakalo Dausa	—	3.48 ± 0.09	—	—	—	—	—	—	—	—	—	—	—	—	—	—	—	—	1.10 ± 1.63	1.26 ± 0.23

(*continued*)

Table 7.2 (*continued*)

District	Name of the Tehsils	Name of the Villages	Crops and Vegetables (mg kg^{-1})																			
		Malarana/ Dausa	—	9.11 ± 0.04	—	—	—	—	—	—	—	—	—	—	—	—	—	—	—	—	2.15 ± 1.21	2.9 ± 0.11
		Jhonpuria	—	10.72 ± 0.06	—	—	—	—	—	—	—	—	—	—	—	—	—	—	—	—	4.6 ± 0.18	2.92 ± 1.31
		Jog Sahaipura/ Dausa	—	3.56 ± 0.01	—	—	—	—	—	—	—	—	—	—	—	—	—	—	—	—	1.5 ± 0.29	1.27 ± 0.25
		Khandelwal/ Dausa	—	14.3 ± 0.26	—	—	—	—	—	—	—	—	—	—	—	—	—	—	—	—	4.19 ± 1.74	2.92 ± 0.23
Tonk	Newai	Raholi	—	6.76 ± 0.06	12.53 ± 0.21	—	—	—	45.97 ± 1.34	32.30 ± 1.45	—	—	—	83.23 ± 0.61	—	—	—	—	—	—	—	—
		Chatarbhu- jpura	39.09 ± 0.98	4.76 ± 0.04	14.56 ± 0.83	—	—	—	47.58 ± 1.67	48.42 ± 2.84	—	—	—	81.09 ± 0.86	—	—	—	—	—	—	—	—
		Sunara	37.01 ± 0.69	3.89 ± 0.06	10.23 ± 0.45	—	—	—	42.72 ± 1.72	—	—	—	—	—	—	—	—	—	—	—	—	—
		Banasthali	33.02 ± 0.61	5.86 ± 0.05	12.36 ± 0.72	—	—	—	42.68 ± 1.67	44.08 ± 1.96	—	—	—	82.03 ± 1.23	—	—	—	—	—	—	—	—
		KVK Farm	41.04 ± 1.07	7.98 ± 0.09	13.61 ± 0.62	—	—	—	51.97 ± 3.63	64.09 ± 2.12	—	—	—	98.42 ± 2.81	—	—	—	—	—	—	—	—
		Mashi Dam	35.78 ± 1.29	4.34 ± 0.08	11.26 ± 0.45	—	—	—	39.72 ± 1.25	30.41 ± 1.13	—	—	—	54.23 ± 1.54	—	—	—	—	—	—	—	—
		Newai	37.44 ± 1.37	6.42 ± 0.02	12.40 ± 0.56	—	—	—	42.17 ± 2.12	38.92 ± 1.72	—	—	—	71.00 ± 1.67	—	—	—	—	—	—	—	—
		Chainpura	—	6.31 ± 0.05	13.75 ± 0.19	—	—	—	44.27 ± 2.76	—		—	—	78.92 ± 1.31	—	—	—	—	—	—	—	—
		Mundiya	34.00 ±0.51	4.52 ± 0.45	12.45 ± 0.24	—	—	—	41.84 ± 2.18	37.82 ± 2.10		—	—	80.27 ± 4.30	—	—	—	—	—	—	—	—
		Gunshi	34.87 ±0.81	6.82 ± 0.08	11.22 ± 0.85	—	—	—	37.84 ± 1.37	35.03 ± 1.64		—	—	67.21 ± 2.42	—	—	—	—	—	—	—	—
		Baroni	36.09 ±1.21	—	13.87 ± 0.98	—	—	—	35.48 ± 1.79	41.38 ± 2.63		—	—	—	—	—	—	—	—	—	—	—
		Siras	31.08 ±0.67	—	10.75 ± 0.21	—	—	—	—	43.93 ± 2.14		—	—	76.51 ± 0.72	—	—	—	—	—	—	—	—
		Jamdoji	37.97 ±0.64	5.93 ± 0.04	8.96 ± 0.17	—	—	—	47.97 ± 2.31	—		—	—	79.86 ± 0.47	—	—	—	—	—	—	—	—
		Sirohi	36.07 ±0.32	6.21 ± 0.02	12.78 ± 0.52	—	—	—	43.86 ± 2.89	40.71 ± 2.67		—	—	72.01 ± 2.24	—	—	—	—	—	—	—	—
		Karera	—	7.32 ± 0.04	—	—	—	—	41.97 ± 1.78	38.96 ± 0.34		—	—	67.62 ± 1.53	—	—	—	—	—	—	—	—
		Datwas	34.98 ±0.57	—	13.79 ± 0.43	—	—	—	—	38.09 ± 1.32		—	—	57.09 ± 2.73	—	—	—	—	—	—	—	—

Table 7.3 Fluoride accumulation scenario in crops and vegetables of different endemic areas of Uttar Pradesh, India. This table represents accumulation of F by crops and vegetables grown on two potentially fluoride contaminated sites such as brick kiln area and F-rich sodic area with a comparison on normal area (F uncontaminated). Sampling sites were within the periphery of 5 km from the brick kiln. In this table nd stands for not detected and values are indicated as mean ± SD. Data was taken from Jha *et al.* (2011) with permission from the publisher.

Vegetables and Cereal Crops	Concentration of F(mg $kg^{-1}{}_{dwt}$)							
	Brick kiln sites				Sodic sites			
	Bhaulikhera	Chaida	Hindukhera	Average	Nawabganj	Newai	Maurawan	Average
Vegetables								
Spinach	44.8 ± 3.9	38.9 ± 2.3	55.2 ± 3.7	46.30 ± 3.3	29.2 ± 3.1	30.4 ± 3.8	28.8 ± 2.7	29.47 ± 3.2
Ridge gourd	20.3 ± 1.0	26.6 ± 1.5	24.8 ± 1.7	23.90 ± 1.4	17.5 ± 1.1	16.8 ± 1.5	18.1 ± 1.9	17.47 ± 1.5
Lady's finger	26.2 ± 1.7	22.1 ± 4.1	25.6 ± 1.4	24.77 ± 2.4	18.8 ± 1.6	16.1 ± 2.5	16.4 ± 1.6	17.10 ± 1.9
Menthe	44.4 ± 3.1	54.6 ± 3.6	49.8 ± 3.4	49.60 ± 3.4	nd	34.5 ± 2.3	37.4 ± 1.6	23.97 ± 1.3
Tomato	7.6 ± 0.8	nd	6.3 ± 0.7	4.63 ± 0.5	5.5 ± 0.3	6.0 ± 0.8	5.8 ± 0.75	5.77 ± 0.6
Pumpkin	2.8 ± 0.5	3.2 ± 0.8	nd	2.00 ± 0.4	2.1 ± 0.4	2.5 ± 0.6	2.2 ± 0.4	2.27 ± 0.5
Bottle gourd	2.4 ± 0.6	2.8 ± 0.5	2.0 ± 0.4	2.40 ± 0.5	1.9 ± 0.4	2.2 ± 0.4	2.0 ± 0.1	2.03 ± 0.3
Cauliflower	2.7 ± 0.6	3.0 ± 0.2	2.6 ± 0.5	2.77 ± 0.4	1.7 ± 0.3	2.2 ± 0.4	2.0 ± 0.3	1.97 ± 0.3
Cabbage	3.4 ± 0.4	4.1 ± 0.5	3.8 ± 0.4	3.77 ± 0.4	2.8 ± 0.5	2.5 ± 0.5	1.9 ± 0.3	2.40 ± 0.4
Carrot	2.7 ± 0.4	3.0 ± 0.3	2.6 ± 0.1	2.77 ± 0.3	1.7 ± 0.3	2.2 ± 0.4	2.0 ± 0.3	1.97 ± 0.3
Cereal crops								
Paddy	11.2 ± 0.5	12.0 ± 0.9	10.3 ± 0.8	11.17 ± 0.7	8.0 ± 0.4	8.4 ± 0.6	9.1 ± 0.4	8.50 ± 0.5
Wheat	15.3 ± 0.6	14.8 ± 0.7	14.4 ± 0.5	14.83 ± 0.6	12.8 ± 0.6	11.9 ± 0.5	12.1 ± 0.4	12.27 ± 0.5

agricultural crops are known to have F ranging from 0.2 to 11.0 mg kg^{-1} with the exception of coconut water where no traces of F were detected.

7.4 Fluoride Accumulations in Crops and Vegetables of Birbhum District, West Bengal – A Case Study

Study of accumulation of F in crops and vegetables was conducted in Junidpur village (24° 06′07.5″N & 87° 46′ 54.7″E) of Rampurhat block of Birbhum district, where fluorosis has been prevailing for some 6–8 years (Figure 7.4). To control the prevalence of fluorosis, the local villagers started to drink defluoridated water supplied from a nearby water treatment plant. In contrast, Moula village of the same block was taken as a control area. Both the villages under its administration have been put on the list of economically and socially backward, with farming industry being predominant and almost no industrial enterprise. The distance between the two villages is about 6–8 kilometers. They both grow rice as their main foodstuff and there are no factories around. Hence, almost no man-made pollution source exists around the two villages. Apart from an F-contaminated source of irrigation water there is no significant difference in both natural and social backgrounds between the two villages. In addition to Junidpur another study area named Bakrashwar (23° 52′45.4″N & 87° 22′42.9″E) in the Dubrajpur block of Birbhum district was also selected for the present research work (Figure 7.4). In this area hot spring water, having higher F content, first stored in ponds and tanks and thereafter used for irrigation purposes. Among these two study areas Junidpur is the ideal one for carrying out this experimental work because people of this area not only take F-contaminated drinking water but also consume crops/vegetables cultivated in their own agricultural field (irrigated with the aid of F-contaminated irrigation water) as food items.

7.4.1 Fluoride in Irrigation Water

In Junidpur F concentrations in the irrigated water varies from 0.6 to 4.06 mg L^{-1} whereas in Bakreshwar F concentration in irrigation water ranges from 7 to 5 mg L^{-1} with a mean of 6 mg L^{-1}. In the control area F content in irrigation water ranges from 0.11 to 0.26 mg L^{-1}.

7.4.2 Fluoride in Irrigated Soil

Both $F_{(H_2O)}$ and $F_{(CaCl_2)}$ varied between 1.45 and 3.80 mg kg^{-1} (mean 2.91 mg kg^{-1}) and between 0.85 and 6.75 mg kg^{-1} (mean 5.16 mg kg^{-1}), respectively, in the Junidpur area and between 2.80 and 2.95 mg kg^{-1} (mean 2.90 mg kg^{-1}) and between 4.03 and 4.68 mg kg^{-1} (mean 4.52 mg kg^{-1}). In this study $F_{(CaCl_2)}$ is higher than the (1:1) $F_{(H_2O)}$, in agreement with earlier research showing that a solution of $CaCl_2$ is generally more efficient than water in extracting F from soil (Supharungsun and Wainwright, 1982). On the other hand, F_{Total} in both

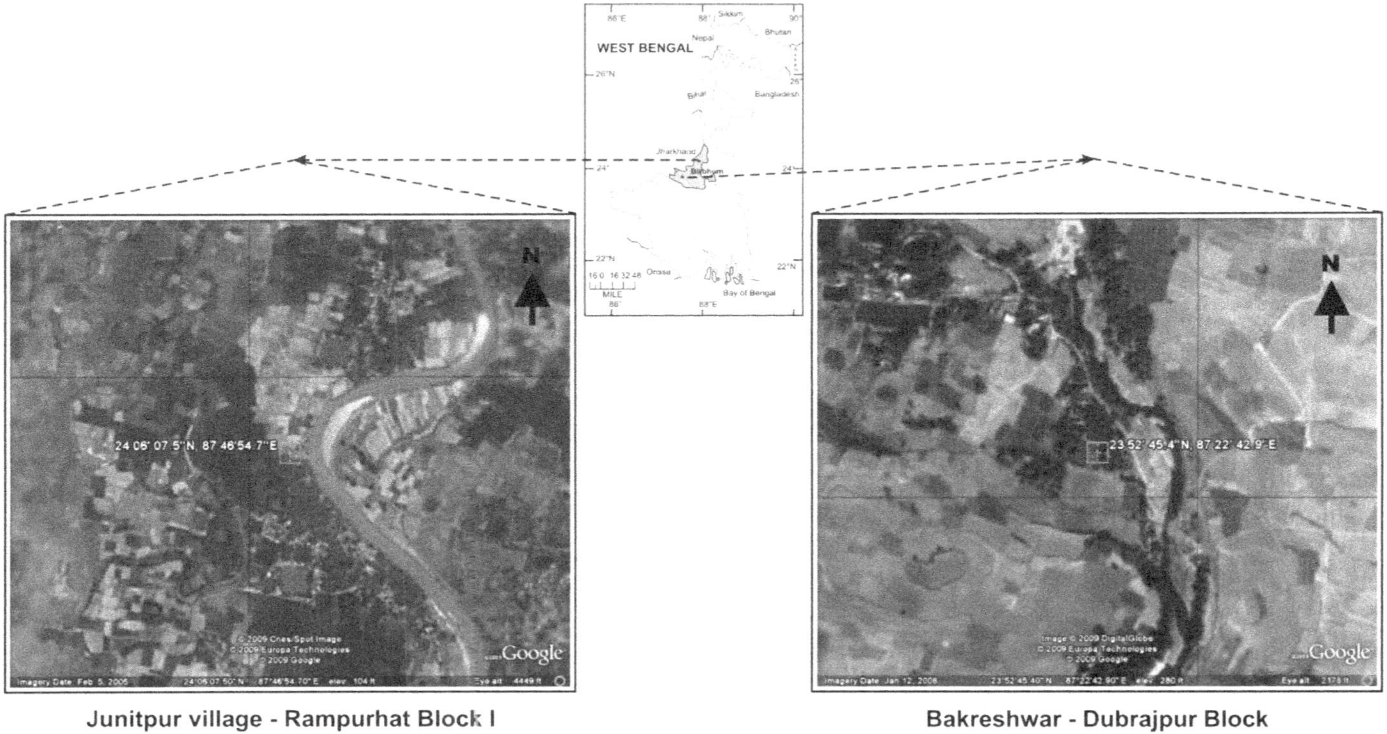

Figure 7.4 Study area location. This figure indicates location details of two endemic areas, namely Junidpur and Bakreshwar situated in the Rampurhat and Dubrajpur Block of Birbhum district, West Bengal, India. (Source: Google image and own art work.)

the areas varies between 143 and 144 mg kg^{-1} (mean 143 mg kg^{-1}) and 24 to 68 mg kg^{-1} (mean 50 mg kg^{-1}) suggesting a significant contribution towards its accumulation in paddy from all these sources. Only water-soluble F in soils can be absorbed directly by animals and plants (Brewer, 1966).

7.4.3 Fluoride in the Harvested Crops and Vegetables of the Study Area

From Table 7.4 it is found that the contents of F in paddy and vegetables in the endemic fluorosis area are higher than those in the fluorosis-free area (control area). For the convenience of discussion, the enrichment multiples here means that the contents of F in the samples from a fluorosis-diseased area are a multiple of those of the samples from a fluorosis-free area. The contents of fluoride in paddy are above the National Standard (1 mg kg^{-1}) for both Junidpur and control villages. The enrichment multiple of paddy seed in Junidpur area is 2.36 and 1.54 with respect to F_{Total} and $F_{(H_2O)}$, respectively. No multiple in Bakreshwar area is calculated because no paddy cultivation is done by the use of F-contaminated hot spring water. Spinach, coriander, cauliflower, mercellia, cabbage, raddish, potato, tomato, brinjal, onion and bean are the main edible vegetables for the local villagers of the study areas. The main edible parts of spinach are stem and leaves and combined enrichment multiples with respect to F_{Total} and $F_{(H_2O)}$ in both Junidpur and Bakreshwar areas are 2.37, 1.23, 4.68 and 4.80, respectively. In the case of coriander, the leaf is the main edible part and enrichment multiples with respect to F_{Total} and $F_{(H_2O)}$ concentration in Junidpur area are 3.23 and 3.06, respectively. No coriander vegetables are cultivated in Bakreshwar area. The edible part of cauliflowers (stalk) collected from Bakreshwar area shows the multiples of 2.99 and 8.05, respectively. Enrichment multiples of mercellia are alarming, which show 81.23 and 41.63 in Junidpur and 89.23 and 57.25 in Bakreshwar with respect to F_{Total} and $F_{(H_2O)}$. The edible part of cabbages (stalk) collected from Bakreshwar area shows multiples of 111.48 with respect to $F_{(H_2O)}$. The multiples of stalk and leaves of radish, collected from Bakreshwar area are 1.48 and 1.23 and 10.46 and 0.36, respectively. The stem part of potato is the main edible part and enrichment multiples are 1.52 and 14.50 and 1.81 and 40.75, respectively, from both the study areas. With respect to F_{Total} enrichment multiples of tomatoes are 2.18 and 2.11, respectively in Junidpur and Bakreshwar. Brinjals are collected from both the areas and the multiples are 4.11 and 1.57 and 3.75 and 4.44, respectively. Onions are collected only from the Junidpur area and have a multiple of 1.24 and 9.25 with respect to F_{Total} and $F_{(H_2O)}$. Apart from all the above-mentioned vegetables alarming levels of F accumulation are also found in beans, mustard seed and poppy seeds (Table 7.4). In these vegetables and crops multiples cannot be calculated as F content in all the above-mentioned crops/vegetable, collected from the control area, has F concentration below the detection limit (BDL). In the same environmental background, the contents of F vary from one kind of vegetable to

Table 7.4 Water-soluble and total fluoride content in different parts of crops and vegetables (mg kg^{-1}) ($n = 4^a$) in endemic areas of Birbhum district, West Bengal. Fluoride accumulation in crops and vegetables of different endemic areas along with control site of Birbhum district, West Bengal has been studied by Gupta and Banerjee (2009, 2011). Each sample analyzed in parallel 4 times. The average value and standard deviation were taken, calculated in terms of the dry mass of plants.

	$F_{(Total)}$	$F_{(H_2O)}$	$F_{(Total)}$	$F_{(H_2O)}$	$F_{(Total)}$	$F_{(H_2O)}$	Multiples of enrichment			
Crops/ Vegetables	Junidpur	Junidpur	Bakrashwar	Bakrashwar	Control	Control	Junidpur $F_{(Total)}$	Junidpur F_{H_2O}	Bakrashwar $F_{(Total)}$	Bakrashwar $F_{(H_2O)}$
Paddy seeds	12.31 ± 0.64	2.88 ± 0.04	*	*	5.22 ± 0.24	1.87 ± 0.13	2.36	1.54	—	—
Paddy leaf	50.30 ± 2.40	5.88 ± 0.04	*	*	23.10 ± 0.12	3.75 ± 0.33	2.18	1.57	—	—
Paddy steam	19.57 ± 1.15	3.26 ± 0.01	*	*	2.74 ± 0.09	1.82 ± 0.18	7.14	1.79	—	—
Paddy root	80.83 ± 2.6	6.28 ± 0.01	*	*	52.73 ± 1.02	4.80 ± 0.11	1.53	1.31	—	—
Poppy	*	*	12.52 ± 1.02	*	BDL	BDL	—	—	—	—
Mustard seed	4.40 ± 0.21	0.87 ± 0.08	21.67 ± 1.07	*	BDL	BDL	—	—	—	—
Spinach leaf	11.37 ± 0.32	4.70 ± 0.17	22.47 ± 0.32	18.29 ± 0.19	4.80 ± 0.22	3.81 ± 0.18	2.37	1.23	4.68	4.8
Coriander leaf	26.94 ± 0.16	11.97 ± 0.26	*	*	8.33 ± 0.11	3.91 ± 0.07	3.23	3.06	—	—
Cauliflower	*	*	11.95 ± 0.19	5.07 ± 0.14	3.99 ± 0.32	0.63 ± 0.03	—	—	2.99	8.05
Marcellia	24.37 ± 1.12	3.33 ± 0.16	26.77 ± 1.15	4.58 ± 0.08	0.3 ± 0.12	0.08 ± 0.01	81.23	41.63	89.23	57.25
Cabbage	*	*	3.01 ± 0.08	2.81 ± 0.05	0.03 ± 0.003	BDL	—	—	111.48	—
Radish leaf	*	*	7.93 ± 0.23	2.1 ± 0.05	5.37 ± 0.23	1.71 ± 0.05	—	—	1.48	1.23
Radish	*	*	11.93 ± 0.19	0.87 ± 0.01	1.14 ± 0.32	0.31 ± 0.01	—	—	10.46	2.81
Potato	4.01 ± 0.19	0.58 ± 0.02	4.77 ± 0.03	1.63 ± 0.12	2.64 ± 0.42	0.04 ± 0.01	1.52	14.50	1.81	40.75
Tomato	8.75 ± 0.12	0.28 ± 0.01	8.5 ± 0.02	2.48 ± 0.16	4.02 ± 0.19	BDL	2.18	—	2.11	—
Bringel	14.46 ± 1.03	1.29 ± 0.09	13.21 ± 0.12	3.64 ± 0.09	3.52 ± 0.32	0.82 ± 0.01	4.11	1.57	3.75	4.44
Onion	9.19 ± 0.19	2.59 ± 0.06		*	7.40 ± 0.26	0.28 ± 0.02	1.24	9.25	—	—
Bean	15.26 ± 0.32	3.07 ± 0.15	*	*	BDL	BDL	—	—	—	—

*Mark in the table indicates crops/vegetables were not cultivated in the corresponding areas; BDL means below detection limit. Data was taken from Gupta and Banerjee (2009, 2011) with permission from the publisher.

another, indicating that each sort of vegetable has its own ability of enriching F. The distance between the two villages (one represents a fluorosis-diseased area and the other represents a fluorosis-free area) is only 6–8 km, so no difference is recognized in weather, landform, farming style and geological background between the two villages. Thereby, the enrichment of F in foodstuff and vegetables should be attributed to irrigating water with high F and soils with relatively high water-soluble F.

In order to determine the F accumulation scenario in different parts of crop plants separate analyses were carried out in root, stem, and leaf and seed part of paddy, collected both from the endemic area and the control area. A substantial accumulation of both $F_{(H_2O)}$ and F_{Total} were found throughout the plant body, namely, root, stem, leaf, and seeds (Table 7.4). Maximum accumulation of F (mg kg^{-1} dry wt.) occurred in the root (mean $F_{(H_2O)}$ 6.28 mg kg^{-1} and mean F_{Total} 80.83 mg kg^{-1}), followed by the leaf (5.88 mg kg^{-1} and 50.30 mg kg^{-1}), stem (3.26 mg kg^{-1} and 19.57 mg kg^{-1}) and seeds (2.88 mg kg^{-1} and 12.31 mg kg^{-1}). The translocation of F ion from soil to plant parts (transfer factor, TF) was calculated to determine the relative uptake of $F_{(H_2O)}$ ion by the plants with respect to the presence of F in soil solution. The ratio of F concentration between soil and plant parts (TF) is an important criterion for the contamination assessment of soils with a high level of ionic F. The ratio ">1" means hyperaccumulation of F in plant parts than soil. In the Junidpur study area, except seeds and all the paddy parts (*i.e.*, root, stem and leaf) with respect to $F_{(H_2O)}$ in soil the ratio was >1, suggesting hyperaccumulation of F in the sequence: TF_{root} (2.15) >TF_{leaf} (2.02) >TF_{stem} (1.12) >TF_{seeds} (0.98). A similar trend was also seen in the F_{CaCl_2} in the sequence: TF_{root} (1.22) >TF_{leaf} (1.13) >TF_{stem} (0.63) >TF_{seeds} (0.56). This observation is very similar to the study of Lan De *et al.* (2008). As the contents of F in soil were so high the contents of F in the roots increased sharply. Those very small clay minerals attached to the roots still contain some F. Such a small amount of F-bearing clay may also be harmful to human.

7.4.4 Quantitative Estimation of Fluoride in the Food Chain

Fluoride concentration in the environment is closely associated with human health. Total F content taken up into a human body from the environment is calculated according to epidemiological investigations of endemic fluorosis in experimental areas in Junidpur and Bakrashwar. It is assumed that 500 g of meal natural ingredient, 500 g of vegetable, 5 L of water (amount of drinking water for each resident in this region is more than that of other regions because of tropical climatic conditions), the proportion of F content in other pathways (such as tea and meat) to total daily intake of F is very small, so that has to be neglected in the calculation of the total daily intake of F. The meal is mainly rice. The total F concentration in meal is 12.31 mg kg^{-1}. Average values of F concentration in eight vegetables (*i.e.*, spinach, coriander leaf, marcellia, potato, tomato, onion, brinjel and bean) are

Table 7.5 Quantification of F in the meals and vegetables along with absorption capacity to human body. This table represents that each of the selected adult family members required an average of 500 g each of meal ingredient and vegetable food and 5 L water per day. No quantitative data on F absorption was determined from Bakreshwar due to nonavailability of data on rice and drinking water. Data was taken from Gupta and Banerjee (2009, 2011) with permission from the publisher.

Foods	Crops/ Vegetables	Study Area	Uptake Amount (g)	F Concentration (mg kg^{-1})	F Absorption (mg)
Meal	Paddy seeds	Junidpur	500	12.31	6.16
		Bakrashwar	—	—	—
Vegetable	Spinach	Junidpur	500	11.37	5.69
		Bakrashwar	500	22.47	11.24
	Coriander leaf	Junidpur	500	26.94	13.47
		Bakrashwar	—	—	—
	Cauliflower	Junidpur	—	—	—
		Bakrashwar	500	11.95	5.98
	Marcellia	Junidpur	500	24.37	12.19
		Bakrashwar	500	26.77	13.39
	Cabbage	Junidpur	—	—	—
		Bakrashwar	500	3.01	1.51
	Radish	Junidpur	—	—	—
		Bakrashwar	500	11.93	5.97
	Potato	Junidpur	500	4.01	2.01
		Bakrashwar	500	4.77	2.39
	Tomato	Junidpur	500	8.75	4.38
		Bakrashwar	500	8.5	4.25
	Onion	Junidpur	500	9.19	4.60
		Bakrashwar	—	—	—
	Brinjal	Junidpur	500	14.46	7.23
		Bakrashwar	500	13.21	6.61
	Poppy	Junidpur	—	—	—
		Bakrashwar	500	12.52	6.26
	Mustard seed	Junidpur	500	4.4	2.20
		Bakrashwar	500	21.67	10.84
	Bean	Junidpur	500	15.26	7.63
		Bakrashwar	—	—	—
Water		Junidpur	5 L	2.06(mg L^{-1})	10.3
		Bakrashwar	—	—	—

14.29 mg kg^{-1} and average value of F concentration in drinking water is 2.06 mg L^{-1} (Table 7.5) in Junidpur. Total F intake amounts from meal, vegetable and drinking water is equal to 23.61 mg day^{-1}. Fluoride content from the meal, vegetables and drinking water accounted for 26.09%, 30.28% and 43.63% of the total F amount, respectively. Therefore, it was found that the F amount in drinking water was not only the main source of F taken up by the human body but a major portion of F has been taken up by the meal and vegetables. The sum of the total percentage from the latter is 56% which is greater than the 44% intake from the water.

7.4.5 Salient Outcomes of the Case Study

- The fluoride contents of rice grains and vegetables (irrigated with F-contaminated groundwater and hot spring water) are obviously higher in fluorosis-diseased areas than in fluorosis-free areas (irrigated with groundwater having permissible levels of F). It is shown that water-soluble F in soil and F in irrigating water tend to be enriched in rice grains and vegetables. In order to diminish the risk of human exposure to F, it is of great importance to prevent crops from being irrigated by the F-contaminated irrigation water. It is advisable to grow crops that have relatively lower F fixation ability or species.
- The tissues of the plants that are of strong metabolism have higher F contents. The fluoride contents of roots and leaves are apparently higher than those in stalks and seeds. The fluoride contents of rice grains vary slightly with increasing water-soluble F contents in soil. The seeds of crops have a weak ability to enrich F; 80–95% of F in foods will be absorbed by a human body (Cerklewski, 1997). Therefore, growing the crops whose stalks and seeds are used as the main edible parts can help to decrease the intake of F by human beings.
- The input of high-F irrigation water/hot spring water into the environment will lead to the increase of the environmental F background value, resulting in a higher F level of drinking water and soil. Thus, the risk of human exposure to F will be enhanced. Only such measures as water-improving and defluoridation can essentially suppress the prevalence of fluorosis. Some sort of integrated controlling measures to avoid the excessive intake of F should be taken.

Summary Points

- This chapter focuses on F accumulation in crops and vegetables in India.
- The amount of plant-available F in the soil solution depends strongly on the solubility of mineral phases, soil type, and pH of the soil and on the presence of aluminum and calcium in soil.
- Plant species also play a critical role in F uptake.
- In spite of different soil types in India *viz.* alkaline soil in Rajasthan, acidic soil in Andhra Pradesh and alluvial soil in Uttar Pradesh crops and vegetables grown on different kind of soils show a more or less common accumulation trend.
- Within the F-contaminated irrigated states in India the most commonly grown vegetables *Chenopodium album, Spinacea oleracea and Mentha arvensis* is more accumulator, whereas within cereals *Pennisetum glaucum, Brassica juncea and Triticum sativam* are the dominant ones.
- Fluoride content in roots and leave are apparently higher than those in stalks and seeds, except for tea.

- The seeds of crops have a weak ability to enrich F; 80–95% of F in foods will be absorbed by a human body. Therefore, growing the crops whose stalks and seeds are used as the main edible parts can help to decrease the intake of F by human beings.
- A case study shows that in some endemic areas of West Bengal, India ingestion of F through crops and vegetable is substantially higher than through drinking water intake.

Key Facts of Fluoride Accumulation in Crops/ Vegetables

1. Fluoride has beneficial effects on teeth at low concentrations, but excessive exposure to F in drinking water or in combination with other sources can give rise mild dental fluorosis to crippling skeletal fluorosis as the level and period of exposure increases.
2. In India, about 20 and 70 million people suffer from dental and skeletal fluorosis.
3. The source of F in soil is the parent rock itself. The average crustal abundance is 300 mg kg^{-1}.
4. Fluoride is found in all natural waters at concentrations depending on the geology, chemistry, physical characteristics and climate of the area.
5. Although F in drinking water is usually the largest contributor to daily intake, there are other sources of F that may in certain cases be significant.
6. In some regions of India F-contaminated water is the major source of irrigation. So in this case substantial amounts of F may be ingested by crops and vegetables and thereafter enters the human body through diet.

Definitions and Explanations of Key Terms

Fluoride: Fluoride is the 13th most abundant element on the Earth's crust. It is a chemical ion of the element fluorine – F that has one extra electron that gives it a negative charge. Fluoride is found naturally in soil, water, foods, and several minerals, such as fluorapatite and fluorite.

Fluorosis: Excessive exposure to high concentrations of F during tooth development (during childhood) can result in tiny white streaks or specks in the enamel of the tooth in mild cases of dental fluorosis. Too much consumption of F may cause bone disease. In severe cases, there is damage to bones and joints, as well as pain.

Adsorption: Accumulation of sorbed substance (here F) on the surface of clay particle.

Desorption: The process of removing or changing of sorbed substance (here F) by the reverse of adsorption or absorption.

Xylem: The xylem is a transport tissue. Its basic function is to transport water and some nutrients through the plant.

Casparian strip: The casparian strip is a band of cell-wall material in the wall of the endodermis, which is chemically different from the rest of the cell wall and it is used to block passive flow of material from the root to the shoot.

Point of zero charge (PZC): Point of zero charge is the phenomenon of adsorption, and it describes the condition that the electrical charge density on a surface has a net neutral charge. The significance of this kind of plot is that a given mineral surface will have a positive charge at solution pH values less than the PZC and thus be a surface on which anions may adsorb. On the other hand, that mineral surface will have a negative charge at solution pH values greater than the PZC and thus be a surface on which cations may adsorb.

Apoplast: Apoplastic or nonliving pathway provides a route toward the vascular stele through free spaces and cell walls of the epidermis and cortex. An additional apoplastic route that allows direct access to the xylem and phloem is along the margins of secondary roots.

Symplast: Symplastic or living pathway route to the vascular stele involves cell-to-cell transport by plasmodesmata. Plasmodesmata are channels of cytoplasm lined by plasma membrane that transverse cell walls.

Transfer factor (TF): This is a ratio between the concentrations of $F_{(H_2O)}$ in the plant body and in the soil at that site. A ratio greater than 1 means higher accumulation of F^- in the plant parts than is present in the soil.

Bioconcentration factor (BCF): This is a common parameter for estimating the F concentration in vegetables and subsequently human exposure through consumption of vegetables, which is defined as the ratio of between the concentrations of F in the edible part of the vegetable to F concentration in soil.

Estimated daily intake (EDI): This is the quantitative health risk assessment due to consumption of various F-contaminated vegetables and cereal crops by using a generic equation (USEPA, 1992).

List of Abbreviations

BIS	Bureau of Indian Standards
BDL	Below Detectable Level
BCF	Bioconcentration Factor
CGWB	Central Ground Water Board
EDI	Estimated Daily Intake
PZC	Point of Zero Charge
SAR	Sodium Adsorption Ratio
TF	Transfer Factor
USEPA	United States Environmental Protection Agency
WHO	World Health Organisation

References

Amini, M., Mueller, K., Abbaspour, K. C., Rosenberg, T., Afyuni, M., Moller, K. N., Sarr, M. and Johnson, C. A., 2008. Statistical modeling of global geogenic fluoride contamination in groundwaters. *Environmental Science and Technology*. **42**: 3662–3668.

Arnesen, A. K. M., 1997. Availability of fluoride to plants grown in contaminated soils. *Plant and Soil*. **191**: 13–25.

Arnesen, A. K. M. and Krogstad, T., 1998. Sorption and de-sorption of fluoride in soil polluted from the aluminium smelter at Ardal in Western Norway. *Water Air and Soil Pollution*. **103**: 357– 373.

Ayoob, S. and Gupta, A. K., 2006. Fluoride in drinking water: a review on the status and stress effects. *Critical Reviews in Environmental Science and Technology*. **36**: 433–487.

Barrow, N. J., 1986. Testing a mechanical model 1. The effect of time and temperature on the reaction of fluoride and molybdate with a soil. *Journal of Soil Science*. **37**: 267–275.

Bear, F. E., 1954. Trace elements: progress report on research, with particular reference to New Jersey soils. *Journal of Agricultural Food Chemistry*. **2**: 244–251.

Bureau of Indian Standards (BIS), 2003. *Drinking Water Specification, IS:10500*. Bureau of Indian Standards, New Delhi.

Brahmbhatt, N. H. and Patel, R. V., 2013. Effect of sodium flouride and magnesium chloride and its interaction on seedling growth of *Medicago Sativa*, Anand, Gujarat. *Asian Journal of Plant Science and Research*. **3**: 80–84.

Brewer, R. F., 1966. Fluorine. *In Diagnostic Criteria for Plants and Soils*. Ed. H D Chapman. Univ. California Press, California, pp. 80–196.

Cerklewski, F. L. 1997. Fluoride bioavailability—nutritional and clinical aspects. *Nutrition Research*. **17**: 907–929.

Gupta, S. and Banerjee, S. 2009. Fluoride accumulation in paddy (*Oryza sativa*) irrigated with fluoride-contaminated groundwater in an endemic area of the Birbhum district, West Bengal. *Fluoride*. **42**: 170–173.

Gupta, S., and Banerjee, S. 2011. Fluoride accumulation in crops and vegetables and dietary intake in a fluoride-endemic area of West Bengal. *Fluoride*, **44**:153–157.

Huang, P. M. and Jackson, M. L., 1965. Mechanism of reaction of neutral fluoride solution with layer silicates and oxides of soil. *Soil Science Society of America, Proceedings*. **29**: 661–665.

Istas, J. R. and Alaerts, G., 1974. The effects of fluorine and its compounds on plants. *Revue de l'Agriculture*. **27**: 487–496.

Jha, S. K., Nayak, A. K. and Sharma, Y. K., 2011. Site specific toxicological risk from fluoride exposure through ingestion of vegetables and cereal cropsin Unnao district, Uttar Pradesh, India. *Ecotoxicology and Environmental Safety*, **74**:940–946.

A. Kabata-Pendias. 2001. *Trace Elements in Soils and Plants*, 3rd edn. Florida: CRC Press LLC.

Kumar, S., Rai, D., Rai, P. K., Shukla, A. B. and Saxena, A., 2004. Fluoride pollution in ground water in parts of Unnao district: impact and distribution, extended summaries. In: *Proceedings of the International Conference on Sustainable Management of Sodic Lands, Lucknow*, p. 474.

Lan De, Daishe, W. U., Ping, L. I., Tengsheng, W., Chengguang, C. and Wuyi, W., 2008. Influence of high-fluorine environmental background on crops and human health in hot spring-type fluorosis-diseased areas. *Chinese Journal of Geochemistry*. **27**: 335–341.

Loganathan, P., Hedley, M. J., Grace, N. D., Lee, J., Cronin, S. J., Bola, N. S. and Zanders, J. M., 2003. Fertiliser contaminants in New Zealand grazed pasture with special reference to cadmium and fluorine: a review. *Australian Journal of Soil Research*. **41**: 501–532.

Madhavan, N. and Subramanian, V., 2006. Environmental impact assessment including evolution of fluoride and arsenic contamination process in ground water and remediation of contaminated ground water system. In: M. Thangarajan (ed.) *Sustainable Development and Management of Ground Water Reserve*. Capital Publi. Co., New Delhi, pp. 128–155.

McLaughlin, M. J., Simpson, P. G., Fleming, N., Stevens, D. P., Cozens, G. and Smart, M. K., 1997. Effects of fertilizer type on cadmium and fluorine concentrations in clover herbage. *Australian Journal of Experimental Agriculture*. **37**: 1019–1026.

Miller, G. W. 1993. The effect of fluoride on higher plants with special emphasis on early physiological and biochemical disorders. *Fluoride*. **26**: 3–22.

Muller, K. 2005. *Fluorid im Grundwasser: Mogliche Ursachen und Aufbereitungstechniken in Entwicklungslandern*. Diplomarbeit ETH Zurich .

Oelschlager, W., 1974. Fluoride-containing mineral supplement in agriculture. *Fluoride*. **2**: 84–88.

Raju, K. C. B., 1998. Importance of recharging depleted aquifers: state of the art of artificial recharge in India. *Journal of Geological Society of India*. **5**: 429–454.

Saini P., Khan S., Baunthiyal M., and Sharma V., 2013. Mapping of fluoride endemic area and assessment of F^{-1} accumulation in soil and vegetation. *Environmental Monitoring Assessment*. **185**: 2001–2008.

Sheldrake, R., George, E. D., Leigh, E., St John Jr, E. and Donald, J. L., 1978. Lime and charcoal amendments reduced fluoride absorption by plants cultured in perlite peat medium. *Journal of the American Society for Horticulture Science*, **103**: 268–270.

Singh, V., Gupta, M. K., Rajwanshi, P., Srivastava, S. and Dass, S., 1993. Studies on ingestion of fluoride through tobacco pan masala and toothpaste. *Indian journal of Environmental Health*. **35**: 215– 220.

Singh, V., Gupta, M. K., Rajwanshi, P., Mishra, S., Srivastava, S., Srivastava, R., Srivastava, M., Prakash, S. and Dass, S., 1995. Plant uptake of fluoride in irrigation water by lady finger (*Abelmoschus Esculentus*). *Food and Chemical Toxicology*. **33**: 399–402.

Supharungsun, S. and Wainwright, M., 1982. Determination, distribution, and adsorption of fluoride in atmospheric-polluted soils. *Bulletin of Enviornment Contamination and Toxicology*. **28**: 632–636.

Stevens, D. P., McLaughlin, M. J. and Alston, A. M., 1995. Limitation of acid digestion techniques for the determination of fluoride in plant material. *Communication of Soil Science and Plant analysis*. **26**: 1823–1842.

Susheela, A. K. (ed.), 1993. *Prevention and Control of Fluoride in India (Health Aspects)*, Rajiv Gandhi National Drinking Water Mission, New Delhi. **1**: 20–22.

Vike, E. and Habjorg, A., 1995. Variation in fluoride content and leaf injury on plants associated with three aluminium smelters in Norway. *Science of the Total Environment*. **163**: 25–34.

Wenzel, W. W. and Blum, W. E. H., 1992. Fluorine speciation and mobility in contaminated soils. *Soil Science*. **153**: 357–364.

World Health Organisation (WHO), 1984. *Guidelines for Drinking Water Quality. Values 3; Drinking Water Quality Control in Small Community Supplies*. WHO, Geneva.

World Health Organisation (WHO), 2008. *Guidelines for Drinking-water Quality; Third Edition Incorporting the First and Second Addenda*. Geneva: WHO library cataloguing-in-publication data.

Yadav, R. K., Sharma, S., Bansal, M., Singh, A., Panday, V. and Maheshwari, R., 2013. Effects of fluoride accumulation on growth of vegetables and crops in Dausa district, Rajasthan, India. *Advances in Bioresearch*. **3**: 14–16.

CHAPTER 8

Fluoride Levels in Herbal and Tea Infusions

EBRU EMEKLI-ALTURFAN*[a], AYŞEN YARAT [a], AND SERAP AKYUZ[b]

[a]Department of Biochemistry, Faculty of Dentistry, Marmara University, Guzelbahce Buyukciftlik Sok, No:6 Nisantasi, Istanbul, Turkey; [b]Department of Pediatric Dentistry, Faculty of Dentistry, Marmara University, Nisantasi, 34365 Istanbul, Turkey

*E-mail: ebruemekli@yahoo.com

8.1 Tea

Tea has been consumed for over 4000 years and it is the most consumed beverage all over the world after water (Gardner *et al.*, 2007). Tea has a slightly bitter, cooling and relaxing flavor that many people enjoy (Dufresne and Farnworth, 2001). Drinking tea can be considered as a social habit beyond the demand for it. There is even a concept called a tea ceremony practiced in the East Asian cultural sphere by the Chinese and the Japanese. Tea consumption has a unique place, especially due to the rituals in China, Japan, and England taking place in a relaxed atmosphere in the afternoon between the major meals (Dufresne and Farnworth, 2001). People used to drink tea traditionally for improving blood flow, eliminating toxins, and for improving resistance to diseases (Balentine *et al.*, 1997). The tea plant *Camellia sinensis* can be cultivated in many regions from sea level to high mountains having fair temperature, high humidity, and acidic soils (Balentine *et al.*, 1997). When the enzymatic oxidation of leaf enzymes are inactivated or controlled Oolong

Food and Nutritional Components in Focus No. 6
Fluorine: Chemistry, Analysis, Function and Effects
Edited by Victor R Preedy

Published by the Royal Society of Chemistry, www.rsc.org

Table 8.1 Key facts about tea. Black tea originated from the tea plant *Camellia sinensis* is prepared by pouring hot or boiling water over cured leaves of the tea plant and can be in granular, stick-shaped, paper-bagged powder and canned or bottled beverage forms.

1. Tea is the most consumed beverage all over the world after water and in many countries drinking tea is considered as a social habit.
2. The tea plant *Camellia sinensis* can be cultivated in many regions having high humidity, fair temperature, and acidic soils.
3. When the enzymatic oxidation of leaf enzymes are inactivated or controlled Oolong or green tea are produced.
4. Tea is usually prepared by pouring hot or boiling water over cured leaves of the tea plant, *Camellia sinensis.*
5. One of the major tea types consumed worlwide is the black tea prepared as granular, stick shaped, paper bagged powder and canned or bottled beverage forms. In western countries black tea bags are usually used, whereas in Asia green tea and Oolong tea are more popular.

or green tea are produced (Balentine *et al.*, 1997). Tea is usually prepared with hot or boiling water and cured leaves of the tea plant, *Camellia sinensis.* One of the major tea types consumed worlwide is the black tea prepared as granular, stick-shaped, paper-bagged powder and canned or bottled beverage forms (Balentine *et al.*, 1997). In western countries black tea bags are usually used, whereas in Asia green tea and Oolong tea leaves are more popular (Cao *et al.*, 2004). Herbal infusions do not contain black tea leaves but contain dried fruit products, a mixture of dried berries or dried fruit leaves together with other fillers and flavorings. In recent years the number of people preferring herbal infusions instead of tea is on the rise and accordingly the popularity of herbal infusions is increasing day by day. However in general black tea is consumed more than other types of tea such as the green and the Oolong tea (Table 8.1).

8.1.1 Composition of Tea

Camella sinensis is a rich source of fluoride and other components such as aluminum catechins and flavonols (Dufresne and Farnworth 2001; Cao *et al.*, 2004). The chemical composition of tea has been widely studied and special attention has been paid to the non-nutritive components. Among these are the polyphenols especially the flavonoids that have a high antioxidant capacity of protecting cells from the reactive oxygen species (Dufresne and Farnworth, 2001). Catechins are a group of active flavonoids present in tea. Moreover, tea has beeen shown to be the best dietary source of catechins. The major catechin found in tea is epigallocatechin gallate and on a dry weight basis epigallocatechin gallate occupies more than ten percent (Hollman *et al.*, 1996, Rice-Evans *et al.*, 1997). Epicatechin gallate, gallocatechin gallate and epicatechin are the other catechins found in the black and the green tea (Dufresne and Farnworth, 2001). The important flavonols

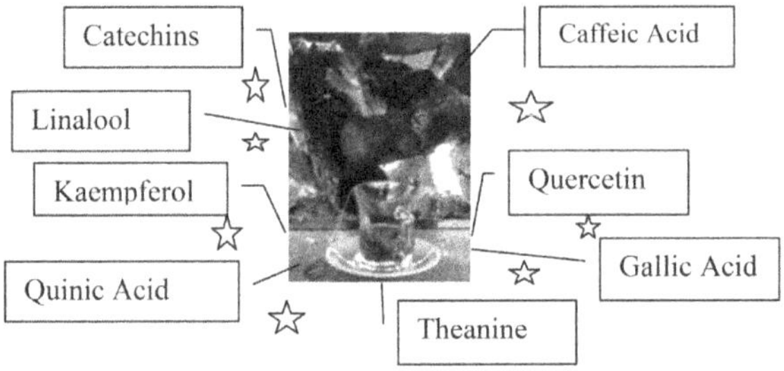

Figure 8.1 Chemical composition of black tea. The chemical composition of tea consists of polyphenols (quercetin, kaempferol and rutin), catechins (such as epigallocatechin gallate, gallocatechin gallate and epicatechin), phenolic acids (mainly caffeic, quinic and gallic acids), amino acid named theanine and linalool as a flavor compound.

present in tea are quercetin, kaempferol and rutin, and phenolic acids are mainly caffeic, quinic and gallic acids (Rice-Evans *et al.*, 1997; Hakkinen *et al.*, 1999; Dufresne and Farnworth, 2001). Caffeic acid can be found in tea and herbal infusions as it has been shown to be present in most fruits, berries and most vegetables (Hakkinen *et al.*, 1999). Moreover, there is an amino acid named theanine that is found only in tea leaves and tea also contains linalool as a flavor compound (Kubo and Morimitsu, 1995) (Figure 8.1).

8.1.2 The Quality of Black Tea

The quality of black tea is determined by its content. Professional tea tasters consider the characteristics of the tea liquor during the determination of tea quality and tea price (Biswas and Biswas, 1971). The quality of the black tea depends mainly on the components and the color of it therefore developing objective methods to identify the quality of tea chemically or physically is very important and necessary (Biswas and Biswas, 1971). Accordingly, in order to explain the quality of black tea researchers evaluated the chemical components of it. Theaflavin in black tea has been shown to be important to determine its quality (Liang *et al.*, 2003). Black tea is a fermented tea and during this fermentation an enzymatic oxidation of tea polyphenols in particular tea catechins occurs that leads to the formation of some colored chemical compounds including theaflavins and thearubigins (Biswas and Biswas, 1971). Theaflavin content directly affected the valuation of the broker of Central African Black Tea and the total theaflavin content of black tea has been shown to correlate significantly with the value of the tea (Roberts and Smith, 1963; Hilton and Ellis, 1972). The contents of black tea include cathecins like epigallocatechin, gallocatechin, epicatechin, epicatechin gallate, catechin gallate as well as total catechins and caffeine, nitrogen, amino acids, polyphenols, theaflavin and theaflavin gallate. All these content have been shown to be significantly correlated with the total quality score (Figure 8.2). The contribution of theaflavin to the brightness of black tea infusions has been

Figure 8.2 The factors affecting the quality of black tea. Theaflavin in black tea is important to determine its quality. On the other hand, other contents of black tea including caffeine, nitrogen, amino acids, polyphenols, gallocatechin, epigallocatechin, catechin, epicatechin, epicatechin gallate, catechin gallate, total catechins, and theaflavin gallate have also been shown to be significantly correlated with the total quality score.

reported to be greater when compared with the theaflavin gallates that form the tea cream stronger than theaflavin (Hilton and Ellis, 1972). Chemical and physical methods have been developed for identifying tea quality such as capillary electrophoresis, electronic tongue as well as lipid membrane taste sensor (Horie and Kohata, 1998; Ivarsson *et al.*, 2001).

There is still a need for tea quality estimation methods. One of the attemps has been presented by Liang *et al.* who analyzed the chemical composition and color differences of black tea infusions and correlated these with the results of sensory quality attributes assessed by tea tasters (Liang *et al.*, 2003). The authors suggested that the polyphenol level was highest in black tea and next was caffeine. The authors reported in their previous research in 2001 that epigallocatechin gallate was the most important catechin found in the unfermented green tea and they reported gallocatechin as the most important catechin in the black tea in 2003 (Liang and Xu, 2001; Liang *et al.*, 2003).

8.1.3 Fluoride in Tea Plant

Tea contains a significant amount of fluoride and therefore tea consumption can contribute to fluoride intake significantly. Tea trees absorb fluoride selectively from the air and soil and accordingly they accumulate and store fluoride (Hidekazu and Toshiyuki, 1977). By passive diffusion the tea plant

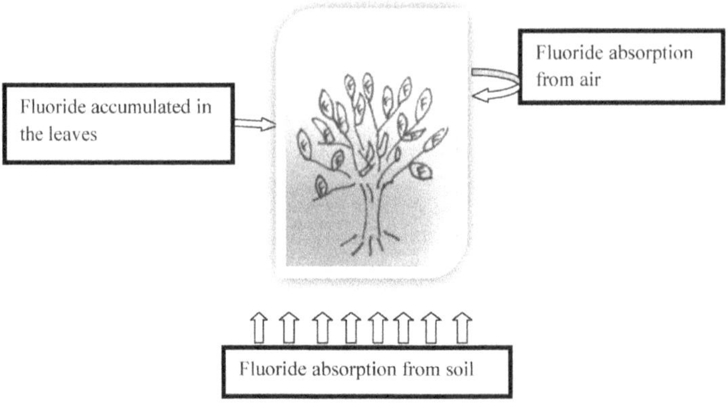

Figure 8.3 Fluoride in tea plants. Tea trees absorb fluoride from the air and soil. This absorbed fluoride is accumulated in the leaves during the plant's life span. The maturation of the leaves affects the fluoride content, as the maturation stage of the leaves increases the fluoride content also increases.

absorbs fluoride from acid soil and this absorbed fluoride is kept in the leaves during the plant's life span (Ruan and Wong, 2001). In tea trees, up to 98% of fluoride has been reported to be stored in the leaves. The maturation of the leaves also affects the fluoride content, as the maturation stage of the leaves increases the fluoride content also increases. For instance 100–430 mg kg^{-1} fluoride content can be found in delicate buds and young leaves, whereas in the older leaves the levels were up to 530–2350 mg kg^{-1} (Lu *et al.*, 2004).

Tea grows in soils that are usually acidic with a pH less than 5.5 and the fluoride uptake mechanism from these soils is well understood (Wong *et al.*, 2003). Some factors, like the quantity of minerals, the pH of the soil, adsorption of positively charged complexes, iron, calcium and aluminum concentration influence the mobility of the fluoride ion in the soil (Fung *et al.*, 1999). The complex of aluminum fluoride halide present in the soil decomposes into aluminum ions and fluoride ions under acidic conditions and this leads to high solubility and therefore the availability of the anion for uptake by the tea plant. Thereafter the free fluoride ion is absorbed by tea roots and transported to and stored in the leaves of the tea plant (Lu *et al.*, 2004) (Figure 8.3).

8.2 Fluoride Content in Tea Commodities, and Herbal Infusions

8.2.1 Fluoride Content in Black Tea

Exposure to high fluoride levels has been shown to cause some symptoms such as teeth discoloration in children and severe osteoarthritis in adults (Cao *et al.*, 1996). Since tea is a potentially speculative source of fluoride, the

Table 8.2 Key facts of fluoride content of black tea. The fluoride content of black tea varies depending on factors such as leaf age, plant genetics, rainfall, altitude, fertilizer and soil type. For instance, the fluoride content has been reported as 0.57–3.72 mg L^{-1} in Turkish teas using the fluoride ion-specific electrode after 5 min of brewing (Emekli-Alturfan *et al.*, 2009).

1. Tea is a potentially speculative source of fluoride therefore the safety of fluoride levels in black tea commodities has attracted attention.
2. More than 90% of total fluoride in tea is the ionic fluoride and about 75–90% of the ionic fluoride ingested each day is absorbed from the alimentary tract.
3. The fluoride content in black tea consumed in Turkey has been reported as 0.57–3.72 mg L^{-1} using the fluoride ion-specific electrode after 5 min of brewing.
4. Black teas originated from Turkey had higher fluoride levels compared with teas originated from Sri Lanka.
5. In literature the concentration range of fluoride content in tea is due to the factors such as leaf age, plant genetics, rainfall, altitude, fertilizer and soil type.

safety of fluoride levels in black tea commodities have attracted the attention of healthcare workers and food-regulation agencies. On the other hand, fluoride levels in different black tea commodities and the problem of safety are still ignored in some countries (Cao *et al.*, 2004) (Table 8.2).

More than 90% of total fluoride in tea is the ionic fluoride and nearly 75–90% of the ionic fluoride ingested is absorbed from the alimentary tract each day (Cao *et al.*, 1998). Only this ingested water-soluble ionic fluoride is suitable for human health (Cao *et al.*, 2004). The ionic fluoride can be detected by the fluoride ion-specific electrode. Our group has investigated the fluoride content in black tea using the fluoride ion-specific electrode. Accordingly, after 5 min of brewing fluoride content in black tea extracts ranged from 0.57 to 3.72 mg L^{-1}. Black teas originated from Turkey had higher fluoride levels when compared with teas originating from Sri Lanka. The highest level was found for the infusion of black tea (Earl Grey) originated from Turkey and the lowest one was found in the extract of Istikan-type Ceylon tea originated from Sri Lanka (Emekli-Alturfan *et al.*, 2009). Concentrations of 55 and 127 μg g^{-1} corresponding to 0.6 to 1.3 μg mL^{-1} and 0.64–3.55 μg mL^{-1} have been reported by Tokalioglu *et al.* in another study performed in Turkish teas (Tokalioglu *et al.*, 2001; Tokalioglu *et al.*, 2004). Kalayci and Somer reported in 2003 that fluoride concentrations were 2.6 and 3.9 μg mL^{-1} in black tea liquors extracted for 5 and 20 min, respectively (Kalayci and Somer, 2003). Hudaykuliev *et al.* reported a range of fluoride concentrations for Turkish teas that corresponded to between 0.8 and 2.6 μg mL^{-1} liquor concentrations, consistent with the findings of Kalayci and Somer (Hudaykuliev *et al.*, 2005).

Cao *et al.* evaluated the fluoride content in black tea in different commodities and packages originated from India, Sri Lanka, China, Japan and UK using a fluoride ion-selective electrode method (Cao *et al.*, 2004). They reported the fluoride content of 5 stick-shaped black teas to be 96.9–148 mg kg^{-1}, and that of 8 granular tea to be between the range of 139–223 mg kg^{-1}. The authors also suggested that bagged tea paper or the addition of milk and sugar did not affect the fluoride content.

Mahvi *et al.* investigated the fluoride content of some commercial tea marketed in Iran. The authors reported the ranges of fluoride content as 35 ± 6 mg kg^{-1} and 182 ± 20 mg kg^{-1} in Sedaghat and Sharzad tea, respectively. The fluoride contents of tea infusions have been evaluated as well and were found to be in the range of 0.53 ± 0.07 and 2.60 ± 0.18 mg L^{-1} (Mahvi *et al.*, 2006).

Moseti *et al.* determined the fluoride content in Kenyan black tea and compared it with black tea from other tea-producing countries in East Africa (Rwanda, Uganda and Tanzania) (Moseti *et al.*, 2013). They prepared 1% (w v^{-1}) tea liquor by dissolving 1 g of black tea in 100 mL boiling water, agitated for 10 min on a mechanical shaker and filtered. The fluoride levels were measured using a potentiometric fluoride-specific electrode in tea liquors. The authors reported a mean fluoride concentration of 0.37 μg mL^{-1} for the 1% tea liquors for the 72 Kenyan tea samples. They suggested that tea samples from the same region, regardless of the grade have the same fluoride levels. In their study, the samples from Uganda and Kenya had the highest mean fluoride concentration (0.66 μg mL^{-1}), whereas the lowest concentration was determined in the samples from Rwanda (0.25 μg mL^{-1} of fluoride) (Moseti *et al.*, 2013).

Hayacibara *et al.* evaluated the fluoride content in 72 samples of Brazilian black tea (bags) and 11 different types and brands of imported tea (bags) purhased in Brazilian supermarkets. Brazilian black teas had higher values of fluoride ranging from 0.08 to 0.40 mg per bag. Green tea had lower fluoride levels, whereas Oolong and black tea showed similar and higher fluoride content. The authors suggested that the concentration range was due to factors such as leaf age, plant genetics, rainfall, altitude, fertilizer and soil type (Hayacibara *et al.*, 2004) (Table 8.3).

8.2.2 Effects of Manufactued Forms of Black Tea on Fluoride Concentrations

Black tea is manufactured as three different forms on today's world market. They can be stick-shaped, granular and black tea bags. The manufacturing techniques of these forms are usually similar. Stick-shaped black tea is made of tender buds and leaves, which is why it keeps the original shapes of the leaves. Granular black tea coming from the small bits ground mechanically is no longer similar to the original shape. On the other hand, black tea bags are composed of granular black tea enclosed in a fitler paper bag. Black tea has been reported to be 80% of all tea commodities on worldwide markets with 90% of it being granular black tea (Cao *et al.*, 2006).

Table 8.3 Results of studies that evaluated fluoride contents in black teas. The fluoride contents of commercial black tea samples have been evaluated in different countries and varying results have been reported.

Authors	Reports
Tokalioglu *et al.* (2001, 2004)	Fluoride content of Turkish black tea extracts have been reported as 55 and 127 μg g^{-1} corresponding to 0.6 to 1.3 μg mL^{-1} and 0.64-3.55 μg mL^{-1}.
Kalayci and Somer (2003)	Fluoride concentrations have been reported as 2.6 and 3.9 μg mL^{-1} in black tea liquors extracted for 5 and 20 min, respectively.
Cao *et al.* (2004)	Fluoride contents of stick-shaped black teas and granular tea have been reported as 96.9–148 mg kg^{-1} and 139–223 mg kg^{-1}, respectively.
Hayacibara *et al.* (2004)	Fluoride content in Brazilian black teas have been reported to range from 0.08 to 0.40 mg per bag.
Hudaykuliev *et al.* (2005)	Fluoride concentrations in Turkish teas have been reported to be between 0.8 and 2.6 μg mL^{-1}.
Mahvi *et al.* (2006)	The fluoride content of commercial tea marketed in Iran have been reported as 35 ± 6 mg kg^{-1} and 182 ± 20 mg kg^{-1} in Sedaghat and Sharzad tea, respectively. The fluoride content of tea infusions were found in the range of 0.53 ± 0.07 and 2.60 ± 0.18 mg L^{-1}.
Emekli-Alturfan *et al.* (2009)	Fluoride content of black tea extracts ranged from 0.57 to 3.72 mg L^{-1} after 5 min of brewing with higher fluoride levels in black teas originated from Turkey compared with teas from Sri Lanka.
Moseti *et al.* (2013)	A mean fluoride concentration of 0.37 μg mL^{-1} for the 1% tea liquors has been reported for the 72 Kenyan tea samples.

We have reported higher fluoride levels in infusions prepared from granulated black tea when compared with stick-shaped black tea. However, fluoride levels in extracts prepared from black tea bags were higher compared with both granular and stick-shaped black teas (Emekli-Alturfan *et al.*, 2009). Cao *et al.* determined the fluoride content in 37 commodities in different manufactured tea forms and suggested that fluoride contents were 0.95–1.41 mg L^{-1} and 0.70–2.44 mg L^{-1} in black tea sticks and black tea granules, respectively. Consistent with the results of our study the authors reported higher fluoride levels in black tea bags (1.15–6.01 mg L^{-1}). Cao *et al.* suggested the older tea leaves content with low cost as the reason for this high fluoride content in black tea bags (Cao *et al.*, 2006). Lu *et al.* reported that there was a large difference in fluoride and polyphenols content in different types of black tea commodities. It is also interesting that a negative correlation has been reported between the fluoride and polyphenol content (Lu *et al.*, 2004). This negative correlation indicates that as the fluoride content increases the polyphenol content decreases, leading to a lower tea quality (Lu *et al.*, 2004; Cao *et al.*, 2006).

8.2.3 Fluoride Levels in Herbal Infusions

There are no black tea leaves in herbal infusions and herbal infusions are preferred by many people instead of tea. They contain dried fruit products, a mixture of dried berries, fruit leaves with some flavorings. We have evaluated the fluoride contents of herbal and fruit infusions (linden, sage, nettle, chamomile, femel, peppermint, cinnamon, anise, melisa, savory, ginseng, echinacea, mandarin-orange, lemon, rosehip, apple, blackberry) and found that herbal and fruit infusions contained low values of fluoride, having minimum and maximum values as 0.02–0.04 mg L^{-1} after 5 min of brewing. In order to evaluate the effect of time of brewing on the fluoride contents in herbal and fruit infusions, the impact of different brewing times (0, 5 and 10 min) on fluoride content was also determined. However, increasing the brewing time led to only slight increases in fluoride contents. On the other hand, increasing the brewing time did not change the fluoride content of anise, rosehip and apple infusions (Emekli-Alturfan *et al.*, 2009). Hayacibara *et al.* investigated the fluoride content of herbal infusions such as chamomile, balm-mint, anise, flowers and fruit (bags) using the ion-specific electrode and reported that herbal infusions showed very low amounts of fluoride. Moreover, most results were even below the limit of quantification for the analysis (Hayacibara *et al.*, 2004).

Malinowska *et al.* evaluated the fluoride content in commercially available herbal infusions by an ion-selective electrode (Malinowska *et al.*, 2008). In aggrement with our results they reported lower values for herbal tea infusions of 0.02–0.09 mg L^{-1}. Also consistent with our results they reported that brewing time (5, 10 and 30 min) did not increase the fluoride content.

8.3 Fluorosis and Tea Drinking

It was shown in epidemiologic observations, case reports and even in some experimental studies that excessive tea drinking may cause fluorosis. The studies concerning tea drinking and fluorosis have been reviewed and suggested that fluorosis was usually misdiagnosed and overlooked and people did not pay attention to the high fluoride content in some tea commodities (Yi and Cao, 2008).

8.3.1 Case Reports

Excessive consumption of tea has been shown to cause fluorosis in some regions. The results of these studies were presented as case reports. The most remarkable case reports on this issue are reviewed. An early case report indicating that fluorosis could be caused by fluoride in tea was about a 57 year-old Englishman diagnosed as Paget's disease with osteoartritis for leg paralysis, flexor spasm and low back pain in 1959 (Yi and Cao, 2008). It was later (in 1963) confirmed that his syptoms were due to his long-term heavy tea-drinking habit. Another case report from Pakistan showed dental fluorosis in a woman who had drunk tea since she was two and never lived where water contained fluoride (Cook, 1968). A brick tea fluorosis case report was from China. A 36 year-old woman with pains from upper limbs to the lumber, both knees and the joints for more than 10 years was diagnosed as hypoparathyroidism renal osteodystrophy. She had a long-term brick-tea-drinking habit and her urine fluoride content increased significantly. She was treated with fluoride inhibitors like Mg and her pain in joints disappeared. She stopped drinking tea and she became normal with only slight limitation of the right elbow (Hayem *et al.*, 2004). Moreover, five patients were identified in France who developed skeletal fluorosis due to drinking Darjeeling tea for 10 to 25 years (Yi and Cao, 2008). Also, an American patient was diagnosed with osteosclerosis and cortical elephantiasis in her spine and after stopping drinking instant tea her symptomps disappeared (Whyte *et al.*, 2005).

Very recently, Kakamanu and Rao reported in 2013 that a 47 year-old woman was referred for bone pain and abnormal findings on radiography, suggesting skeletal fluorosis. The patient reported that she habitually consumed a pitcher of tea for the past 17 years, which is equivalent to an estimated fluoride intake, >20 mg per day. The authors suggested that brewed tea had one of the highest fluoride contents among beverages in the United States. When the patient discontinued tea consumption, her symptoms improved. However, since it would take years to deplete skeletal fluoride, the authors considered whether to increase bone remodeling with the intermittent use of teriparatide to facilitate the elimination of skeletal fluoride (Kakamanu and Rao, 2013).

8.3.2 Epidemiological Reports

Fluorosis has been shown to be caused by long-term consumption of tea in large epidemiological studies conducted in Tibet, Mongolia and west China inhabitants (Cao *et al.*, 1996). Bilbeissi *et al.* showed that in children living in a chronic dental fluorosis area, dental fluorosis was positively associated with the daily tea drinking (Bilbeissi *et al.*, 1998). Sergio *et al.* reported that children living in a fluoridated area had a 22.1% risk of dental fluorosis consuming tea obtained in Valparaiso, Chile (Sergio *et al.*, 1989). A pilot survey was carried out by Cook who investigated drinking habits, tea consumption and fluoride ingestion in 662 children aged 5–15 in Pakistan (Cook, 1969). Tea was found to be the main sorce of fluoride for children. Fluoride intake from tea was 1.26 mg among the total daily intake of 2.24 mg.

8.3.3 Animal Studies

Reid *et al.* fed rats with tea and tea infusions to produce dental fluorosis very early in 1936 (Reid, 1936). In order to test whether low-fluoride brick tea could prevent the occurrence of fluorosis, Cao *et al.* performed a study in which rats had access only to a specially prepared low-fluoride brick tea for 1 year. In the ordinary brick tea group the daily intake of fluoride was reported as 0.3 mg. At the end of the 8th month this group was shown to develope dental fluorosis. Furthermore, white chalky dental fluorosis has also been observed at the end of month 12. On the other hand, the low-fluoride brick-tea-consuming group had a daily fluoride intake as 0.19 mg, and no signs of dental fluorosis developed in this group. Based on their results the authors suggested that brick tea containing low fluoride did not induce fluorosis in rats and therefore can be adviced to be used as an effective control measure for humans (Cao *et al.*, 2001).

8.4 Conclusion

Considering that black tea is the most popular beverage worldwide controlling the tea quality is unavoidable to protect human health against excessive uptake of fluoride from black tea. Accordingly, excessive intake of fluoride with black tea drinking especially in the regions with high fluoride levels in the drinking water, increases the risk of dental fluorosis and even skeletal fluorosis. However, further well-designed epidemiological and animal studies are required to illuminate the effects of tea drinking on human health.

Summary Points

- This chapter focuses on the fluoride content of herbal and tea infusions.
- The fluoride content in tea plants can be affected by many factors.
- People consume a significant amount of fluoride by tea drinking.
- Excessive tea drinking can lead to fluoride toxicity that is an important health problem.
- Herbal infusions are popular beverages and the fluoride content of herbal infusions varies.

Key Facts of Tea

1. Tea is the most consumed beverage all over the world after water and in many countries drinking tea is considered as a social habit.
2. The tea plant *Camellia sinensis* can be cultivated in many regions having high humidity, fair temperature, and acidic soils.
3. When the enzymatic oxidation of leaf enzymes is inactivated or controlled Oolong or green tea are produced.
4. Tea is usually prepared by pouring hot or boiling water over cured leaves of the tea plant, *Camellia sinensis.*
5. One of the major tea types consumed worlwide is the black tea prepared as granular, stick-shaped, paper-bagged powder and canned or bottled beverage forms. In western countries black tea bags are usually used, whereas in Asia green tea and Oolong tea leaves are more popular.

Key Facts of Fluoride Content of Black Tea

1. Tea is a potentially speculative source of fluoride, therefore the safety of fluoride levels in black tea commodities have attracted attention.
2. More than 90% of total fluoride in tea is the ionic fluoride and about 75–90% of the ionic fluoride ingested each day is absorbed from the alimentary tract.
3. The fluoride content in black tea consumed in Turkey has been reported as 0.57–3.72 mg L^{-1} using the fluoride ion-specific electrode after 5 min of brewing.
4. Black teas originated from Turkey had higher fluoride levels compared with teas originated from Sri Lanka.
5. In the literature the concentration range of fluoride content in tea is due to factors such as leaf age, plant genetics, rainfall, altitude, fertilizer and soil type.

Definitions of Words and Terms

Caffeic acid: This is an organic compound found in all plants consisting of both phenolic and acrylic functional groups.

Camellia sinensis: This is the plant that is used to produce tea from its leaves and buds.

Catechins: They are phytochemical compounds found in high concentrations in many plant-based foods and beverages. They are classified as flavanols and they include catechin, epicatechin, epigallocatechin, epicatechin gallate and epigallocatechin gallate.

Darjeeling tea: This is a type of tea from the Darjeeling district in West Bengal, India.

Elephantiasis: This is a disease characterized by the thickening of the skin and underlying tissues, occuring in the legs and male genitals especially.

Epigallocatechin gallate: This is the most abundant catechin with an antioxidant activity found in tea.

Fluorosis: This is a developmental disturbance of dental enamel due to excessive exposure to high concentrations of fluoride during tooth development.

Green tea: This is made from the leaves from Camellia sinensis undergoing minimal oxidation during processing.

Linalool: This is a terpene alcohol that is found in flowers and spice plants.

Oolong tea: This is a traditional type of tea produced through a process including withering under the strong sun and oxidation.

Osteoarthritis: This is a group of mechanical abnormalities consisting of degradation of joints, including articular cartilage and subchondral bone.

Osteosclerosis: This is a term defining the increase in bone density.

Paget's disease: This is a chronic disorder that can lead to enlarged and misshapen bones.

Skeletal fluorosis: This is type of a bone disease due to excessive consumption of fluoride. Skeletal fluorosis may lead to pain and damage in bones and joints.

Quercetin: This is a plant pigment with a molecular structure similar to flavone. It is present in tea, fruits, vegetables, leaves and grains.

Theaflavin: This is an antioxidant polyphenol formed from the condensation of flavanols in tea leaves during the fermentation of black tea.

Theanine: This is an amino acid and a glutamic acid analog that is found primarily in tea.

Thearubigins: They are polymeric polyphenols formed during the fermentation and condensation of two gallocatechins with the participation of polyphenol oxidases.

References

Balentine D. A., Wiseman S. A. and Bouwens L. C., 1997. The chemistry of tea flavonoids. *Critical Reviews in Food Science and Nutrition.* **37**: 693–704.

Bilbeissi, M. W., Fraysse, D., Mitre, D., Kerebel, L. M. and Kerebel, B., 1998. Dental fluorosis in relation to tea drinking in Jordan. *Fluoride.* **21**: 121–126.

Biswas, A. K. and Biswas, A. K., 1971. Biological and chemical factors affecting the valuations of North-East Indian plain teas. I. Statistical association of liquor characteristics with cash valuations of black teas. *Journal of the Science of Food and Agriculture.* **22**: 191–195.

Cao J., Bai, X. X. and Zhao, Y., 1996. The relationship of fluorosis and brick tea drinking in Chinese Tibetans. *Environmental Health Perspectives.* **104**: 1340–1343.

Cao J., Luo S. F., Liu J. W. and Li Y., 2004. Safety evaluation on fluoride content in black tea. *Food Chemistry.* **88**: 233–236.

Cao, J., Zhao, Y. and Liu, J., 2001. Prevention of brick tea fluorosis in rats with low-fluoride brick tea on laboratory observation. *Food and Chemical Toxicology.* **39**(6): 615.

Cao, J., Zhou, Y. and Liu, J. W., 1998. Safety evaluation and fluoride concentration of Pu'er brick tea and Bianxiao brick tea. *Food and Chemical Toxicology.* **36**: 1061–1063.

Cao, J. A., Zhao, Y., Li, Y., Deng, H. J., Yi, J. and Liu, J. W., 2006. Fluoride levels in various black tea commodities: measurement and safety evaluation. *Food and Chemical Toxicology.* **44**: 1131–1137.

Cook, H. A., 1968. Case report-tea and fluorosis. *Pakistan Dental Review.* **18**: 100.

Cook, H. A., 1969. Fluoride and tea. *Lancet.* **2**: 329.

Dufresne, C. J. and Farnworth, E. R., 2001. A review of latest research findings on the health promotion properties of tea. *Journal of Nutritional Biochemistry.* **12**: 404–421.

Emekli-Alturfan, E., Yarat, A. and Akyuz, S., 2009. Fluoride levels in various black tea, herbal and fruit infusions consumed in Turkey. *Food and Chemical Toxicology.* **47**: 1495–1498.

Fung, K. F., Zhang, Z. Q., Wong, J. W. C. and Wong, M. H., 1999. Fluoride contents in tea and soil from tea plantations and the release of fluoride into tea liquor during infusion. *Environmental Pollution.* **104**(2): 197–205.

Gardner, E. J., Ruxton, C. H. S. and Leeds, A. R., 2007. Black tea – helpful or harmful? A review of the evidence. *European Journal of Clinical Nutrution.* **61**: 3–18.

Häkkinen, S., Heinonen, M., Kärenlampi, S., Mykkänen, H., Ruuskanen, J. and Törrönen, R., 1999. Screening of selected flavonoids and phenolic acids in 19 berries. *Food Research International.* **32**: 345–353.

Hayacibara, M. F., Queiroz, C. S., Tabchoury, C. P. M. and Cury J. A., 2004. Fluoride and aluminum in teas and tea-based beverages. *Revista de Saude Publica.* **38**: 100–105.

Hayem, G., Ballard, M., Palazzo, E., Somogyi, F., Roux, F. and Meyer, O., 2004. Insufficiency bone fractures due to fluorosis in heavy tea drinkers. *Annals of the Rheumatic Diseases.* 2004, 63–488.

Hidekazu, Y. and Toshiyuki, F., 1977. Study on the plant fluoride content and relationship with aluminum. *Japanese Journal of Soil Science and Plant Nutrition.* **48**: 253–261.

Hilton, P. J. and Ellis, R. T., 1972. Estimation of the market value of central African tea by theaflavin analysis. *Journal of the Science of Food and Agriculture.* **23**: 227–232.

Hollman P. C. H., Hertog, M. G. L. and Katan, M. B., 1996. Analysis and health effects of flavonoids. *Food Chemistry.* **57**: 43–46.

Horie, H. and Kohata, K., 1998. Application of capillary electrophoresis to tea quality estimation. *Journal of Chromatography A.* **802**: 219–223.

Hudaykuliev, Y., Tastekin, M., Poyrazoglu, E. S., Baspinar, E. and Velioglu, Y. S., 2005. Variables affecting fluoride in Turkish black tea. *Fluoride.* **38**: 38–43.

Ivarsson, P., Homin, S., Hojer, N. E., Krantz-Rulcker, C. and Winquist, F., 2001. Discrimination of tea by means of a voltammetric electronic tongue and different applied waveforme. *Sensors and Actuators B.* **76**: 449–454.

Kakumanu, N. and Rao, S. D., 2013. Skeletal Fluorosis Due to Excessive Tea Drinking. *The New England Journal of Medicine.* **368**: 1140.

Kalayci, S. and Somer, G., 2003. Factors affecting the extraction of fluoride from tea: Application to three samples. *Fluoride.* **36**: 267–270.

Kubo, I. and Morimitsu, Y., 1995. Cytotoxicity of green tea flavor compounds against two solid tumor cells. *Journal of Agricultural and Food Chemistry.* **43**: 1626–1628.

Liang, Y., Lu, J., Zhang, L., Wu, S. and Wu, Y., 2003. Estimation of black tea quality by analysis of chemical composition and colour difference of tea infusions. *Food Chemistry.* **80**: 283–290.

Liang, Y. R. and Xu, Y. R., 2001. Effect of pH on cream particle formation and solids extraction yield of black tea. *Food Chemistry.* **74**: 155–160.

Lu, Y., Guo, W. F. and Yang, X. Q., 2004. Fluoride content in tea and its relationship with tea quality. *Journal of Agricultural and Food Chemistry.* **52**: 4472–4476.

Mahvi, A. H., Zazoli, M. A., Younecian, M., Esfandiari, Y., 2006. Fluoride Content of Iranian Black Tea and Tea Liquor. *Fluoride.* **39**(4): 266–268.

Malinowska, E., Inkielewiczb, I., Czarnowskib, W. and Szefer, P., 2008. Assessment of fluoride concentration and daily intake by human from tea and herbal infusions. *Food and Chemical Toxicology.* **46**: 1055.

Moseti, K. O., Wanyoko, J. K., Kinyanjui, T. and Wachira, F. N., 2013. Extractability of Fluoride into Black Tea Liquors. *Kenya International Journal of Environmental Protection.* **3**(5): 14–19.

Reid, E., 1936. The fluorine content in some Chinese food materials. *Chinese Journal of Physiology.* **10**: 259–272.

Rice-Evans C. A., Miller N. J. and Paganga G., 1997. Antioxidant properties of phenolic compounds. *Trends in Plant Science.* **2**: 152–159.

Roberts, E. A. H. and Smith, R. F., 1963. The phenolic substances of manufactured tea. IX. The spectrophotometric evaluation of tea liquors. *Journal of the Science of Food and Agriculture.* **14**: 689–700.

Ruan, J. Y. and Wong, M. H., 2001. Accumulation of fluoride and aluminium related to different varieties of tea plant. *Environmental Geochemistry and Health.* **23**: 53–63.

Sergio, G. S., Weber C. and Torres, C., 1989. Fluoride content of tea and amount ingested by children. *Odontología Chilena.* **37**: 251–255.

Tokalioglu, S., Kartal, S. and Sahin, U., 2004. Determination of fluoride in various samples and some infusions using a fluoride selective electrode. *Turkish Journal of Chemistry.* **28**: 204–211.

Tokalioglu, S., Sahin, U. and Kartal, S., 2001. Determination of fluoride and some metal ion levels in the drinking waters in Kayseri Province. *Turkish Journal of Chemistry.* **25**: 113–121.

Whyte, M. P., Essmyer, K., Gannon, F. H. and Reinus, W. R., 2005. Skeletal fluorosis and instant tea. *American Journal of Medicine.* **118**(1): 78–82.

Wong, M. H., Fung, K. F. and Carr, H. P., 2003. Aluminium and fluoride contents of tea, with emphasis on brick tea and their health implications. *Toxicology Letters.* **137**: 111–120.

Yi, J. and Cao, J., 2008. Tea and fluorosis. *Journal of Fluorine Chemistry.* **129**: 76–81.

CHAPTER 9

Bioavailability of Fluoride: Factors and Mechanisms Involved

RENÉ A. ROCHA[a], DINORAZ VÉLEZ[b], AND VICENTA DEVESA*[b]

[a]Institute of Biochemistry and Microbiology, Faculty of Science, Campus Isla Teja s/n, Valdivia, Chile; [b]Institute of Agrochemistry and Food Technology (IATA), Spanish National Research Council (CSIC), Av. Catedrático Agustín Escardino, 7, 46980, Paterna (Valencia), Spain
*E-mail: vdevesa@iata.csic.es

9.1 Concept of Bioavailability and Models for its Evaluation

The term bioavailability has many definitions, depending on the field in which it is applied. In nutrition and food toxicology it can be defined as the quantity of a nutrient or food contaminant that after solubilization during gastrointestinal digestion is absorbed across the gastrointestinal epithelium and reaches the systemic circulation (see Figure 9.1). Bioavailability and the factors and mechanisms involved in it can be studied by means of *in vivo*, *in situ* or *in vitro* approaches.

In vivo bioavailability studies are conducted mainly in rodents, pigs, dogs and primates, and in the case of compounds that are not toxic they may also be conducted in humans. There are two possible approaches: evaluation of the appearance of the compound in plasma at different times, and quantification of the compound in excreta (mainly urine and feces). *In situ* models using perfusion

Food and Nutritional Components in Focus No. 6
Fluorine: Chemistry, Analysis, Function and Effects
Edited by Victor R Preedy

Published by the Royal Society of Chemistry, www.rsc.org

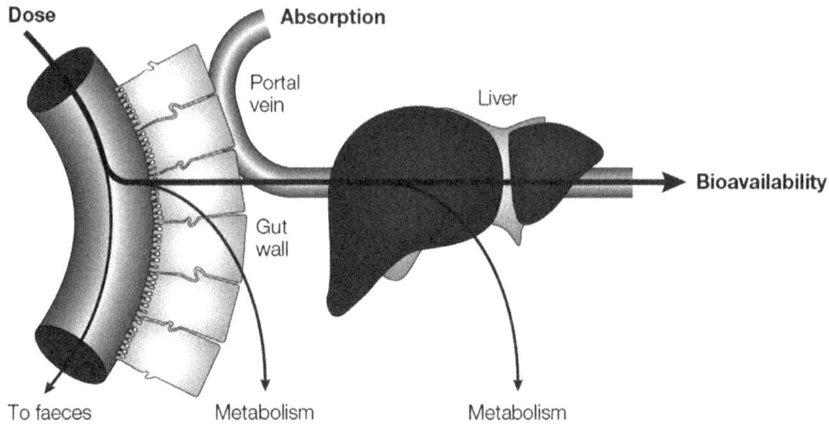

Figure 9.1 Concept of bioavailability. Reproduced from van de Waterbeemd and Gifford (2003) with permission of Nature Publishing Group.

techniques, in which the animal is anaesthetized, allow better control of certain parameters and, often, a better interpretation of results than *in vivo* methods. Although the animal has undergone a surgical process, neural, endocrine, lymphatic, and mesenteric blood supplies are maintained intact and therefore all the transport mechanisms present in a live animal should be functional (Griffin and O'Driscoll, 2008). This kind of approach is used mainly to study factors that may modify the quantity absorbed and to elucidate transport mechanisms.

With regard to *in vitro* approaches, the cell lines derived from colorectal cancers (Caco-2, HT-29, T84) can be considered one of the most suitable models. Some of these cell types preserve parts of the intestine's epithelial differentiation programme, expressing many of the characteristic genes of differentiated epithelium cells, and also retain the ability to form monolayers of polarized cells once they have reached confluence. The Caco-2 human cell line, originally isolated from a colon adenocarcinoma, is the one that is currently most used in intestinal absorption studies. It is also possible to obtain brush border membrane vesicles or everted intestinal ring or sacs of different portions of the gastrointestinal tract of laboratory animals, which are mainly useful for mechanistic studies.

9.2 Fluoride Bioavailability Studies

9.2.1 *In Vivo* Bioavailability Studies

Whitford *et al.* (1991) conducted a comparative study with 5 animal species (young adult dogs, cats, rabbits, rats and hamsters), using estimates of the plasma, renal and extrarenal clearances, and came to the conclusion that the dog was the best model to evaluate the pharmacokinetics of fluoride in humans. However, the rat, the model most used for the study of fluoride, presented renal and extrarenal clearances much greater than those observed for humans. This study, conducted with the use of an intravenous dose, does not show whether in these animal species the absorption of fluoride from the intestine and its arrival in the blood after oral ingestion may be comparable to what happens in humans. Furthermore, it must be borne in mind that a substantial part of the fluoride absorbed is retained in hard tissues, and therefore it is difficult to obtain quantitative pharmacokinetic information from urinary data alone (Ekstrand and Ehrnebo, 1979), which is a common practice in studies of this kind.

Studies have shown that sodium fluoride (NaF) in aqueous solution has a high bioavailability (70–90%, Cerklewski, 1997), and that it quickly appears in the blood, in less than an hour after its ingestion (Nopakun *et al.*, 1989). Absorption takes place mainly in the intestine, although gastric absorption is also important. A study conducted by Nopakun *et al.* (1989) in rats showed that 10 min after application of NaF by gastric intubation approximately 25% of the fluoride had been absorbed in the gastric region.

The form of fluoride most commonly studied is the sodium salt, but there are also data about the bioavailability of other forms. Insoluble sources of fluoride such as calcium fluoride are much less well absorbed (EFSA, 2008). Supplementation with 4 mg of calcium fluoride showed very small fluoride bioavailability (<10%) based on the areas under the plasma fluoride curve measured in volunteers during a 6 h period (Afseth *et al.*, 1987). The same phenomenon was demonstrated by Trautner and Einwag (1987), who evaluated bioavailability by means of an 8 h profile of fluoride concentrations in plasma after ingestion of calcium fluoride in the form of tablets, observing low and very variable absorption (0–47%). This low bioavailability contrasts with what has been observed for disodium monofluorophosphate (Na_2FP_3O, MFP), used as a supplement or in the preparation of toothpaste. van Asten *et al.* (1996) determined the absolute bioavailability of two fluoride formulations (tablets of MFP and enteric-coated tablets of NaF) in 13 healthy volunteers, aged 61–70 years. They concluded that the bioavailability of MFP (102.8%) was greater than that of sodium salt (64.2%). Rigalli *et al.* (1996) arrived at the same conclusion when they studied the area under the curve of total plasma fluoride in human volunteers exposed to oral doses of MFP (1540 ± 117 μmol min L^{-1}) and NaF (811 ± 52 μmol min L^{-1}).

The bioavailability of the fluoride present in toothpaste or dental rinses has also been evaluated. In these cases the bioavailability is determined by evaluating the concentration of fluoride in the saliva and/or in urine or plasma. In an investigation with 20 volunteers, Falcão *et al.* (2013) showed that absorption was greater for the fluoride present in NaF/SiO_2-based toothpastes than in those formulated on the basis of Na_2FPO_3/$CaCO_3$, and that this bioavailability was related to the quantity of soluble fluoride in each toothpaste.

9.2.2 *In Vitro* Bioavailability Studies

In vitro studies on fluoride have been mainly aimed at evaluating transport mechanisms or factors that affect absorption of this element (Parkins *et al.*, 1966; Nopakun and Messer, 1990; He *et al.*, 1998; Rocha *et al.*, 2012, 2013a). Evaluation of fluoride bioavailability in *in vitro* models is limited to one study in which monolayers formed by differentiated Caco-2 cells were used as a model of the intestinal epithelium (Rocha *et al.*, 2013a) and the apparent permeability of NaF was determined.

Many studies have shown that it is possible to predict absorption in humans by calculating the apparent permeability coefficient (P_{app}) in Caco-2, a parameter that evaluates the rate at which a compound passes across a cell monolayer (Yee, 1997). For this type of study, the cells are grown in permeable inserts that separates the well into two compartments: apical (upper) and basolateral (lower). For analyzing the absorption rate, P_{app} is evaluated in the apical-basolateral direction, emulating penetration into the portal circulation and subsequently the systemic bloodstream from intestinal lumen (see Figure 9.2).

On the basis of the results obtained by Rocha *et al.* (2013a), NaF may be considered an element of moderate absorption in saline solution, with an apparent permeability that is influenced by factors such as pH and the presence of chlorides. The prediction obtained by using this cell model was not what was expected from *in vivo* results, where, in the absence of food, absorption is high. The authors explain this by pointing out that the *in vitro* model may lack some components that are present in chyme during digestion (bile salts, *etc.*) and that may facilitate absorption of fluoride. Moreover, as we shall explain in the section on transport mechanisms (Section 9.4), it is possible that the Caco-2 monolayer may not be the best model for predicting the *in vivo* bioavailability of fluoride.

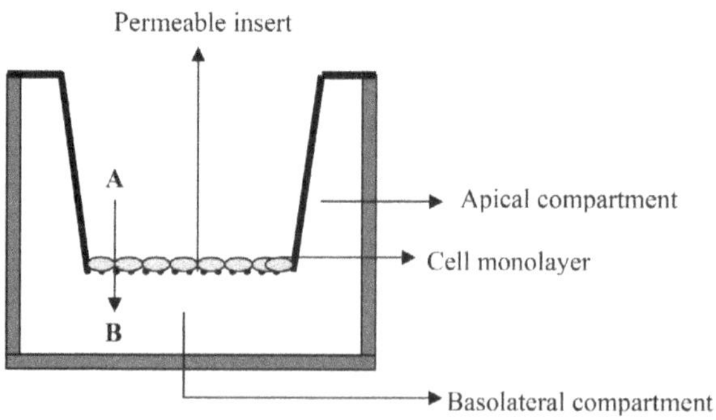

Figure 9.2 Scheme of the transport system for evaluating *in vitro* transport using cells of intestinal origin. Cells are seeded over a permeable insert that separates the well into apical and basolateral compartments. The absorption of a compound is predicted by evaluating the transport through the cell monolayer in the apical-basolateral direction (A–B).

9.3 Factors that Affect the Bioavailability of Fluoride

9.3.1 Components of the Diet that Modulate the Bioavailability of Fluoride

Many studies show that the bioavailability of fluoride from food is lower than what has been observed from administering it in aqueous medium. Hattab (1988) showed that the relative bioavailability of this element is less when it forms part of dry fish. Intake of milk and milk products reduces fluoride bioavailability by 13–50% (Ekstrand and Ehrnebo, 1979; Trautner and Einwag, 1989). Shulman and Vallejo (1990) evaluated the effect of food (lunch or milk) in 10 healthy volunteers exposed to an aqueous solution of NaF. The plasma analysis showed that fluoride bioavailability decreased by 13% in the presence of milk and by as much as 47% when ingested with lunch. The same study showed that the presence of food delays the appearance of fluoride in blood and reduces the peak level in serum. It also showed that milk does not reduce fluoride absorption when taken 15 min prior to fluoride administration.

In general, it seems that the food matrix has a negative effect on absorption of fluoride, although not all studies arrive at the same conclusion. For example, Arnold *et al.* (1989) showed that ingestion of fluoride in tablets with food (standard calcium-rich breakfast or calcium-poor lunch) made it take longer to reach the systemic circulation and reduced the peak serum concentration but did not reduce the final bioavailability. Another study showed that the presence of food increases the quantity of fluoride absorbed (Trautner and Einwag, 1989), as a consequence, according to the authors, of the longer residence time of the element in the intestine.

The decrease in the bioavailability of fluoride in the presence of food may be due to the interaction of fluoride with divalent cations such as calcium, magnesium and aluminum, with which it forms salts that have low solubility. In fact, the presence of Mg in the diet (>200 mg kg^{-1}) leads to a reduction in the absorption of fluoride in rats (Cerklewski, 1987; see Figure 9.3).

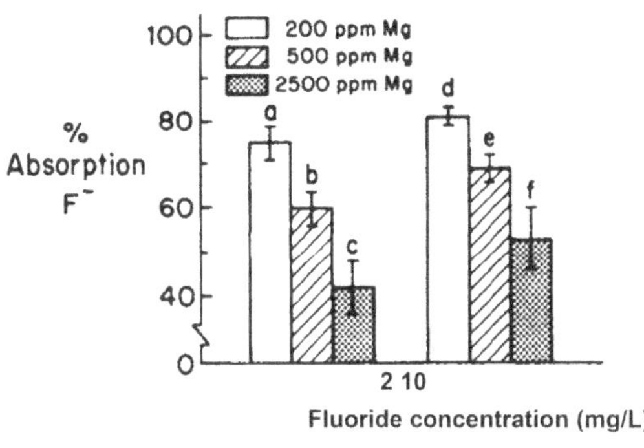

Figure 9.3 Effect of dietary magnesium on fluoride absorption *in vivo*. Values are means ± SD ($n = 8$). Different superscripts indicate a difference ($p < 0.05$) between means. Reproduced from Cerklewski (1987) with permission of the American Society for Nutrition.

Jowsey and Riggs (1978) investigated the effect of free calcium ions on fluoride absorption in humans. Nine female adults received oral fluoride supplements with or without calcium carbonate. The authors were able to show that the serum levels were 22% lower when the fluoride was administered with calcium carbonate. Likewise, Spencer *et al.* (1975) showed that an increase in the quantity of calcium ingested leads to an increase in faecal excretion of fluoride in humans. The same effect was observed after coadministration of fluoride and aluminum hydroxide for several weeks (Spencer *et al.*, 1980).

An *in vitro* study conducted by Rocha *et al.* (2013b) that emulated the process of gastrointestinal digestion showed that in the presence of salts of calcium, magnesium and aluminum the quantity of fluoride solubilized during the digestion of food decreases considerably, especially during the intestinal step (see Figure 9.4). This may be the reason for a reduction in the bioavailability of fluoride in the presence of matrices such as milk (Ekstrand and Ehrnebo, 1979; Trautner and Einwag, 1989), fish (Hattab, 1988) or rice (Rocha *et al.*, 2012), foods rich in these minerals.

Not only has the effect of essential cations on the absorption of fluoride been analyzed, but also research has been conducted on the effect of toxic

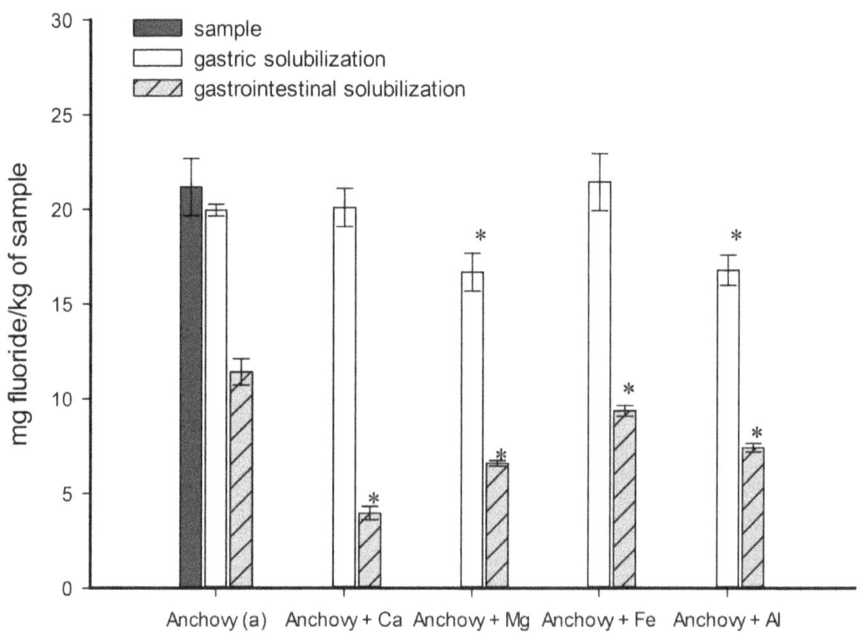

Figure 9.4 Influence of the presence of Ca^{2+}, Mg^{2+}, Fe^{3+} and Al^{3+} on solubilization of fluoride from seafood samples. Solubilization of fluoride expressed as mg of fluoride/kg of sample (mean ± standard deviation, $n = 3$). Asterisks indicate statistically significant differences with respect to the bioaccessible fluoride obtained without addition of cations ($p < 0.05$). Reproduced from Rocha *et al.* (2013b) with permission of Elsevier.

cations such as lead. Cerklewski and Ridlington (1987) evaluated the influence of lead acetate and lead carbonate on the absorption of fluoride in rats and showed that absorption decreased in the presence of low concentrations of these contaminants. It is possible that these conditions lead to the formation of lead fluoride (PbF_2), which is less soluble than sodium fluoride, and that this reduces the absorption of fluoride.

In addition to the cations just mentioned, it has also been observed that the presence of certain anions may affect the bioavailability of fluoride. The data concerning the influence of chloride are contradictory. In a study conducted in rats, Ericsson (1968) showed a reduction in the absorption of fluoride in the presence of 350 mM of NaCl. According to the author, this decrease might be due to the existence of competition between halogens for transport through the gastric or intestinal wall and/or an osmotic movement of fluid toward a high chloride concentration in the gastric or intestinal lumen, particularly immediately following the ingestion, before the chloride has been diluted or absorbed. However, Cerklewski *et al.* (1986) conducted a study in rats in which they evaluated the absorption and retention in bone of fluoride in the presence of chloride and came to the conclusion that the effect of chloride on fluoride bioavailability is due more to an enhancement of fluoride retention by low salt (sodium chloride) diets than to a possible negative effect of a high salt diet on fluoride absorption. The authors indicated that the differences with regard to the study conducted by Ericsson (1968) might be a result of the chloride concentration employed. Cerklewski *et al.* (1986) used concentrations that they described as normal dietary conditions, whereas Ericsson (1968) tested much higher concentrations.

Another anion that seems to reduce the quantity of fluoride that enters the systemic circulation is borate. A study conducted by Xianling *et al.* (1996), using the intestinal perfusion technique *in situ* in rats, showed that the presence of borate in the intestinal tract makes absorption of fluoride more difficult. The authors suggested that this might be due to the formation of complexes of tetrafluoroborate (BF_4^-), which were detected in the intestinal lumen during the tests. Finally, *in vitro* studies using brush border membrane vesicles of rabbits have shown that the presence of nitrates in the medium reduces the transport of fluoride across this intestinal model (He *et al.*, 1998). According to the authors, this seems to suggest that fluoride and nitrates might use common transport mechanisms.

Apart from the mineral components of the diet, there have been reports of interactions of other micro- and macronutrients on the absorption of fluoride. In a study conducted in rats, Chan *et al.* (1992) showed that the presence of vitamins may increase the bioavailability of fluoride. The authors demonstrated that some of the vitamins present in the complex (A, D and C) increased the quantity of fluoride present in plasma after intake. McGown *et al.* (1976) evaluated the effect of fat, observing that an excess of fat increased urinary excretion and reduced faecal excretion. They concluded that the increase in absorption was due to the delaying effect of fat on gastric emptying, thereby allowing more time for the fluoride to be absorbed from the stomach.

In a study in rats, Boyde and Cerklewski (1987) evaluated the effect of proteins on the absorption of fluoride. After a period of 4 weeks of exposure in which the apparent fluoride absorption was evaluated daily, the authors concluded that the greater the quantity of protein in the diet, the greater the absorption of fluoride. This result was not dependent on the quantity of fluoride employed (2 or 10 mg kg^{-1} as NaF) or the type of protein (casein or lactalbumin). The explanation given for this was the possible increase in luminal acidity in diets rich in proteins, which, as we shall show later, may favor the absorption of fluoride, mainly in the gastric region. However, a study conducted in volunteers under strictly controlled dietary conditions, in which fluoride was administered in aqueous solution in the form of NaF or as a fish protein concentrate, showed that the presence of protein reduced the net absorption of fluoride (Spencer *et al.*, 1970). In this case, however, the reduction might have been due, not to the presence of protein as such, but rather to the form in which the fluoride was present in the protein concentrate. In this kind of concentrate, most of the fluoride may come from bones and scales and be in the form of fluorapatite. Finally, Chan *et al.* (1990) suggested that coingestion of fluoride and caffeine or some other methylxanthines increases the bioavailability of fluoride; however, other studies failed to confirm this effect (Chen and Whitford, 1999).

9.3.2 Nondietary Factors that Modulate the Bioavailability of Fluoride

There are some factors that are not provided by the diet but are modulated with it (pH, presence of bile salts, emptying of the stomach) and that may affect the bioavailability of fluoride. One of the factors most studied, with regard to the bioavailability of fluoride, is luminal pH. Many studies show that an acid pH favors absorption of fluoride in the stomach (Whitford and Pashley, 1984; Messer and Ophaug, 1993). However, Messer *et al.* (1989) showed that this factor does not noticeably affect the absorption of this element in the intestine. Most studies agree that this is because at acid pHs fluoride is found in neutral form (HF), and therefore it is more easily absorbable than in its ionic form, which may also interact with certain cations, giving rise to salts with lower solubility. However, He *et al.* (1998) and Rocha *et al.* (2012, 2013b) working with intestinal models *in vitro*, showed that the transport of fluoride in the intestinal region is pH dependent and is greater when there is a H^+ gradient. According to the authors, this might be due to the participation of a transporter dependent on H^+ concentration in the intestinal transport of fluoride.

Another factor that also has a demonstrated influence on the transport of fluoride is bile salts. Rocha *et al.* (2012) showed that in the presence of taurocholic acid, a bile acid that is a conjugate of cholic acid with taurine, transport of fluoride increased in various intestinal epithelium models (monocultures or cocultures of Caco-2 and HT29-MTX). According to the authors, this effect might be related with the ability of this bile acid to bond Ca^{2+} ions (Baruch *et al.*, 1991), which might favor transport of fluoride by eliminating from the medium an element with which it forms insoluble salts (see Figure 9.5).

In fact, another *in vitro* study conducted by the same authors (Rocha *et al.*, 2013b) showed that during gastrointestinal digestion the solubilization of fluoride from seafood products increased as a result of an increase in the concentration of bile salts. However, we cannot rule out the possibility that the effect observed on the bioavailability of fluoride in the cell model may be a result of the opening of intercellular junctions observed during the treatment with bile salts, which, as we shall see in the section on transport mechanisms (Section 9.4), may favor the absorption of this element.

Gastric emptying influences the rate of absorption of fluoride but does not seem to affect the final bioavailability of the element (Nopakun *et al.*, 1989;

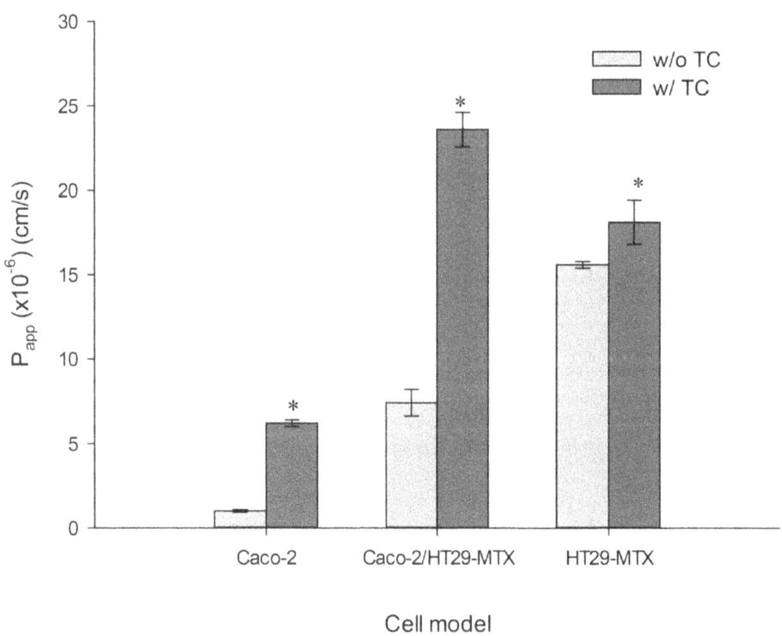

Figure 9.5 Influence of taurocholic acid, a bile salt component, on fluoride transport. Apparent permeability coefficients (P_{app}) of fluoride (20 mg L^{-1}, 1.05 mM) in the presence or absence of taurocholic acid (20 mM TC) in various cell models of intestinal epithelium. Values expressed as cm s^{-1} (mean ± SD, $n = 4$). Significant differences with respect to the control (without TC) are marked with an asterisk (*) ($p < 0.05$). Reproduced from Rocha *et al.* (2012) with permission of Elsevier.

Messer and Ophaug, 1991). Messer and Ophaug (1991) studied this phenomenon in rats treated with pectin (which reduces gastric transit) and untreated rats and showed that the rate of fluoride absorption was initially much slower in the pectin group than in the group given fluoride in water, and plasma concentration increased more slowly and reached a lower maximum value. After 2 h, however, fluoride absorption exceeded 90% in both groups.

An *in vitro* study conducted by Rocha *et al.* (2012) using the HT29-MTX cell line, which differentiates and produces monolayers that secrete mucus, showed that mucus reduces transport of fluoride across the intestinal epithelium. It is considered that the gel formed by mucins, glycosylated proteins that constitute the mucus, acts as a barrier to the diffusion of small molecules such as ions. Therefore, in normal conditions, this layer of glycoproteins could reduce the absorption of fluoride. However, any situation that leads to a thinning of the mucus layer, such as inflammatory processes or treatment with antibiotics, might result in an increase in the absorption of this element.

9.4 Intestinal Transport Mechanisms for Fluoride

The transport of a compound across the intestinal epithelium is typically a very complex process in which various routes or pathways may be involved: paracellular, passive transcellular and carrier-mediated transport, and transcytosis. Knowledge of these mechanisms is very important for characterization of the absorption processes that allow the compound to enter the systemic circulation (see Figure 9.6).

It is considered that the transport of fluoride across biological membranes takes place mainly by diffusion of the undissociated form HF. The classic studies with artificial lipid bilayers and pH electrodes indicated that HF is a highly

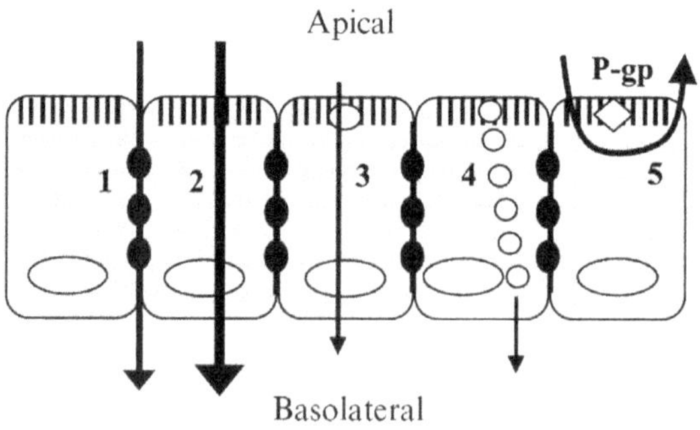

Figure 9.6 Graphical representation of the main physiological transport mechanisms across cells. Compounds can be transported across the intestinal epithelium by (1) paracellular passive diffusion (between cells); (2) transcellular passive diffusion; (3) carrier-mediated transport (active or facilitated) or (4) transcytosis mechanisms. Reproduced from Press (2011) with permission of Springer.

permeable solute with a permeability coefficient similar to that of water. The permeability of this nonionic form is five to seven orders of magnitude greater than the permeability of the dissociated form (Gutknecht and Walter, 1981).

In general, we can say that in many cases there is a lack of agreement about the characteristics of the transport of fluoride, which may be partly due to the use of different animal models and different transport conditions. Most studies indicate that the intestinal absorption of fluoride takes place by simple passive diffusion. In a recent work using the differentiated Caco-2 cell line on permeable porous supports, Rocha *et al.* (2013a) showed that there is a substantial participation of the paracellular pathway (see Figures 9.7 and 9.8).

In view of the hydrophilic nature of fluoride and its salts, it is possible that most of the passive diffusion to which the studies refer is due to transport across intercellular junctions. If so, any compound, micro-organism or situation that produces an opening of intercellular junctions might increase transport of this element from the lumen towards the systemic circulation. This is what was suggested as one of the possible reasons for the increase in absorption in the presence of bile salts (Section 9.3.2). On the other hand, as we commented at the beginning of this chapter (Section 9.2.2), the possible participation of the paracellular pathway in the transport of fluoride means that Caco-2 is not the most suitable model to use for estimating the bioavailability of this element. One of the limitations of this cell model is the intercellular pore size, much smaller than in the human small intestine. This means that the evaluation of the bioavailability of compounds transported mainly by the paracellular pathway is much lower in Caco-2 cells than what is observed in humans.

Passive transport of fluoride seems to coexist with carrier-mediated transport. Most studies show that the entry of fluoride does not increase

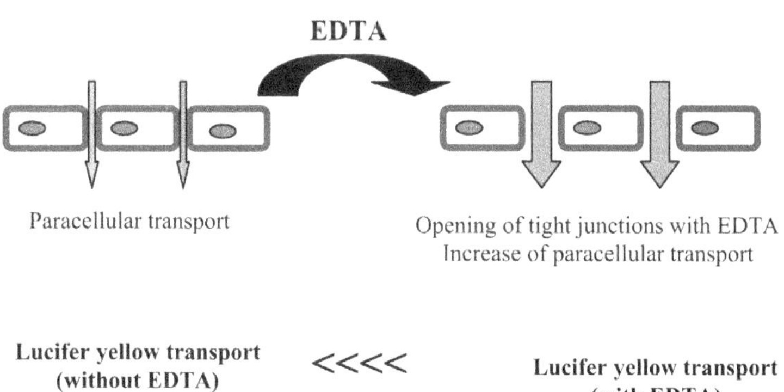

Figure 9.7 Scheme of the effect of EDTA on the paracellular pathway. Preincubation of EDTA before a transport assay produces an opening of the tight junctions, and the subsequent increase in the transport of substances, such as lucifer yellow (LY), which are transported mainly by the paracellular route.

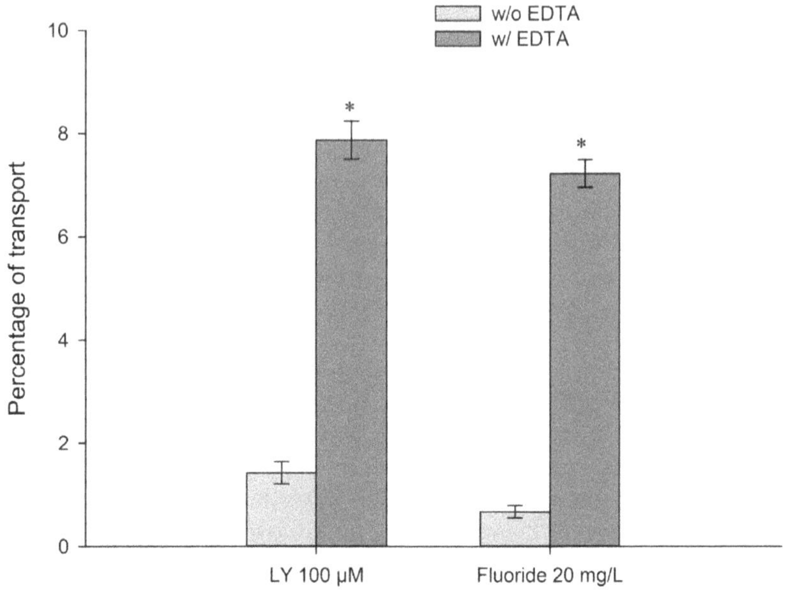

Figure 9.8 Participation of paracellular route in fluoride and lucifer yellow (LY) transport. Percentage of transport of fluoride (20 mg L^{-1}) and the paracellular marker LY (100 μM) in the apical-basolateral direction in Caco-2 cells previously treated with EDTA or without previous treatment with EDTA. Values expressed as mean ± standard deviation ($n = 3$). Significant differences with respect to the control (without EDTA) are marked with an asterisk ($p < 0.05$).

linearly with the fluoride concentration (Sato *et al.*, 1986; He *et al.*, 1998) and that therefore at least part of the transcellular transport is carrier mediated. According to some studies, this transport may have an active component. Parkins *et al.* (1966), using various regions of everted rat intestine, demonstrated the participation of an active component in the jejunum and ileum. Furthermore, in a study using the gastrointestinal reflux method *in situ* on rats, Sato *et al.* (1986) showed that the transport of fluoride is inhibited in the presence of dinitrophenol, an uncoupler of oxidative phosphorylation, suggesting the involvement of an energy-requiring mechanism.

It has also been suggested that fluoride may use the same transport mechanisms as chloride. A reduction in the transport of chloride in the presence of fluoride in the medium has been shown *in vitro* (Simchowitz, 1988). It has also been shown that the presence of chloride reduce the transport of fluoride *in vivo* and *in vitro* (Sato *et al.*, 1986; He *et al.*, 1998; Rocha *et al.*, 2013a). In this respect, however, there are also contradictory data, because Cerklewski *et al.* (1986), in a study in rats, did not observe a reduction in the transport of fluoride in the presence of chloride.

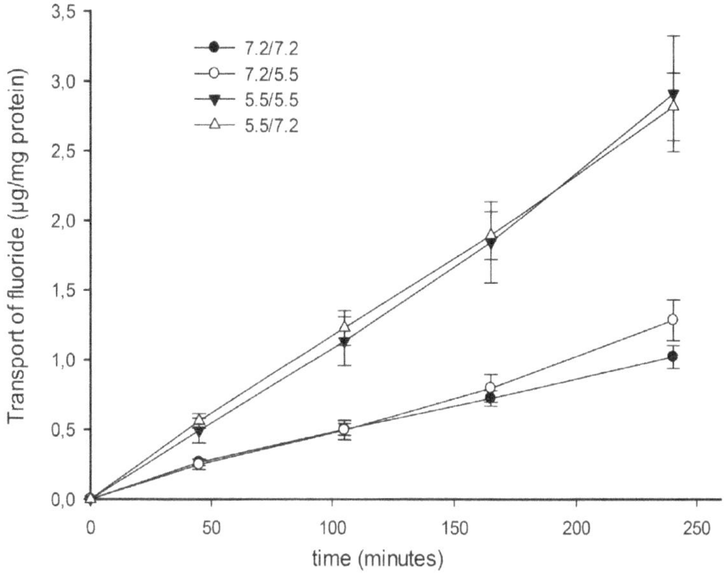

Figure 9.9 Influence of H^+ concentration on fluoride transport. Transport of fluoride (µg fluoride/mg protein) in the apical-basolateral direction in Caco-2 cells exposed to 20 mg L^{-1} of fluoride for 240 min at various pHs in the apical/basolateral sides (7.2/7.2; 7.2/5.5; 5.5/7.2; 5.5/5.5). Values expressed as mean ± standard deviation ($n = 3$). Reproduced from Rocha *et al.* (2013a) with permission of Elsevier.

He *et al.* (1998), using intestinal vesicles of rabbit, and Rocha *et al.* (2013a), using the Caco-2 cell line, concluded that the transport of fluoride is pH dependent and that it is favored by an acid pH (Figure 9.9). The authors suggested that the transport is mediated by a F^-/H^+ transporter or a F^-/OH^- exchanger. The possible participation of a pH-dependent transport has also been shown in the gastric region in rats (Whitford and Pashley, 1984).

So far, however, although some of the characteristics of the transport of fluoride have been studied, it has not been possible to identify the transporters that may be involved in intestinal absorption of this trace element. Rocha *et al.* (2013a) suggested, on the basis of the results obtained (H^+ dependence and inhibition with chloride) that transport of fluoride in the absorptive direction might be partly mediated by the downregulated in adenoma transporter (DRA). DRA (SLC26A3) is a transmembrane glycoprotein that is located in the apical membrane of columnar and goblet cells in the small intestine, especially in the duodenum and the distal ileum. It is considered the major anion exchanger in the duodenum. It mediates Cl^-/CO_3H^- or Cl^-/OH^- electroneutral exchange, and is involved in the transport of sulfates and other anions. However, there is not enough scientific data to confirm this assumption.

Summary Points

- This chapter is a review of the available data on intestinal absorption of fluoride (F).
- The bioavailability of F (fraction absorbed) varies, depending on the form in which it is present.
- Many food components affect the absorption of F (divalent cations, monovalent anions, vitamins and some macronutrients).
- Factors such as luminal pH, residence time in the tract and the layer of mucus on the gastrointestinal wall may affect the bioavailability of F.
- It has been suggested that passive transport may feature among the transport mechanisms, possible across intercellular space.
- Transcellular mechanisms that may be mediated by a carrier have been described, and there is even evidence of an energy-dependent transport.
- Competition with chloride has been observed, which in a way indicates that these two elements may share intestinal transport mechanisms.

Key Facts

Key Features of Fluoride Chemistry

- Fluorine is the most electronegative element of the Periodic Table; as a result it rarely occurs naturally in the elemental state.
- Fluoride is the ionic form of fluorine. It forms salts or complexes with a large number of elements and compounds.
- Some salts of fluoride, such as calcium fluoride, magnesium fluoride and aluminum fluoride, are highly insoluble.
- Fluoride can combine with a proton forming hydrogen fluoride (HF), the conjugate acid of fluoride.
- The pKa of HF is 3.17, so it is in the neutral form at acid pHs.
- All these chemical characteristics affect the bioavailability of this element.

Key Features of Cellular Transport

- The transport of compounds across the intestinal epithelium may take place by paracellular, passive transcellular, carrier-mediated transport, and transcytosis.
- Paracellular route is transport between cells. It is not concentration dependent and it takes place by passive diffusion.
- The transcellular transport that takes place across the cell may be passive, facilitated or active.
- Active transport requires energy, either directly (primary active transport) or else indirectly, by coupling to an energy-producing process (secondary active transport).

- Carrier-mediated transport, either facilitated or active, is concentration dependent.

Definitions of Words and Terms

Bioavailability: the fraction of an administered dose of unchanged compound that reaches the systemic circulation.
Area under curve (AUC): corresponds to the integral of the plasma concentration *versus* an interval of definite time.
Peak plasma concentration (C_{max}): the highest level of a compound that can be obtained in the blood, usually after administration.
Apparent permeability coefficient (P_{app}): initial flux of a compound through a membrane (normalized by membrane surface area and donor concentration).
Fluorapatite ($Ca_5(PO_4)_3F$): Fluorinated calcium phosphate.
Gastric emptying: Physiological process that consists in emptying of the stomach contents after gastric filling towards the duodenum.
Mucus: A layer of water-insoluble gel that forms a continuous cover over the gastrointestinal mucosal surfaces. It is the first line in mucosal defence against the natural aggressors, acid and pepsin, in the lumen.
Paracellular transport: Transfer of substances across an epithelium by passing through the intercellular space between the cells.
Passive transcellular transport: The transport of a molecule across the cell by simple diffusion.
Carrier-mediated transport: The transport of a molecule across the cell *via* a transmembrane carrier protein.
Active transport: The movement of all types of molecules across a cell membrane against its concentration gradient (from low to high concentration), requiring the consumption of energy.

List of Abbreviations

DRA	Downregulated in adenoma, anion transporter
MFP	Sodium monofluorophosphate
P_{app}	Apparent permeability coefficient

References

Afseth, J., Ekstrand, J. and Hagelid, P., 1987. Dissolution of calcium fluoride tablets in vitro and bioavailability in man. *Scandinavian Journal of Dental Research.* **95**: 191–192.

Arnold, P., Wermeille, M., Chapuy, M. C., Biollaz, J., Grandjean, E. M., Schelling, J. L. and Meunier, P. J., 1989. Influence of food and antacid administration on fluoride bioavailability from enteric-coated sodium fluoride tablets. *Bone.* **10**: 401–407.

van Asten, P., Duursma, S. A., Glerum, J. H., Ververs, F. F., van Rijn, H. J. and van Dijk A., 1996. Absolute bioavailability of fluoride from disodium monofluorophosphate and enteric-coated sodium fluoride tablets. *European Journal of Clinical Pharmacology.* **50**: 321–326.

Baruch, E., Lichtenberg, D., Barak, P. and Nir, S., 1991. Calcium binding to bile salts. *Chemistry and Physics of Lipids.* **57**, 17–27.

Boyde, C. D. and Cerklewski, F. L., 1987. Influence of type and level of dietary protein on fluoride bioavailability in the rat. *Journal of Nutrition.* **117**: 2086–2090.

Cerklewski, F. L., Ridlington, J. W. and Bills, N. D., 1986. Influence of dietary chloride on fluoride bioavailability in the rat. *Journal of Nutrition.* **116**: 618–624.

Cerklewski, F. L., 1987. Influence of dietary magnesium on fluoride bioavailability in the rat. *Journal of Nutrition.* **117**: 496–500.

Cerklewski, F. L. and Ridlington, J. W., 1987. Influence of dietary lead on fluoride bioavailability in the rat. *Biological Trace Element Research.* **14**: 105–113.

Cerklewski, F. L., 1997. Fluoride bioavailability - Nutritional and clinical aspects. *Nutrition Research.* **17**: 907–929.

Chan, J. T., Qui, C. C., Whitford, G. M. and Weatherred, J. G., 1990. Influence of coffee on fluoride metabolism in rats. *Proceedings of the Society for Experimental Biology and Medicine.* **194**: 43–47.

Chan, J. T., Stark, C. M., Wild, T. W. and Jeske, A. H., 1992. Influence of vitamins and iron on plasma fluoride levels in rats. *Pediatric Dentistry.* **14**: 37–40.

Chen, X. and Whitford, G. M., 1999. Effects of caffeine on fluoride, calcium and phosphorus metabolism and calcified tissues in the rat. *Archives of Oral Biology.* **44**: 33–39.

EFSA, European Food Safety Authority., 2008. Sodium monofluorophosphate as a source of fluoride added for nutritional purposes to food supplements. *The EFSA Journal.* **886**: 1–18.

Ekstrand, J. and Ehrnebo, M., 1979. Influence of milk products on fluoride bioavailability in man. *European Journal of Clinical Pharmacology.* **16**: 211–215.

Ericsson, Y., 1968. Influence of sodium chloride and certain other food components on fluoride absorption in the rat. *Journal of Nutrition.* **96**: 60–68.

Falcão, A., Tenuta, L. M. and Cury, J. A., 2013. Fluoride gastrointestinal absorption from $Na_2FPO_3/CaCO_3^-$ and NaF/SiO_2-based toothpastes. *Caries Research.* **47**: 226–233.

Griffin, B. T. and O'Driscoll, C. M., 2008. Models of the small intestine In: Kim, K. J. and Ehrhardt, C. (ed.) *Drug Absorption Studies: In Situ, In Vitro and in Silico Models.* Springer, NY, USA.

Gutknecht, J. and Walter, A., 1981. Hydrofluoric and nitric acid transport through lipid bilayer membranes. *Biochimica et Biophysica Acta.* **644**: 153–156.

Hattab, F. N., 1988. Pharmacokinetics of fluoride absorbed from dried seafoods by healthy adults. *European Journal of Clinical Pharmacology.* **34**: 489–493.

He, H., Ganapathy, V., Isales, C. M. and Whitford, G. M., 1998. pH-dependent fluoride transport in intestinal brush border membrane vesicles. *Biochimica et Biophysica Acta.* **1372**: 244–254.

Jowsey, J. and Riggs, B. L., 1978. Effect of concurrent calcium ingestion on intestinal absorption of fluoride. *Metabolism.* **27**: 971–974.

McGown, E. L., Kolstad, D. L. and Suttie, J. W., 1976. Effect of dietary fat on fluoride absorption and tissue fluoride retention in rats. *Journal of Nutrition.* **106**: 575–579.

Messer, H. H., Nopakum, J. and Rudney, J. D., 1989. Influence of pH on intestinal fluoride transport in vitro. *Journal of Dental Association of Thailand.* **39**: 226–232.

Messer, H. H. and Ophaug, R. H., 1991. Effect of delayed gastric emptying on fluoride absorption in the rat. *Biological Trace Element Research.* **3**: 305–315.

Messer, H. H. and Ophaug, R. H., 1993. Influence of gastric acidity on fluoride absorption in rats. *Journal of Dental Research.* **72**: 619–622.

Nopakun, J. and Messer, H. H., 1990. Mechanism of fluoride absorption from the rat small intestine. *Nutrition Research.* **10**: 771–779.

Nopakun, J., Messer, H. H. and Voller, V., 1989. Fluoride absorption from the gastrointestinal tract of rats. *Journal of Nutrition.* **119**: 1411–1417.

Parkins, F. M., Hollifield, J. W., McCaslin, A. J., Wu, S. M. and Faust, R. G., 1966. Active transport of fluoride by the rat intestine in vitro. *Biochimica et Biophysica Acta.* **126**: 513–524.

Press, B., 2011. Optimization of the Caco-2 permeability assay to screen drug compounds for intestinal absorption and efflux. *Methods in Molecular Biology.* **763**: 139–154.

Rigalli, A., Morosano, M. and Puche, R. C., 1996. Bioavailability of fluoride administered as sodium fluoride or sodium monofluorophosphate to human volunteers. *Arzneimittelforschung.* **46**: 531–533.

Rocha, R. A., Vélez, D. and Devesa, V., 2012. In vitro evaluation of intestinal fluoride absorption using different cell models. *Toxicology Letters.* **210**: 311–317.

Rocha, R. A., Devesa, V. and Vélez, D., 2013a. *In vitro* study of intestinal transport of fluoride using the Caco-2 cell line. *Food and Chemical Toxicology.* **55**: 156–163.

Rocha, R. A., de la Fuente, B., Clemente, M. J., Ruiz, A., Vélez, D. and Devesa. V., 2013b. Factors affecting the bioaccessibility of fluoride from seafood products. *Food and Chemical Toxicology.* **59**: 104–110.

Sato, T., Yoshitake, K. and Hitomi, G., 1986. Mechanism of fluoride absorption from the gastrointestinal tract in rats. In: Tsunoda, H. and Yu, M.-H. (ed.) *Fluoride Research 1985, Studies in Environmental Science.* vol. 27. Elsevier Science Publishers B.V., Amsterdam, pp. 325–332.

Shulman, E. R. and Vallejo, M., 1990. Effect of gastric contents on the bioavailability of fluoride in humans. *Pediatric Dentistry.* 1990, **12**: 237–240.

Simchowitz, L., 1988. Interactions of bromide, iodide, and fluoride with the pathways of chloride transport and diffusion in human neutrophils. *The Journal of General Physiology.* **91**: 835–860.

Spencer, H., Osis, D., Wiatrowski, E. and Samachson, J., 1970. Availability of fluoride from fish protein concentrate and from sodium fluoride in man. *Journal of Nutrition.* **100**: 1415–1424.

Spencer, H., Osis, D., Kramer, L., Wiatrowski, E. and Norris, C., 1975. Effect of calcium and phosphorus on fluoride metabolism in man. *Journal of Nutrition.* **105**: 733–740.

Spencer, H., Kramer, L., Norris, C. and Wiatrowski, E., 1980. Effect of aluminum hydroxide on fluoride metabolism. *Clinical Pharmacology and Therapeutics.* **28**: 529–535.

Trautner, K. and Einwag, J., 1987. Factors influencing the bioavailability of fluoride from calcium-rich, health-food products and CaF_2 in man. *Archives of Oral Biology.* **32**: 401–406.

Trautner, K. and Einwag, J., 1989. Influence of milk and food on fluoride bioavailability from NaF and Na_2FPO_3 in man. *Journal of Dental Research.* **68**: 72–77.

van de Waterbeemd, H. and Gifford, E., 2003. ADMET in silico modelling: towards prediction paradise? *Nature Reviews Drug Discovery.* **2**: 192–204.

Whitford, G. M. and Pashley, D. H., 1984. Fluoride absorption: the influence of gastric acidity. *Calcified Tissue International.* **36**: 302–307.

Whitford, G. M., Biles, E. D. and Birdsong-Whitford, N. L., 1991. A comparative study of fluoride pharmacokinetics in five species. *Journal of Dental Research.* **70**: 948–951.

Yee, S., 1997. In vitro permeability across Caco-2 cells (colonic) can predict in vivo (small intestinal) absorption in man – fact or myth. *Pharmaceutical Research.* **14**: 763–766.

Xianling, L., Zengguang, X. and Rong'an, C., 1996. Effects of boron on intestinal fluoride absorption in rats. *Journal of Yunyang Medical College.* 1996–2003.

CHAPTER 10

Fluoride in Saliva and its Impact on Health

SERAP AKYUZ*[a], AYSEN YARAT[b], EBRU EMEKLI ALTURFAN[b], AND SARP KAYA[a]

[a]Department of Pediatric Dentistry, Faculty of Dentistry, Marmara University, Nisantasi, 34365, Istanbul, Turkey; [b]Department of Biochemistry, Faculty of Dentistry, Marmara University, Nisantasi, 34365, Istanbul, Turkey
*E-mail: sakyuz@marmara.edu.tr

10.1 Saliva

Saliva is a clear and slightly acidic (pH 6–7) fluid, is the common product of three major and several minor salivary glands, consists of water, electrolytes and protein. Saliva has numerous qualities aiding different oral functions. Water (approximately 99%) is the prevailing compound of saliva, followed by ions (Table 10.1). The ionic composition of saliva differs from the plasma although it is actually derived from it. An important component of saliva is fluoride. Saliva delivers fluoride regularly to the tooth surface and due to this ability salivary fluoride gains an important role in caries protection by promoting remineralization and concurrently reducing demineralization (Smith and Ekstrand, 1996; Dowd, 1999; Mese and Matsuo, 2007; Carpenter, 2013).

In addition to the above-stated roles, saliva also has the following important qualities: (i) buffering the acid, (ii) providing minerals in order to replace the minerals dissolved from the tooth during demineralization (Table 10.2). Saliva constantly contacts enamel surface that is saturated with certain calcium phosphate salts and helps to maintain the integrity of the enamel

Food and Nutritional Components in Focus No. 6
Fluorine: Chemistry, Analysis, Function and Effects
Edited by Victor R Preedy

Published by the Royal Society of Chemistry, www.rsc.org

Table 10.1 Major organic and inorganic salivary components.

Organic	Inorganic
Amylase	Sodium
Lipase	Calcium
Mucins	Chloride
Lactoferrin	Potassium
Lysozyme	Bicarbonate
Peroxidases	Phosphate
Immunoglobulins	Fluoride
Histatins	Magnesium
Cystatins	Thiocyanate
Agglutinin	Hydrogen

Table 10.2 Salivary functions.

Masticatory properties	Lubrication Taste Formation of Bolus Digestive Enzymes
Anticaries properties	Remineralization Oral Clearance Inhibition of Demineralization
Antimicrobial properties	Antibacterial Antifungal Antiviral

surface. Bounded with physiological pH limits, it was suggested that, the salivary composition of calcium and inorganic phosphate was qualified to supersaturate the saliva with respect to hydroxylapatite. The protective factors, including salivary calcium, phosphate, proteins, salivary flow, and fluoride in saliva, can balance, maintain, avoid or reverse dental caries (Rosin-Grget *et al.*, 2013).

10.2 Fluoride

Fluoride is an inorganic anion of fluorine and its chemical formula is F^-. Fluoride is the most electronegative element, having atomic weight 19 and atomic number 9. From the biological viewpoint, fluoride is mainly classified under trace elements. Fluoride can be found densely in nature and therefrom it is dispersed in the lithosphere, biosphere, hydrosphere and the atmosphere (Smith and Ekstrand, 1996).

During the 1930s it was observed that children living in areas with naturally high fluoride concentrations experienced less tooth decay than their peers living in areas without fluoride in water (see Figure 10.1). High fluoride in drinking water was altering enamel appearance, depending on the dose consumed, from white striations to mottling spectrum; however, it was also making the enamel resistant to caries attack. This has led to water fluoridation at 1 ppm being

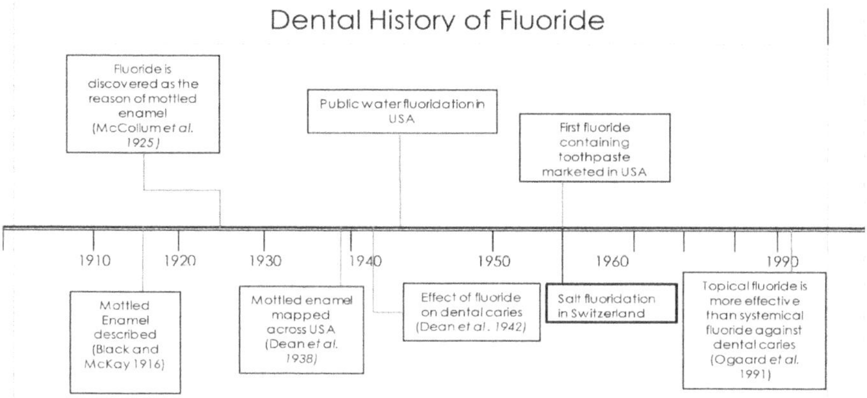

Figure 10.1 Dental history of fluoride. Progress in the understanding of fluoride's effect on dental caries is shown.

accepted as an optimal level in USA based on epidemiological studies in order to avoid dental caries without esthetical alterations (Harrison, 2005). Many countries adopted programmes to establish fluoridated water in order to reduce dental caries and today 39 countries use artificially fluoridated water (Browne *et al.*, 2005). Fluoride is considered as one of the best preventive health applications and the most effective public dental health measure in the 20th century.

Today, people are exposed to fluoride in two major ways. Systemic fluoride exposure is *via* water, foods, beverages and artificial fluoride supplements and topical fluoride exposure is through dental hygiene products; toothpastes, mouth rinses, professional fluoride gels, foams and varnishes (Figure 10.2). However, taking into consideration that people travel around the world and food and beverages are circulated worldwide, it is hard to estimate one's own chronic exposure level (Ozsvath, 2008).

Ever since fluoride has been embraced by the medical community for its effectiveness against dental caries, cumulative fluoride ingestion sources has multiplied, besides the systemic and topical sources some are also exposed due to their occupation or as part of their osteoporosis/osteosclerosis treatment. Besides the willful ingestion of fluoride, people are also unknowingly exposed to it from beverages and food prepared with fluoride content (Harrison, 2005). Although estimation of the average amount of fluoride a person ingests daily depends on many individual factors, 0.03–0.04 mg kg^{-1} for non-fluoridated and 0.04–0.06 mg kg^{-1} have been reported for mean intake values (Harrison, 2005).

After ingestion, some fluoride ions can be retained in the oral cavity and incorporated to the teeth but most of it is absorbed by the stomach and small intestine and distributed to soft and hard tissues. Ten to fifty percent

SYSTEMICAL FLUORIDE SOURCES	TOPICAL FLUORIDE SOURCES
• Water • Foods and beverages • Fluoride supplements (drops, tablets, lozenges) • Ingested fluoride from topical sources	• Toothpaste • Mouthrinse • Gels, foams, varnishes • Fluoride-eleasing dental restoratives

Figure 10.2 Systemical and topical fluoride sources. Systemical and topical fluoride sources are listed.

of the ingested fluoride can be retained in the body and the unabsorbed fragment is excreted through feces and sweat. Ninety nine percent of the fluoride in the body is deposited in the calcified tissues from where it can be dispersed into blood plasma, and the remaining small amount of fluoride in the soft tissues are in an equilibrium with their extracellular compartments. The majority of the absorbed fluoride not incorporated to the calcified tissues is excreted *via* urine and the remainder is excreted with sweat and feces (Buzalaf *et al.*, 2011).

Fluoride is dispersed through the body with blood plasma after its absorption. Blood plasma fluoride levels reach their peak levels depending on the amount and type of the fluoride 20 to 60 min after its ingestion (Whitford, 1989).

10.3 Fluoride in Saliva

Whole salivary fluoride levels vary from 0.01 to 0.1 ppm depending on the water, diet and topical fluoride use, it also depends on the blood plasma fluoride concentration of the individual (Toumba and Curzon, 2001). Gingival crevicular fluid fluoride level is higher and glandular saliva fluoride is lower than blood plasma. Glandular saliva to plasma fluoride concentrations has been reported to be 0.9 for submandibular and 0.8 for parotid glands; the whole saliva to blood plasma fluoride ratio is 1.10 (Buzalaf and Whitford, 2011). Glandular saliva fluoride content is not effected by flow rate unlike whole saliva, which decreases in fluoride content as the flow rate increases (Doull *et al.*, 2006).

Fluoride content in glandular saliva from systemic exposure in single-dose ingestions follows blood plasma levels and peaks in about 30 min and remains elevated for 8 h. But from drinking water, it remains elevated throughout the day with frequent peaks depending on drinking pattern (Whitford, 1989). Although both parotid and submandibular saliva reflect plasma fluoride levels, the submandibular gland is more stable in this than parotid (Rugg-Gunn *et al.*, 2011).

Fluoride content in saliva from topical exposure, decreases biphasic exponentially with time (ten Cate, 1997). Numerous studies reported that topical fluoride is cleared from the oral cavity within two hours after its application (Oliveby *et al.*, 1990; Whitford *et al.*, 2002; Naumova *et al.*, 2012).

Fortunately, fluoride's bioavailability in the oral fluids; saliva and biofilm is extended by the its reservoir sources; mineral CaF_2 and biologically/bacterially bound calcium fluoride deposits (Vogel, 2011). Some of the dental materials may also serve as a fluoride reservoir that increases the fluoride level in saliva and biofilm, which helps to prevent or reduce the carious process (Wiegand *et al.*, 2007).

Saliva can be used as a fluoride level biological marker that can be collected less invasively then blood. Glandular saliva is highly correlated to plasma fluoride, rather than whole saliva. However, saliva collection directly from a gland is challenging especially for the submandibular gland (Rugg-Gunn *et al.*, 2011).

10.4 Effects of Fluoride on Health

The majority of the fluoride entering the body is deposited in the calcified tissues; to teeth depending on the stage of dental development and bones. Major health impacts of fluoride are observed in these tissues.

10.4.1 Toxic Effects

10.4.1.1 Acute Toxicity

Acute fluoride toxicity occurred from accidental ingestion of pesticides in the past. Today, accidental overdoses from fluoride-containing substances is a risk especially for children. Depending on the amount ingested and absorbed, symptoms of vomiting, cramping of the extremities, bronchospasm, hyperkalemia, hypocalcemia, ventricular fibrillation, cardiac arrest can be observed and even in certain cases death may follow.

Oral ingestion of 5 mg kg^{-1} bodyweight fluoride (11 mg kg^{-1} NaF) is accepted as a possible toxic dose; which could cause serious life-threatening systemic signs and symptoms requiring immediate hospitalization (Buzalaf *et al.*, 2011).

Today fluoride is rarely used in pesticides but overdoses from fluoride dietary supplements, mouthrinses, toothpastes and professional gels can be possible risk sources without proper supervision, especially for young children.

10.4.1.2 Chronic Toxicity

Excess nonlethal fluoride consumption over extended periods may result in a serious health problem called fluorosis. Fluorosis exhibits dental and skeletal manifestations. Dental fluorosis (enamel fluorosis) is characterized with

Table 10.3 Fluoride toothpaste recommendation by EAPD. Fluoride toothpaste regimen recommended by the European Academy of Paediatric Dentistry.

Age	Fluoride concentration	Daily use	Amount to be used daily
6 months–2 years	500 ppm	Twice	Pea size
2–6 years	1000 ppm	Twice	Pea size
6 years and over	1450 ppm	Twice	1–2 cm

hypocalcified enamel, caused by excess fluoride intake during dental formation that varies from mild opaque lines to pitting and staining of enamel. Skeletal fluorosis is a variant of chronic fluoride toxicity that goes usually undetected without radiography. Symptoms are increased bone mass in milder cases with pain and calcification of spinal ligaments in more severe stages (Sampaio and Levy, 2011).

Fluoride's effect on dental caries used to be perceived from incorporation into enamel structure from its systemic ingestion. This approach has shown to be problematic in the past due to difficulties in estimating a person's regular fluoride exposure and accidental prescription of fluoride supplements, risking fluorosis. The paradigm shift in the scientific community defending minimal and frequent use of fluoride provided the benefits against dental caries without enamel fluorosis risk. For this reason systemical fluoride delivery methods are losing their support against topical fluoride use (Ellwood *et al.*, 2008).

The European Academy of Paediatric Dentistry (EAPD) suggests that the daily use of fluoride should be a significant part of protective program for the control of dental caries in children. However, the intake of fluoride must be well balanced between the accurate estimation of caries risks and the possible risks for toxic effects of the fluoride (EAPD, 2009) (Table 10.3).

10.4.2 Dental Caries

Dental caries describes progressive mineralized tooth surface dissolution caused by the acid attack from biofilm. It is one of the most common diseases affecting humans (Selwitz *et al.*, 2007). Nutrition, oral hygiene habits, host immune capabilities, salivary factors and local dental properties play critical roles in its progression. Biofilm is a complex interplay of acid-producing micro-organisms, constantly formed on the teeth following routine oral hygiene procedures. This structure protects its resident microbiota from the host immune system, antimicrobial agents and provides a thriving environment for multiplication. Acid-producing bacteria; *Streptococcus mutans*, *Streptococcus sobrinius*, *Lactobacillus* spp. metabolize fermentable carbohydrates ingested and produce acetic, butyric, formic, lactic, and propionic acids and cause the dissolution of the mineral structure. Dental caries is the result of imbalance from constant acid challenge and its counter-reparative remineralization in the oral cavity.

Table 10.4 Effects of fluoride on dental caries. The effect of both topical and systemic fluoride on dental caries is shown.

Effects of fluoride on dental caries	Mechanism
Improved enamel structure	Formation of fluoroapatites during enamelogenesis
Promoting remineralization and improved crystal precipitation	Lowering critical pH for enamel dissolution, creating supersaturated environment for calcium, phosphate
Reducing biofilm microorganism's ability to produce acids	Inhibition of glycolysis enzyme enolase

The carious process starts with the formation of biofilm on teeth after routine oral hygiene practices, complying different types of micro-organisms (filamentous, veilonella, *etc.*) throughout its existence. At neutral oral conditions (pH 7) biofilm, saliva calcium and phosphate levels are supersaturated against teeth (enamel). Micro-organisms like *S. Mutans* and *Lactobacilli* in the biofilm ferment carbohydrates and produce organic acids following a carbohydrate diffusion to its midst. This carbohydrate challenge lowers biofilm pH starting dissolution of enamel crystal structure: demineralization. Enamel becomes 10 times more soluble for each pH unit decrease in the biofilm due to the lowering of available calcium and phosphate levels (Buzalaf *et al.*, 2011). Fortunately this action is self-limiting usually due to carbohydrate depletion, acid diffusion and salivary buffering in the biofilm.

The critical pH level for enamel is 5.5, which signifies the level at which no enamel crystal precipitation nor dissolution takes place, from this level forward more acidic levels cause dissolution and base levels cause precipitation. Following acid attack, the pH returns to neutral and calcium and phosphate precipitate to repair damaged crystal structure; in this way remineralization takes place. Tipping of this equilibrium in the demineralization direction causes dental caries.

Fluoride has been proven to be effective against caries since Dean's discovery but the proposed mechanism was assumed to be its inclusion in enamel structure and formation of acid-resistant fluorapatite. However, this theory was abolished with the demineralization studies that proved that the effects of demineralization on human enamel can be decreased when fluoride is present in the oral environment (Ogaard *et al.*, 1991). This finding changed the recommended mode of application of fluoride against caries in the dental community. Effects of fluoride against caries are promotion of remineralization, improvement of crystal structure and reducing plaque microorganisms to produce acid (Table 10.4).

10.4.3 Bone Effects

The effects of fluoride on the maintenance of skeleton are unclear. It is one of the few agents that can stimulate osteoblast proliferation and increase mineral deposition in cancellous bone. Since 1960s, fluoride has been used in the treatment of osteoporosis with calcium and vitamin D at high dose levels. Fluoride incorporation into the bone structure alters its crystal structure, making it resistant to osteoclastic attack. This feature has resulted in efforts to use fluoride against the treatment of osteoporosis. Even though the artificially fluoridated water is not sufficient to implement osteoporotic fractures, the effects of high fluoride levels on bone fractures are inconclusive (Palmer *et al.*, 2012).

10.4.4 Cancer

Controversy of alleged risk of fluoridation to increase cancer development is not new. But considering the time between factors causing cancer, its clinical onset and different types of cancers make inquiry of this problem complex. The relation of kidney, bladder, stomach, thyroid and especially bone cancer to fluoride has been given attention in the literature with inconclusive evidence in humans (Harrison, 2005; Ozsvath, 2008). The World Health Organization also stated that the weight of evidence does not support the hypothesis that fluoride causes cancer in humans and many of the findings suggesting a correlation suffer from methodological flaws (WHO, 2002).

10.4.5 Reproductive Effects

The majority of the studies investigating effects of fluoride on the reproductive system has been conducted with high doses and short durations on animals. This difference in dose and duration makes the extrapolation of the results doubtful (Harrison, 2005).

Human studies are limited on this subject, Freni (1994) reported a correlation between fluoride levels higher than 3 mg L^{-1} and decreasing mean birth rates in a population. Numerous detrimental effects of ingesting high fluoride on males have been reported; sperm morphology/mobility, on testosterone levels, follicle stimulating hormones and inhibitin-B (Ozsvath, 2008). On the other hand a systematic review conducted by York University on the systemic effects of public water fluoridation reported no evidence on reproductive toxicity (NHS and CRD, 2000).

10.4.6 Renal Effects

Kidneys are exposed to higher fluoride concentrations due to their roles in excreting most of the excess fluoride. Decreased filtration in renal insufficiency or diabetes-related kidney failure may cause fluoride retention. This puts kidneys at higher risk for fluoride toxicity than other soft tissues. There is no evidence published showing that chronic fluoride ingestion at 1 ppm level can affect kidneys (Doull *et al.*, 2006). Singh's study from

India links higher fluoride levels in drinking water to kidney stones but the evidence is not conclusive, because the subjects of the study were already at greater risk of forming kidney stones due to malnutrition (Singh *et al.*, 2001).

10.4.7 Endocrine Effects

Fluoride is reported to affect endocrine function in numerous ways; decreased thyroid function, increased calcitonin activity, secondary hyperparathyroidism and impaired glucose tolerance to different extents between individuals (Ozsvath, 2008).

10.4.8 Gastrointestinal Effects

Fluoride forms hydrofluoric acid upon its contact with hydrochloric acid in the stomach. Hydrofluoric acid is irritating to the gastric mucosa, especially at toxic dose levels. Low level fluoride exposure is reported not to cause gastrointestinal symptoms (Harrison, 2005; Doull *et al.*, 2006).

10.4.9 Neurological Effects

Numerous Chinese studies suggested a possible effect of higher drinking-water fluoride levels on the intelligence of children (Zhao *et al.*, 1996; Lu *et al.*, 2000; Xiang *et al.*, 2003; Doull *et al.*, 2006). According to these studies children from high fluoride areas performed lower scores in the intelligence tests applied. Tang *et al.* (2008) reviewed this possible effect in a meta analysis and reported higher odds for lower IQ levels for children living in high-fluoride areas. The effects of high levels of fluoride on this outcome have not been explained clearly. Different theories on fluoride influence of lead and aluminum uptake, possible effects on brain enzymatic systems have been proposed (Ozsvath, 2008). Further research with strong methodology on this critical claim is necessary to draw substantial conclusions (Doull *et al.*, 2006).

10.4.10 Birth Defects

Fluoride can cross the placenta and incorporate into the tissues of the fetus (Harrison, 2005). Placenta fluoride level is 60% of the maternal blood plasma. The relation of fluoride with human developmental disorders like spina bifida occulta, sudden infant death syndrome and Down's syndrome has been investigated in the literature but the results were inconclusive (Doull *et al.*, 2006). High fluoride levels in water did not result in developmental defects in human and animal studies with the exception of ossification problems (Harrison, 2005).

10.5 Conclusions

Fluoride exposure is very common from numerous sources with a wide range of effects, some of which are well documented and some not. The majority of the information on the health effects of fluoride is on its efficacy against dental caries. Besides its preventive potential, fluoride is a substance that should be handled with caution because its excess can cause toxic results. The body of knowledge on the effects of fluoride on soft tissues and health effects besides fluorosis is limited, warranting further research.

Summary Points

- This chapter focuses on the effects of fluoride on general and dental health.
- Fluoride is regarded as the best preventive public dental health measure of the 20th century.
- Effects of fluoride on soft tissues are not well established, however, its effects on hard tissues are well documented in the literature.
- The only established clinical use of fluoride is against dental caries.
- Systemical use of fluoride is claimed to be associated with a series of side effects depending on the dose, whereas topical use of fluoride is enough to benefit from its established efficacy against caries.
- Systemic fluoride use is losing support in the world.

Key Facts of Dental Caries

1. Dental caries is a common preventable infectious disease.
2. Progression of dental caries may cause tooth loss in children and adults.
3. Biofilm habitates acid-producing bacteria (*Streptococcus mutans*, *Streptococcus sobrinius*, *Lactobacillus* spp.) that initiate enamel demineralization.
4. The efficacy of fluoride against dental caries was discovered by Dean in the 1940s.
5. Fluoride in the oral cavity (saliva, biofilm) is the only effective form of exposure against caries throughout the whole human life span.

Definitions of Words and Terms

Biofilm: A complex interplay of acid-producing micro-organisms, formed on the teeth following routine oral hygiene procedures.

Critical pH concept: Signifies the pH level (5.5) from which the enamel crystalline structure begins to dissolve.

Demineralization: The dissolution of enamel crystalline structure caused by the decrease of pH surrounding enamel.

Dental caries: Multifactorial disease characterized with the dissolution of enamel, dentin and cementum caused by the micro-organisms (*Streptococcus mutans* and other acid-producing bacteria) in the biofilm.
Dental fluorosis: Also known as enamel fluorosis. Caused by excess fluoride intake during dental formation characterized with hypocalcified enamel, which varies from mild opaque lines to pitting and staining of enamel.
Enamel: Outermost calcified structure of teeth visible in mouth.
Fluoride toxicity: The clinical manifestation of doses that can implicate acute or chronic health effects on the organism.
Hydroxylappatite ($Ca_{10}(PO_4)_6(OH)_2$): The most basic component of the crystalline structure of enamel and bone in the human body, consisting of calcium, phosphate and hydroxyl ions.
Remineralization: Precipitation of calcium and phosphate ions in non-acidic conditions into the enamel.
Saliva: Composite fluid from salivary glands and gingival crevicular fluid that mainly consists of water; responsible for masticatory, anticaries and immunological maintenance of oral health.

Abbreviations

ppm	Part per million
IQ	Intelligence quotient
WHO	World Health Organization
NaF	Sodium fluoride
spp	Species
EAPD	European Academy of Paediatric Dentistry

References

Black G. V. and McKay F. S., 1916. Mottled enamel. An endemic developmental imperfection of the teeth, heretofore unknown in the literature of dentistry. *Dental Cosmos.* **58**: 129–156.

Browne, D., Whelton H. and O'Mullane D., 2005. Fluoride metabolism and fluorosis. *Journal of Dentistry.* **33**: 177–186.

Buzalaf, M. A., Pessan, J. P., Honório, H. B. and Ten Cate, J. M., 2011. Mechanisms of action of fluoride for caries control. In: Buzalaf, M. A. (ed.) *Fluoride and the Oral Environment*. Karger, Basel, Switzerland; New York, pp. 97–114.

Buzalaf, M. A. and Whitford, G. M., 2011. Fluoride metabolism. In: Buzalaf, M. A. (ed.) *Fluoride and the Oral Environment*. Karger, Basel, Switzerland; New York, pp. 20–36.

Carpenter, G. H., 2013. The secretion, components, and properties of saliva. *Annual Review of Food Science and Technology.* **4**: 267–276.

ten Cate, J. M., 1997. Review on fluoride, with special emphasis on calcium fluoride mechanisms in caries prevention. *European Journal of Oral Sciences.* **105**: 461–465.

Dean H. T., Arnold F. A. and Elvove E., 1942. Domestic water and dental caries. V. Additional studies of the relation of fluoride domestic waters to dental caries experience in 4,425 white children aged 12–14 years of 13 cities in 4 states. *Public Health Reports.* **57**: 1155–1179.

Doull, J., Boekelheide, K., Farishian, B., Isaacson, R., Klotz, J. and Kumar, J. E. A., 2006. Fluoride in drinking water: a scientific review of EPA's standards. Washington, DC: Committee on Fluoride in Drinking Water, Board on Environmental Studies and Toxicology, Division on Earth and Life Sciences, National Research Council of the National Academies.

Dowd, F. J., 1999. Saliva and dental caries. *Dental Clinics of North America.*, **43**: 579–597.

European Academy of Paediatric Dentistry, 2009. Guidelines on the use of fluoride in children: an EAPD policy document. *European Archives of Paediatric Dentistry.* **10**: 129–135.

Ellwood, R., Fejerskov, O., Cury, J. A. and Clarkson, B., 2008. Fluorides in caries control. In: Fejerskov, O. and Kidd, E. A. M. (ed.) *Dental Caries: The Disease and Its Clinical Management.* second ed. Blackwell Munksgaard, Oxford; Ames, Iowa, pp. 330–352.

Freni, S. C., 1994. Exposure to high fluoride concentrations in drinking water is associated with decreased birth rates. *Journal of Toxicology and Environmental Health.* **42**: 109–121.

Harrison, P. T. C., 2005. Fluoride in water: A UK perspective. *Journal of Fluorine Chemistry.* **126**: 1448–1456.

Lu, Y., Sun, Z., Wu, L., Wang, X., Lu, W. and Liu, S., 2000. Effect of high-fluoride water on intelligence in children. *Fluoride.* **33**: 75–78.

McCollum E. V., Simmonds N., Becker J. E. and Bunting R. W., 1925. The effect of additions of fluoride to the diet of rats on the quality of their teeth. *Journal of Biological Chemistry.* **63**: 553–562.

Mese, H. and Matsuo, R., 2007. Salivary secretion, taste and hyposalivation. *Journal of Oral Rehabilitation.* **34**: 711–723.

Naumova, E. A., Kuehnl, P., Hertenstein, P., Markovic, L., Jordan, R. A., Gaengler, P. and Arnold, W. H., 2012. Fluoride bioavailability in saliva and plaque. *BioMed Central Oral Health.* **12**: 3.

National Health Service (NHS), NHS Centre for Reviews and Dissemination (CRD)., 2000. *A Systematic Review of Public Water Fluoridation. NHS Centre for Review and Dissemination York,* UK, University of York.

Ogaard B., Rolla G., Dijkman T., Ruben J. and Arends J., 1991. Effect of fluoride mouthrinsing on caries lesion development in shark enamel; an in situ caries model study. *Scandinavian Journal of Dental Research.* **99**: 372–377.

Oliveby, A., Weetman, D. A., Geddes, D. A. and Lagerlof, F., 1990. The effect of salivary clearance of sucrose and fluoride on human dental plaque acidogenicity. *Archives of Oral Biology.* **35**: 907–911.

Ozsvath, D. L., 2008. Fluoride and environmental health: a review. *Reviews in Environmental Science and Bio/Technology.* **8**: 59–79.

Palmer, C. A. and Gilbert, J. A., 2012. Academy of Nutrition and Dietetics. Position of the Academy of Nutrition and Dietetics: the impact of fluoride on health. *Journal of the Academy of Nutrition and Dietetics.* **112**: 1443–1453.

Rosin-Grget, K., Peros, K., Sutej, I. and Basic, K., 2013. The cariostatic mechanisms of fluoride. *Acta Medica Academica.* **42**: 179–188.

Rugg-Gunn, A. J., Villab, A. and Buzalaf, M. A., 2011. Contemporary biological markers of exposure to fluoride. In: Buzalaf, M. A. (ed.) *Fluoride and the Oral Environment.* Karger, Basel, Switzerland; New York, pp. 37–51.

Sampaio, F. C. and Levy, S. M., 2011. Systemic fluoride. In: Buzalaf, M. A. (ed.) *Fluoride and the Oral Environment.* Karger, Basel, Switzerland; New York, pp. 133–145.

Selwitz, R. H., Ismail, A. I. and Pitts, N. B., 2007. Dental caries. *Lancet.* **369**: 51–59.

Singh, P. P., Barjatiya, M. K., Dhing, S., Bhatnagar, R., Kothari, S. and Dhar, V., 2001. Evidence suggesting that high intake of fluoride provokes nephrolithiasis in tribal populations. *Urological Research.* **29**: 238–244.

Smith, F. and Ekstrand, J., 1996. The occurrence and chemistry of fluoride. In: Fejerskov, O., Ekstrand, J. and Burt, B. (ed.) *Fluoride in Dentistry.* second ed. Munksgaard, Cophenagen.

Tang, Q. Q., Du, J., Ma, H. H., Jiang, S. J. and Zhou, X. J., 2008. Fluoride and children's intelligence: a meta-analysis. *Biological Trace Element Research.* **126**: 115–120.

Toumba, K. J. and Curzon, M. E. J., 2001. Fluoride concentrations in saliva related to dental caries prevalence in primary teeth. *European Journal of Paediatric Dentistry.* **2**: 15–19.

Vogel, G. L., 2011. Oral fluoride reservoirs and the prevention of dental caries. In: Buzalaf, M. A. (ed.) *Fluoride and the Oral Environment.* Karger, Basel, Switzerland; New York, pp. 146–157.

Whitford, G. M., 1989. *The Metabolism and Toxicity of Fluoride,* Karger, Basel; New York.

Whitford, G. M., Wasdin, J. L., Schafer, T. E. and Adair, S. M., 2002. Plaque fluoride concentrations are dependent on plaque calcium concentrations. *Caries Research.*, **36**: 256–265

Wiegand, A., Buchalla, W. and Attin, T., 2007. Review on fluoride-releasing restorative materials–fluoride release and uptake characteristics, antibacterial activity and influence on caries formation. *Dental Materials.* **23**: 343–362.

World Health Organization (WHO)., 2002. *Fluorides.* Environmental Health Criteria 227. Geneva.

Xiang, Q., Liang, Y., Chen, L., Wang, C., Chen, B., Chen, X. and Zhou, M., 2003. Effect of fluoride in drinking water on children's intelligence. *Fluoride.* **36**, 84–94.

Zhao, L., Liang, G., Zhang, D. and Wu, X., 1996. Effect of a high fluoride water supply on children's intelligence. *Fluoride.* **29**: 190–192.

CHAPTER 11

Teeth–Saliva Migration of Fluoride Ions and Health Implications

BILJANA M. KALIČANIN*[a], DRAGAN S. VELIMIROVIĆ[a], AND ALEKSANDRA N. PAVLOVIĆ[b]

[a]Department of Pharmacy, Faculty of Medicine, University of Niš, dr Zoran Đinđić Boulevard 81, 18000 Niš, Serbia; [b]Department of Chemistry, Faculty of Sciences and Mathematics, University of Niš, Višegradska 33, 18000 Niš, Serbia
*E-mail: bkalicanin@yahoo.com

11.1 Physiological and Biochemical Properties of Fluorine

The main biological role of fluorine in the human body is its participation in bone, tooth enamel and dentin formation. The bone mineralization process occurs together with the generation of fluoroapatite, which prevents the release of calcium and other minerals deposited in the bones. The small amount of fluoride has a beneficial effect on tooth enamel, which becomes more resilient against decay. Fluorine does not make your teeth stronger; rather it has a preventive role in caries, particularly in children. The protective effect of fluoride in caries is explained by the inhibitory effect of fluoride on enzymes in glycolysis and the citric acid cycle, through binding F^- ions to their activators – other trace elements. For their metabolic processes, the bacteria of the normal flora of the oral cavity use carbohydrates (especially sucrose), which they decompose to lactic acid during the process

Food and Nutritional Components in Focus No. 6
Fluorine: Chemistry, Analysis, Function and Effects
Edited by Victor R Preedy

Published by the Royal Society of Chemistry, www.rsc.org

Table 11.1 Average intake of fluorides from toothpaste by children of different ages.

Ages	Average intake of fluorides (mg) two daily *via* tooth washes with toothpaste (1100 ppm F^- ion)
2	0.73
2.5	0.59
2–3	0.62
2–4	0.60
3	0.40
3–6	0.84
4	0.86
4	0.29
5	0.44

of glycolysis. The released lactic acid can act as an agent of tooth enamel demineralization since it is responsible for the decomposition of hydroxy-apatite crystals. This adverse effect of bacteria can be prevented through the usage of fluoride since fluorides very effectively inhibit enolase, an enzyme of glycolysis.

In addition to food and beverages, fluoride can be introduced into the human body through a variety of dental products, such as different types of toothpaste, mouthwash and gels for local application. This method of intake involves only a local increase in the content of fluoride in the oral cavity. However, in children aged 2–6, due to lack of the controlled swallowing reflex, an excess of fluoride intake in the body may occur (Table 11.1), which in turn may have an adverse effect on their health (Levy and Guha-Chowdhury, 1999).

In recent years, fluoride has been added to the dental materials used for filling or sealing deep dental alveoli and fissures, in order to prevent secondary and recurrent caries. Thus, for example, glass ionomer cements, as dental fillings, release significant amounts of fluoride to oral media, which have a positive impact on oral health.

11.2 Teeth Structure

Teeth consist of hard and soft tissues. Hard dental tissues are enamel, dentin and cement, while dental pulp represents the tooth's soft tissue (Figure 11.1).

Enamel (*Substantia Adamantina*) is the hardest tissue in human body. This tissue covers the anatomical crown of the tooth, but its thickness is not the same in all parts of the crown. Enamel exhibits the maximum thickness at the incisal edges and cusps of the teeth where the enamel thickness is 2–3 mm, while at the neck of the tooth the thickness decreases and can be 1 mm. The hardness of the enamel is a very important feature of this tissue, because the enamel covers and protects the dentin that lies beneath it. The enamel of the primary teeth is much thinner than the enamel of the permanent teeth and its maximum thickness is 1 mm.

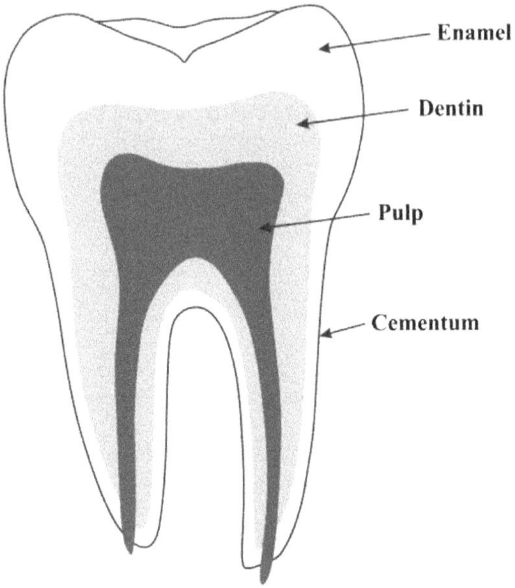

Figure 11.1 Tooth structure.

The chemical structure of tooth enamel consists mainly of inorganic matter (96–97%), while the content of the organic matter is a lot lower (3–4%). From the mineralogical point of view, the enamel can be classified as hydroxylapatite ($Ca_{10}(PO_4)_6(OH)_2$).

Structurally speaking, enamel is composed of millions of calcified enamel prisms extending from the enamel–dentin border to the tooth surface. Between the enamel prisms there is an interprismatic substance, which is not as mineralized as the prisms and contains a higher percentage of organic matter. At the cross section enamel prisms have a hexagonal shape and, compared to enamel prisms dentin they are spread mostly vertically.

Dentin (tooth bone) (*Substantia Eburnea*) is a calcified connective tissue that occupies the bulk of the tooth. Enamel covers the dentine in the crown section, but at the root it is covered by the cement of the tooth. Dentine is much softer compared to the enamel, but is harder than the root cement or the alveolar bone.

In chemical terms, 70% of the dentin is composed of inorganic matter, and about 30% of organic matter. A high percentage of organic matter is one of the reasons caries spreads so rapidly through dentin.

Dentin is entirely permeated by, so-called, dentinal tubules where extensions of odontoblasts, the so-called Thomson's fibers, are located. Thomson's fibers and nerves are located in the dentinal tubules, immersed in liquid fluid that consists of interstitial fluid. The canalicular structure of dentin also contributes to the rapid spread of decay and allows for faster penetration of bacteria into the tooth pulp.

Three different types of dentin are created in the course of human life and these are: the primary dentin (at the 4th month of fetal life), the secondary dentin (the physiological or functional dentin) and the tertiary dentin (reactive, *i.e.* stimulatory dentin).

Cement (*Substantia Ossea*) covers the tooth root, and, compared with enamel and dentin, it is much softer. By its composition it is similar to bones, which is the reason it is named the *Substantia Ossea*. Cementogenesis, *i.e.* the formation of cement, and dentinogenesis take place over the entire life course.

In chemical terms, about 50% of cement is composed of inorganic and 50% of organic matter (Brand and Isselhard, 2013).

11.3 Fluorine in Saliva and Dental Tissue

The amount of fluoride in the so-called "pure" saliva (sample obtained directly from the excretory ducts of the salivary glands) depends on the concentration of fluoride in the blood. The concentration of fluoride in oral fluids of the oral cavity varies considerably, although it is about 5–10 times higher than the concentration in the "pure" saliva. Some studies have shown that after local application of fluoride in the form of mouthwash, toothpaste, *etc.*, the concentration of fluoride in the saliva increased by a hundred or even a thousand times. It has to be noted that this high concentration of fluoride decreased rapidly. After one hour of application, the concentration of fluoride is reduced to a few ppm, and for the next few hours (3–6 h) the amount of this ion in saliva, is returned to the initial level. The maximum concentration of fluoride occurs after one hour of introduction of these ions into the human body, gradually decreasing, along with that of the blood of patients.

The concentration of fluoride in plaque is much higher than in saliva. The reason for this may be due to slower elimination of fluoride ions through the membrane plaques, than its binding in the form of calcium fluoride for different compounds of plaque content. The content of fluoride in plaque depends on its amount in the saliva and the surface layers of the enamel, as well as on the amount of dental plaque. The inner plaque layers contain higher amounts of fluoride compared to the outer layers, which is the result of the presence of higher concentrations of F^- ions in the surface layers of enamel.

Fluoride ions express a strong affinity towards the calcified tissue and, about 99% of fluoride that is introduced into the human body resides in the teeth and bones (Table 11.2). The total fluoride content in bone tissue increases with time due to its accumulation. In the surface layers of calcified tissue, the fluoride content is higher than in the inner layers, since the surface layer is in constant contact with tissue fluids that supply them with fluoride.

The concentration of detected fluoride in the hard dental tissue and bone is an indicator of the amount of fluoride ingested during life. The concentration of fluoride is the highest in the enamel, in its surface layers, and it

Table 11.2 The content of fluorides in some human tissues and organs.

Tissue/Organ	Content of F^- ion (ppm)
Blood	0.02–0.06
Urine	0.78
Saliva	0.00874
Enamel	740–2100
Hair	2.505
Brain	0.0312
Body fat	0.0285
Liver	0.0196
Muscles	0.040
Lunge	0.071
Thymus	9.52
Aorta	4.331
Kidneys	0.159
Nails	1.60–8.80

Table 11.3 The content of fluorides in enamel, dentin and cementum in primary and permanent teeth.

		The content of fluorides (ppm)		
Tissue layer	Teeth	Enamel	Dentin	Cementum
Interior	Primary	39.70	87.00	142.70
	Permanent	49.80	68.00	80.00
Exterior	Primary	101.70	111.70	253.30
	Permanent	448.00	890.00	1144.00

decreases towards the line between the enamel and dentin, while the amount of this element in dentin is about 3 times higher than in the tooth enamel. The layers closest to the predentin (pulp) have the highest fluoride content, as the new amount of fluoride is continuously delivered through the blood. The cement tissue, which is most similar to the bone tissue, contains fluoride at a concentration that is higher compared to the content of this element in the enamel and dentin (Table 11.3). This is due to the fact that the cement is exposed to the influence of tissue liquids, which are rich in fluoride, but also because of the small thickness of this tissue. As with enamel, the amount of fluoride in cement decreases from the outer to the inner parts and increases with age (Gajić, 2002).

11.4 The Cariostatic Effect of Fluoride from the Glass Ionomer Fillings – An *In Vitro* Study

Glass ionomer cements (GICs) are materials that are currently used in dental orthopedic practice with the aim of filling or sealing deep alveoli and fissures of teeth. One of the requirements in the manufacturing of these materials is that, among other things, they should contain around 23% of fluoride components. This enables the glass ionomer cements to have an anticariogenic effect, through releasing fluoride in enamel–dentin substance and saliva. The fluorides released from this material have a preventive role in secondary caries, through certain biological and physicochemical processes. The cariostatic potential of glass ionomer cement depends on both the amount of fluoride released, and the duration of the process of releasing fluoride ions from these materials. Numerous studies have shown that the glass ionomer cements are the materials that, in comparison to other dental and prosthetic materials, provide the easiest release of the present fluorides. Also, these studies have shown that glass ionomer cements tend to bind fluoride from the oral medium. This glass ionomer feature allows for their reflouridation using different means of maintaining oral hygiene.

The large number of studies indicated the possible fluoride migration from GICs into the tooth structure (Kovarik *et al.*, 2005). On the other hand, fluoride levels in the oral fluids affect the demineralization and remineralization processes by shifting the balance to a less cariogenic condition. According to Williams *et al.* (2003) low fluoride concentration in saliva can lead towards remineralization of dentin. Van Duinen *et al.* (2004) demonstrated that glass ionomer adjacent to tooth structure and in contact with the oral fluids releases calcium, phosphate and fluoride content. Davidson (2006) said that glass ionomer can be called intelligent because it releases fluoride proportionally to the acidity. In general, a direct relationship exists between the amount of fluoride present in the cement and the fluoride released (Francci *et al.*, 1999; Perrin *et al.*, 1994). Also, the results obtained from Bayrak *et al.* (2010) are in line with the observation by Xu and Burgess that the materials with higher initial fluoride release have higher recharge capacity. Resin-based restorative materials such as resin-modified GICs, compomers and fluoride-releasing composites are capable of recharge (Attar and Onen, 2002). Previous studies analyzed fluoride release and recharge by restorative materials. Itota *et al.* (1999) analyzed the amount of fluoride release from materials in distilled water and in acid buffer solution until 10 weeks. After the high rates of fluoride release during the first week, the release subsided to a low constant level. In the case of acid buffer solution, rates of fluoride release markedly increased. Results obtained by Preston *et al.* (1999) for the conventional GICs and the RMGIC showed that the early release rates were relatively high, but rapidly decreased over the first 48 h. Bayrak *et al.* (2010) was measures fluoride release during 42 days. The greatest amount of fluoride release occurred during the first day. Further, fluoride release decreased with time, but continued throughout the entire 21 day test period up until recharge. The other studies (Gandolfi *et al.*, 2006; Mousavinasab and Meyers, 2009; Suprastiwi *et al.*, 2010) also demonstrating the mode of fluorine release by GICs, characterized at first by an initial rapid release, and by rapid reduction of fluorine release afterwards. Moreover, the ionic composition of saliva can have an impact of the migration of fluorine ions. Wiegand *et al.*, (2007) reported two mechanisms of fluorine release from GICs. The uptake of fluoride from GICs into adjacent dental tissue is also well documented (Tsanidis and Koulourides, 1992; Murai *et al.*, 1993).

Apostolović *et al.* have, in their research, followed the cariostatic potential of fluoride from the glass ionomer fillings in the extracted teeth, through the content of the released fluorides in artificial saliva (Apostolović *et al.*, 2013). Moreover, the possibility of reflouridation of these fillings had also been followed, by treating the above-mentioned teeth samples with a low concentrated fluoride solution (334 ppm) of a mouthwash.

The study was perform in three phases: teeth extraction and preparation; teeth incubation and fluoridation, and determination of fluorine ion concentration.

After extraction, the surfaces of all the teeth were cleaned, roots were cut off with a metal cutter at the level of the enamel–cement borderline, and the remaining pulp tissue was removed.

From the vestibular aspect, class V cavities were prepared, 3 × 2 × 2 mm in size, using a diamond drill, and all the teeth were restored with GIC (Fuji IX, GC, Japan) following the manufacturer's guidelines. After GIC binding, the excess material was removed and the teeth were washed under the current of distilled water and placed in 100 mL of artificial saliva.

The pH value of artificial saliva was around 6.7, as the closest approximation to physiologic values for saliva in the mouth cavity. The fluorine ion concentration in the artificial saliva was 0.071 ppm.

All glass ionomer cement restored teeth (GICrT) were divided into two groups with six teeth each. The first group was treated daily with low concentration fluoride, while in the second group there was no fluoridation. Both groups were further divided into three subgroups with two teeth each, according to the experiment duration (7, 14, and 21 days).

The prepared teeth samples were treated with the solution of artificial saliva incubated at 37 °C during the aforementioned periods of time.

The first group of GICrT (six samples) was fluoridated with the fluoride solution (concentration of 334 ppm), composed of 10 mL of low concentration fluoride solution and 5 mL of artificial saliva solution. The teeth in this group were fluoridated daily, being submersed in this solution for 1 min, and then washed for 5 s with distilled water, being returned after drying into the artificial saliva. After the planned treatment periods, the samples of artificial saliva were analyzed for fluorine content.

The content of fluorine ions was determined by way of automated potentiometric titration using an fluoride ISE (Ma-5705, Iskra, Slovenia).

The potentiometric analysis of fluoride using this electrode is characterized by a simple, rapid and accurate determination.

The pH value of the solutions in which the fluoride content is determined using ISE, range of 5 to 8, and a temperature range of 0 to 80 °C. Considering the fact that the temperature change of the test systems influences the potential of the electrode, the calibration solutions and the samples have to be thermostated. Electrode response to the content of fluoride ions in the analyzed solution is from 20 to 30 s, depending on the concentration of the

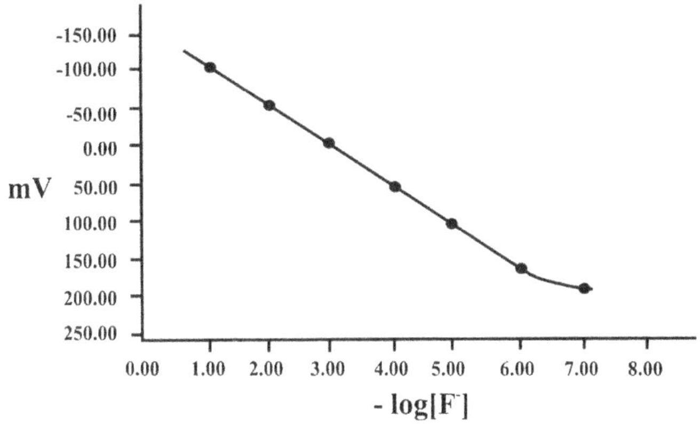

Figure 11.2 Calibration curve for determination of fluoride concentration with fluoride ion-selective electrode in tested samples. The concentration of fluoride in tested samples is determined using a calibration curve method. Standard solutions of fluoride are prepared, and their concentrations are measured using the fluoride ion-selective electrode. The negative logarithm of concentration for prepared F^- ion standard solutions are presented on the *x*-axis, and the corresponded potential (mV) for each standard solution is shown on the *y*-axis.

fluoride present. The content of fluoride ions, which can be determined by this electrode, ranges from 1×10^{-1} to 1×10^{-6} mol L^{-1}.

The determination of fluoride content is interfered by the hydroxide ions in alkaline solutions (pH ≥ 9), and at concentrations that are greater than one tenth of the concentration of fluoride ions present. Some cations such as aluminum(III), silicon (IV) and iron (III), also interfere with the determination of fluoride content, since they build the corresponding complexes with it. These interferences (presence of OH^- ions and certain cations) can be solved by adding a TISAB (*total ionic strength adjustment buffer*) buffer. The addition of this buffer adjusts the pH of the analyzed solution to 5.0–5.5, and the presence of sodium citrate, forming complexes with aluminum or iron, prevents the construction of complex fluorides with interfering metal ions (Singer and Armstrong, 1968).

The fluoride content in the analyzed samples is determined by using a calibration curve, as a linear dependence of the measured potential on the negative logarithm of the fluoride ions concentration (Figure 11.2).

The results of this study have shown that the highest amount of fluoride released from the sampled teeth appeared after 7 days, in both study groups, while the content of the newly released F^- ions was lower 14 and 21 days after the teeth restoration (Table 11.4). These results confirmed that the glass ionomer cement initially releases the highest amount of fluoride, while the content of the newly released F^- ions from this material in any medium becomes lower with time. Based on this, it can be concluded that these materials show the strongest anticariostatic effects in the first week of tooth restoration. The samples of teeth with glass ionomer filling, treated with a low concentrated

Table 11.4 The concentration of newly released fluorine ions from two tested groups in artificial saliva medium. Comparing the concentration of released fluorine ions in the medium within the groups, it can see that the highest ion concentrations were released in the artificial saliva medium in the first week, in both fluoridated and nonfluoridated teeth (0.883 and 0.664 ppm, respectively). During the second week the observed release trend continued, although with a lower concentration of newly released fluorine ions compared to the first week in both studied groups (0.220 and 0.204 ppm, respectively). In the third week, compared to the second week, the trend of reduction of concentration of newly released ions of fluorine continued (0.156 and 0.033 ppm, respectively). These data demonstrated that the release was still present, with the lowest concentration of newly released fluorine ions in the third week, and with a lower reduction rate in fluoridated teeth.

	Concentration (ppm)	
	fluoridated GICrT	**nonfluoridated GICrT**
7 days	0.833	0.644
14 days	0.220	0.204
21 days	0.156	0.033

fluoride, released higher contents of F^- ions into the artificial saliva solution, compared to the group of teeth that was not treated with this solution. This confirmed the existence of a synergistic anticariostatic effect of the glass ionomer cement and mouthwash. The results of this study have also shown that there is a statistically significant difference in the contents of the detected fluorides ($p < 0.05$) 14 and 21 days after the restoration, indicating that a decrease in the concentration of F^- ions in glass ionomer increases the ability of this material to bind fluoride from the solution. As the amount of fluorides released from the glass ionomer decreases with time, reflouridation of these materials is recommended 2 to 3 weeks after the tooth restoration. These results demonstrate that a mouthwash with a low concentration of fluoride may be used for a reflouridation of the glass ionomer fillings.

11.5 Health Effects of Fluorides

Hydroxyapatite is the base of the bone tissue (Figure 11.3). Fluorides are incorporated into the crystal lattice of hydroxyapatite in the process of mineralization and remineralization, resulting in the formation of fluoroapatite. Incorporation of fluorine in the bone tissue is carried out in three stages. In the first stage, fluoride ions reach the hydrated layer, replacing those ions whose bond is unstable. In the second stage there is a replacement of the hydroxyl, carbonate or bicarbonate groups on the hydroxyapatite surface by the fluorine ions from the hydrated layer.

In the final, third stage, the fluoride ions from the surface parts of crystals migrate into the deeper layers of the bone. Most of the fluorides that formulate the crystal are probably located in the areas of the replaced OH^- groups.

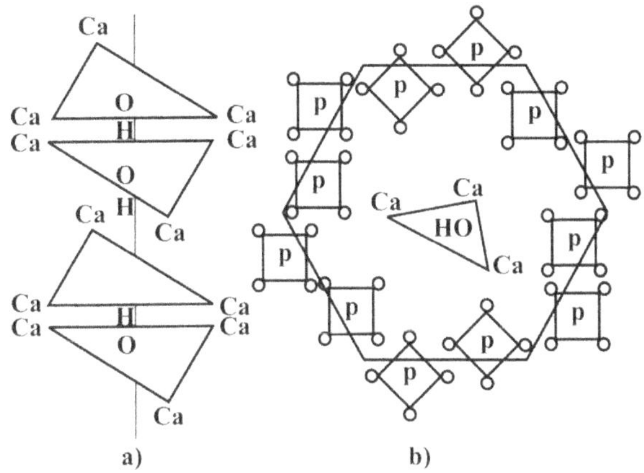

Figure 11.3 Structure of calcium hydroxyapatite ($Ca_{10}(PO_4)_6(OH)_2$): (a) Ca^{2+} ions are axially distributed with centally aligned OH^- groups; (b) coulmnar and axial arrangement ratio of Ca^{2+} and PO_4^{3-} ions in hydroxyapatite.

This provides for the generation of a fluoroapatite that is more stable than hydroxyapatite, which in turns makes the bone tissue more resistant to the onset and progression of the disease (caries, osteoporosis, osteomalacia).

For prevention and treatment of dental caries, fluoride deficit is compensated by using various oral products such as toothpastes, mouthwashes, gels. Fluorides ingested through these products, when in contact with the tooth enamel, react with calcium ions in the hydroxyapatite, resulting in the generation of the calcium fluoride which is deposited on the surface of the enamel:

$$Ca_{10}(PO_4)_6(OH)_2 + 20F^- \rightarrow 10CaF_2 + 6PO_4^{3-} + 2OH^- \qquad (11.1)$$

Fluoride ions from calcium fluoride penetrate the deeper layers of enamel, where, as noted above, fluorides replace OH^- groups of enamel apatite that leads to the formation of a stable fluoroapatite. In this way the fluoride ions can penetrate the tooth enamel up to 50 μm in depth, but this migration into the interior of the enamel takes place very slowly. It is currently considered that calcium fluoride, which covers the outer enamel layers after the local application of concentrated fluoride solutions, has a very important role in the prevention of dental caries, as it represents fluorine and calcium depots to be activated in the conditions of reduced biofilm pH. Namely, when the pH value of the biofilm drops below 5.5, a dissolution of calcium fluoride, which is deposited on the surface of the tooth enamel, occurs, resulting in the release of fluoride, its increased concentration affecting the favorization of the remineralization process in relation to the demineralization.

Fluorides have a cariostatic effect (reducing the risk of dental caries) in the period before the formation of the teeth, and in the later period of a tooth's existence. This effect arises as a consequence of the lower solubility of fluoroapatite in the acidic environment in relation to hydroxiapatite, as the acidic environment is generated in the dental plaque during the metabolism of the bacteria. The acidic environment in the oral cavity can occur as a result of intake of food or drink with a sour taste, which creates an additional effect on the appearance of caries.

Fluorides (if entered in larger amounts) can also stimulate the regeneration process of the damaged bones, and are used as an experimental drug for the treatment of osteoporosis. Recent studies have shown that there is a positive clinical effect on the bones when fluoride is added (23 mg per day) in the form from which it can be gradually released, as compared to the forms in which it is quickly absorbed by the gastrointestinal tract (Pak *et al.*, 1997).

Summary Points

- This chapter focuses on the chemistry of physiological and biochemical properties of fluorine, migration of fluoride ions from teeth and saliva and its implications on human health.
- The main biological role of fluorine in the human body is its participation in bone, tooth enamel and dentin formation. The bone mineralization process occurs together with the generation of fluoroapatite, which prevents the release of calcium and other minerals deposited in the bones.
- An insufficient fluoride intake at any age increases the risk of dental caries.
- Fluorides have a cariostatic effect (reducing the risk of dental caries) in the period before the formation of the teeth, and in the later period of a tooth's existence.
- Glass ionomer cements are materials that are currently used in dental orthopedic practice and one of the requirements in the manufacturing of these materials is that, among other things, they should contain around 23% of fluoride components.
- The fluorides released from glass ionomer cements have a preventive role in secondary caries, through certain biological and physicochemical processes.
- The cariostatic potential of glass ionomer cement depends on both the amount of fluoride released, and the duration of the process of releasing fluoride ions from these materials.
- Reflouridation of glass ionomer fillings can be increased by treating the restored teeth with a low concentrated fluoride solution of a mouthwash.

Key Facts of the Mechanism of Fluoride Effects in Bone Tissue

- Hydroxyapatite represents mineral that forms the base of the bone tissue.
- Fluorides can be incorporated into the crystal lattice of hydroxyapatite in the process of mineralization and remineralization, resulting in the formation of fluoroapatite.
- Most of the fluorides that formulate the crystal are probably located in the areas of the replaced OH^- groups. This provides for the generation of

a fluoroapatite that is more stable than hydroxyapatite, which in turns makes the bone tissue more resistant to the onset and progression of the disease (caries, osteoporosis, osteomalacia).

- The cariostatic effect of fluoride arises as a consequence of the lower solubility of fluoroapatite in the acidic environment in relation to hydroxiapatite, as the acidic environment is generated in the dental plaque during the metabolism of the bacteria.
- For prevention and treatment of dental caries, fluoride deficit is compensated by using various oral products such as toothpastes, mouthwashes, gels.
- Calcium fluoride, which covers the outer enamel layers after the local application of concentrated fluoride solutions, has a very important role in the prevention of dental caries, as it represents fluorine and calcium depots to be activated in the conditions of reduced biofilm pH.
- When the pH value of the biofilm drops below 5.5, a dissolution of calcium fluoride, which is deposited on the surface of tooth enamel, occurs, resulting in the release of fluoride, its increased concentration affecting the favorization of the remineralization process in relation to the demineralization.

Definitions of Words and Terms

Mineralization: The breakdown of organic residues from matrix, by oxidation in order to get inorganic substances for chemical analysis.

Calibration curve: This is a general method for determining the concentration of a substance in an unknown sample by comparing the unknown to a set of standard samples of known concentration.

Dental caries: Dental caries is an infection, bacterial in origin, that causes demineralization and destruction of the hard tissues of the teeth (enamel, dentin and cementum).

Secondary caries: Secondary caries is a caries that appears at a location with a previous history of caries. This is frequently found on the margins of fillings and other dental restorations.

Cariostatic potential: Capacity of used dental material or procedure that prevents or retards the formation and progression of dental caries.

Dental fluorosis: This is a developmental disturbance of dental enamel caused by excessive exposure to high concentrations of fluoride during tooth development.

Osteofluorosis fluorosis: Osteofluorosis fluorosis is a bone disease caused by excessive consumption of fluoride. In advanced cases, skeletal fluorosis causes pain and damage to bones and joints.

Glass ionomer cement: This is a dental restorative material used in dentistry for filling teeth and luting cements. These materials are based on the reaction of silicate glass powder and polyalkenoic acid.

Osteoporosis: This is a progressive bone disease that is characterized by a decrease in bone mass and density that can lead to an increased risk of fracture. In osteoporosis, the bone mineral density (BMD) is reduced, bone microarchitecture deteriorates, and the amount and variety of proteins in bone are altered.

Osteomalacia: This is the softening of the bones caused by defective bone mineralization secondary to inadequate amounts of available phosphorus and calcium, or because of overactive resorption of calcium from the bone as a result of hyperparathyroidism.

List of Abbreviations

ISE	Ion-selective electrode
$[F^-]$	Fluoride ions concentration
TISAB	Total ionic strength adjustment buffer
GICs	Glass ionomer cements
GICrT	Glass ionomer cement restored teeth
OH^-	Hydroxide ion

References

Apostolović, M., Kaličanin, B., Igić, M., Tričković-Janjić, O., Šurdilović, D., Kostadinović, Lj., Stojković, B. and Velimirović, D., 2013. Migration of fluoride ions from the permanent teeth into saliva in children with glass ionomer cement restorations: an in vitro study. *Vojnosanitetski Pregled*. **70**(3): 279–283.

Attar, N. and Onen, A., 2002. Fluoride release and uptake characteristics of aesthetic restorative materials. *Journal of Oral Rehabilitation*. **29**(8): 791–798.

Bayrak, S., Tunc, E. S., Aksoy, A., Ertas, E., Guvenc, D. and Ozer, S., 2010. Fluoride release and recharge from different materials used as fissure sealants. *European Journal of Dentistry*. **4**(3): 245–250.

Brand R. W. and Isselhard D. E., 2013. Anatomy of Orofacial Structures: A Comprehensive Approach (7th edn), Elsevier Health Sciences, USA.

Davidson, C. L., 2006. Advances in glass-ionomer cements. *Journal of Applied Oral Science*. **14**(SPE): 3–9.

Francci, C., Deaton, T. G., Arnold, R. R., Swift, E. J., Perdigao, J. and Bawden, J. W., 1999. Fluoride release from restorative materials and its effects on dentin demineralization. *Journal of Dental Research*. **78**(10): 1647–1654.

Gajić, M., 2002. Fluoride in Preventive Dentistry (2nd edn) [Fluoridi u preventivnoj stomatologiji], Nauka, Beograd, Srbija, 15 pp. (In Serbian).

Gandolfi, M. G., Chersoni, S., Acquaviva, G. L., Piana, G., Prati, C. and Mongiorgi, R., 2006. Fluoride release and absorption at different pH from glass-ionomer cements. *Dental Materials*. **22**(5): 441–449.

Itota, T., Okamoto, M., Sato, K., Nakabo, S., Nagamine, M., Torii, Y. and Inoue, K., 1999. Release and recharge of fluoride by restorative materials. *Dental Materials Journal*. **18**(4): 347–353.

Kovarik, R. E., Haubenreich, J. E. and Gore, D., 2005. Glass ionomer cements: a review of composition, chemistry, and biocompatibility as a dental and medical implant material. *Journal of Long-term Effects of Medical Implants*. **15**(6): 655–671.

Levy, S. M. and Guha-Chowdhury N., 1999. Total fluoride intake and implications for dietary fluoride supplementation. *Journal of Public Health Dentistry*. **59**(4): 211–223.

Mousavinasab, S. M. and Meyers, I., 2009. Fluoride release and uptake by glass ionomer cements, compomers and giomers. *Research Journal of Biological Sciences*. **4**(5): 609–616.

Murai, M., Ikeda, M., Yanagihara, T., Hara, G., Kato, K., Nakagaki, H. and Robinson, C., 1993. Fluoride uptake in human dentine from glass-ionomer cement in vivo. *Archives of Oral Biology*. **38**(12): 1093–1098.

Pak, C. Y., Sakhaee, K., Rubin, C. D. and Zerwekh, J. E., 1997. Sustained-release sodium fluoride in the management of established postmenopausal osteoporosis. *The American Journal of the Medical Sciences*. **313**(1): 23–32.

Perrin, C., Persin, M. and Sarrazin, J., 1994. A comparison of fluoride release from four glass-ionomer cements. *Quintessence International*. **25**(9): 605–608.

Preston, A. J., Mair, L. H., Agalamanyi, E. A. and Higham, S. M., 1999. Fluoride release from aesthetic dental materials. *Journal of Oral Rehabilitation*. **26**(2): 123–129.

Singer, L. and Armstrong, W. D., 1968. Determination of fluoride in bone with the fluoride electrode. *Analytical Chemistry*. **40**(3): 613–614.

Suprastiwi, E. and Npa, D. A., 2010. Fluoride released from GIC and RMGIC in saliva and dentino-enamel substance. *MAKARA of Health Series*. **13**(2): 53–58.

Tsanidis, V. and Koulourides, T., 1992. An in vitro model for assessment of fluoride uptake from glass-ionomer cements by dentin and its effect on acid resistance. *Journal of Dental Research*. **71**(1): 7–12.

Van Duinen, R. N., Davidson, C. L., De Gee, A. J. and Feilzer, A. J., 2004. In situ transformation of glass-ionomer into an enamel-like material. *American Journal of Dentistry*. **17**(4): 223–227.

Williams, J. A., Briggs, E., Billington, R. W. and Pearson G. J., 2003. The effects of adding fluoride compounds to a fluoride-free glass-ionomer cement on subsequent fluoride and sodium release. *Biomaterials*. **24**(7): 1301–1308.

Wiegand, A., Buchalla, W. and Attin, T., 2007. Review on fluoride-releasing restorative materials—fluoride release and uptake characteristics, antibacterial activity and influence on caries formation. *Dental Materials*. **23**(3): 343–362.

CHAPTER 12

Effect of Fluoride on Bone Metabolism, Structure and Remodeling

BRENDA LORENA FINA[a] AND ALFREDO RIGALLI*[a]

[a]Bone Biology Laboratory, School of Medicine, Rosario National University, Santa Fe 3100, 2000, Rosario, Argentina
*E-mail: arigalli@conicet.gov.ar

12.1 Bone Tissue

Bone is a specialized connective tissue consisting essentially of an extracellular matrix, which has the ability to produce a hard mineralized tissue and to provide mechanical support and protection to the body organs. This dynamic tissue further comprises an exquisite cell population: osteoblasts, osteocytes and osteoclasts, which are required to sustain the structural integrity of bone mechanics, biochemistry and fulfil its central role in the mineral homeostasis (Ross and Paulina, 2006). Bone is a unique material that serves conflicting needs: to be resistant to deformation, yet flexible (to withstand the loads to which it is subjected without breaking) and light (to allow movement) (Seeman and Delmas, 2006). Because of its mineral content in the form of hydroxyapatite crystals, bone also serves as a natural reserve of calcium and phosphate.

From the morphological point of view, the bone tissue is classified into two types: cortical bone, which represents 80% of the skeleton forming the outer surface of bones, and trabecular bone, which is located in the inner part of bones and constitutes the remaining 20%. In cortical bone, the mineralized

Food and Nutritional Components in Focus No. 6
Fluorine: Chemistry, Analysis, Function and Effects
Edited by Victor R Preedy

Published by the Royal Society of Chemistry, www.rsc.org

extracellular matrix is arranged in concentric lamellae around Haversian canals inside which there are nerves and blood vessels. Meanwhile, the trabecular bone is composed of interconnected trabeculae forming a network in which spaces bone marrow is located. The location and proportion of cortical and trabecular bone tissue vary according to the shape of the bone (Baron, 2006).

Bone tissue is constantly undergoing remodeling processes due to the action of external forces that induce the tissue to be redistributed or to be repaired when microfractures appear. Bone remodeling consists of the combined action of osteoclasts and osteoblasts separated in time but located in the same place. The process can be initiated by need for redistribution of material, microfractures repair or mineral needs. Some of the mentioned causes can start bone remodeling by the activation of osteoclasts. The osteoclasts start the resorption of bone tissue and consequently release cytokines that activate the differentiation and proliferation of osteoblasts. The osteoblasts initiate the formation of new bone that may fill and replace the bone removed by the action of osteoclasts. Conversely, when parathyroid hormone acts on bone, it stimulates the release of cytokines by osteoblasts that activate the differentiation and proliferation of osteoclasts and thus bone resorption (Figure 12.1).

The process of remodeling and repair can be affected by several pharmacological factors. Fluoride (F^-) is an agent with this capability. Fluoride increases osteoblast proliferation and differentiation from progenitor cells. However, fluoride is a double-edged weapon for human beings. On one hand, its chronic administration has a mitogenic action on osteoblasts (Caverzasio *et al.*, 1998),

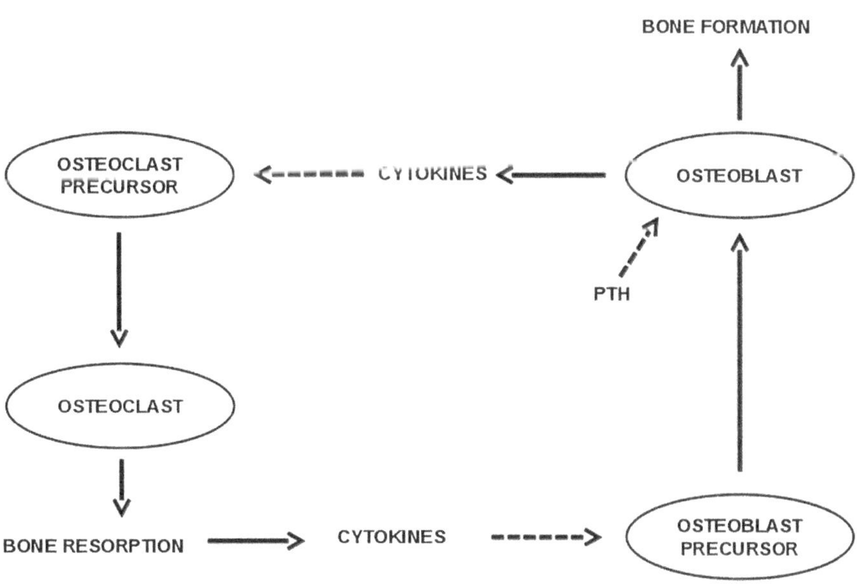

Figure 12.1 Bone remodeling. Schematic figure depicting the process of bone remodeling. Arrows indicate biochemical processes, dashed arrows indicate stimulation.

but on the other hand, fluoride chronic exposure has been demonstrated to be toxic and to cause fluorosis. Fluorosis is clinically characterized by alterations in teeth, musculoskeletal, endocrine and nervous systems (Everett, 2011).

12.2 Effects of Fluoride on Bone Formation and Repair

The effects of fluoride on bone mass began to be studied when it was observed that people living in areas with fluoride in the drinking water had higher bone mineral density (BMD) (Bernstein *et al.*, 1966). In a more recent work, the BMD of lumbar spine and femoral neck of people living in geographic areas with endemic fluorosis were 113% and 111% greater than the values found in a population of the same age and sex living in areas without fluorosis (de la Sota *et al.*, 1997). These results demonstrate a beneficial effect on bones in regard to their BMD and possibly their resistance. Moreover, the inhabitants of endemic fluorosis areas had no increase in their plasma fluoride levels. These results led to the belief that fluoride treatment could be used to increase bone mass. However, when administered as tablets, an increase in plasma levels of fluoride 30–90 min after dosing occurs, returning to baseline within 4 h (Waterhouse *et al.*, 1980; Rigalli *et al.*, 1994a). This phenomenon probably was one of the major causes of adverse effects found after prolonged fluoride administration.

In places where the drinking water contains fluoride 1 ppm (mg L^{-1}), femoral fractures frequency was significantly lower than in populations with water without fluoride (Simonen and Laitinen, 1985). This phenomenon was observed in adult individuals after fluoride was ingested for several years. This suggests that fluoride would be more active on bone tissue early in life rather than in adulthood, where its efficacy in the treatment of osteoporosis could not be confirmed.

Since then, studies to find the appropriate dose or pharmaceutical preparation for administering fluoride or salts containing it, have been unremitting. The most widely used salts are sodium fluoride (NaF) and sodium monofluorophosphate (MFP). There are diverse opinions about the beneficial or deleterious effects of fluoride on bone. Simultaneously, the study of adverse effects has shown great contradictions.

In order to find the appropriate dose for therapeutic treatment with fluoride, there are many published papers where the use of NaF or MFP is compared. Both treatments increased bone mass, but when equivalent dosages of fluoride were used, the bone fluorine content was higher in animals treated with MFP compared to animals treated with NaF (Rigalli *et al.*, 1994a). Administration of 1.52 mg F^- per day as MFP by orogastric tube to rats, produced a faster reparation of a bone defect in the proximal epiphysis of the tibia compared to the control group (Mejia Delgado and Rigalli, 2013). This difference was not observed in the group treated with the same dose of F^- as NaF. The difference between the two treatments may be due to the different bioavailability and pharmacokinetics between MFP and NaF (Rigalli *et al.*, 1994a). The increase in the rate of repair of the defect in the MFP-treated group was accompanied with an increase in serum alkaline phosphatase. The mineral density of the defect was significantly higher 10 days after the defect was made in the MFP-treated group compared to the control group. Histological analysis of bone tissue in the defect area of MFP group showed an organized lamellar bone without the presence of cartilage, 30 days after the defect was made. The presence of hyaline cartilage was maintained in control and NaF groups. These results would indicate that the administration of MFP during bone repair of a noncritical defect can be shortened by 10 days.

Moreover, using a half-dose of MFP, similar results to NaF treatment are achieved. This would indicate that MFP has the double bioavailability of fluoride (Rigalli *et al.*, 1994a, 1996). This difference in bioavailability between MFP and NaF is attributed to the gastric absorption of MFP that does not undergo hydrolysis and generates chemical fluorine compounds in plasma bound to proteins not existing in NaF treatments (Rigalli *et al.*, 1994b). Treatment with NaF (30 mg F^- per day) and MFP (15 mg F^- per day) in women with postmenopausal osteoporosis showed a significant increase in BMD of lumbar spine L2-L4. MFP treatment yielded a value of 60 ± 15 mg Ca cm^{-2}, whereas in the case of NaF treatment BMD was 71 ± 12 mg Ca cm^{-2}, both significantly higher than baseline values 18 months earlier when treatments started. However, no significant differences between the treatments were found. These results confirmed that the bioavailability of fluoride from MFP is twice that of NaF. Furthermore, administration of NaF has adverse gastrointestinal effects such as gastritis, nausea and dyspepsia that impede treatment adherence.

The combination of these effects with the advent of new therapies such as bisphosphonates, caused the decrease in the use of NaF for osteoporosis. However, fluoride given as low doses has proven to be effective without adverse effects. The administration of low doses of NaF (Rubin *et al.*, 2001) or MFP (Reginster *et al.*, 1998) with calcium significantly decreased vertebral fractures risk and increased spinal bone mass in women with established osteoporosis. A meta-analysis of about 2028 references revealed that a daily dose lower or equal to 20 mg F^- (152 mg MFP or 44 mg NaF) significantly reduced vertebral (OR = 0.3, 95% CI: 0.1–0.9) and nonvertebral (OR = 0.5, 95% CI: 0.3–0.8) fracture risk. However, a daily dose greater than 20 mg F^- did not significantly reduce vertebral (OR = 1.3, 95% CI: 0.8–2.0) and nonvertebral (OR = 1.5, 95% CI: 0.8–2.8) fracture risk (Vestergaard *et al.*, 2008). Also, MFP and calcium combined with hormone replacement therapy was demonstrated to be better than MFP alone in protecting against new vertebral fractures in postmenopausal osteoporosis (Ringe and Setnikar, 2002).

Recently, the beneficial effects of MFP on bone mass and strength have been achieved by combining the use of MFP with other treatments that modify bone remodeling. There is evidence that the effect of fluoride on osteoblast proliferation and differentiation depends on the presence of growth factors (Farley *et al.*, 1990). This could be the cause of the failures listed in improving bone mass. The bone matrix is a reservoir of growth factors that are released when bone resorption occurs. Therefore, fluoride would have more effect on a high rate of bone resorption than on a low rate of it. The hypothesis that increased bone remodeling combined with the administration of MFP may have a better effect on bone was tested in ovariectomized rats which were given a hypocalcic diet (0.2 g calcium/100 g diet) for 30 days. After this period the rats showed an increase in deoxypyridinoline (bone resorption marker) compared to rats fed with a normal calcium diet. These animals were then treated with 1.52 mg F^- per day as MFP by orogastric tube and with a hypercalcic diet (2 g calcium/100 g diet) for 60 days. At the end of this period the animals received an antiresorptive treatment with zoledronic acid (Z). At the end of the treatment, bone mass was higher and biomechanical properties were better than the control group with 1 g calcium/100 g diet. Comparison of the experimental group and the groups with pharmacological treatment with MFP or Z, or with groups that had only changes in dietary calcium showed that increases in bone mass and strength were due to a combination of drug treatment and diet, and that could not be attributed to any of them separately (Brance *et al.*, 2014). Briefly, the treatment consisted of administering a hypocalcic diet that increased bone resorption and cytokine release, making the environment conductive for the osteogenic stimulus of the MFP (Figure 12.2). Thus, bone formation and mineralization was enhanced by administering a hypercalcic diet and finally, zoledronic acid was administered to decrease osteoclastic action.

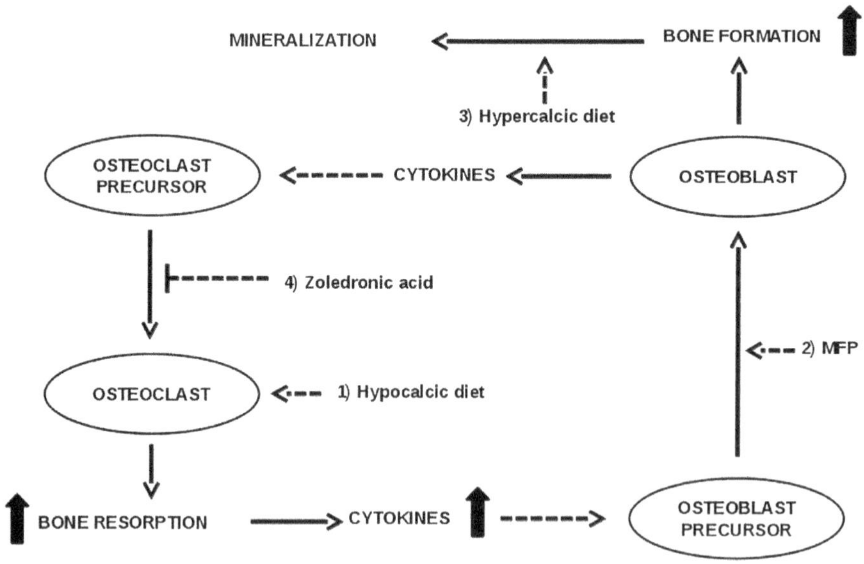

Figure 12.2 Sequential pharmacological treatment to increase bone mass. Treatment to increase bone mass by applying a sequence of pharmacological actions that are numbered in the order of application. Arrows indicate processes, dashed arrows indicate stimulation and dashed-lines with a hyphen indicate inhibition. Thick black arrows indicate increase of a concentration or a process.

As explained before, low doses of fluoride increased osteoblast proliferation and significantly reduced bone fracture risk. However, many studies demonstrated side effects of fluoride when administered chronically. Fluoride treatment decreased nitric oxide response to mechanical loading and affected the arrangement and amount of filamentous-actin in MC3T3-E1 osteoblasts. Also, fluoride treatment resulted in more elongated and smaller osteocytes in hamster mandibles *in vivo* (Willems *et al.*, 2011). Age-related fractures are determined not only by bone mass, but also by biomechanical bone properties depending on microstructure parameters such as structure of type I collagen, osteocytes network and hydroxyapatite mineral properties (Guo, 2008).

To study in more detail the mechanism by which fluoride may be acting on the bone tissue, different doses of NaF were administered for 30 days to four groups of growing rats by orogastric tube (0, 3.8, 7.6, 15.2 mg F^- kg^{-1} bw per day). Histomorphometric, biomechanical and metabolic bone parameters were evaluated. In cortical bone, the fracture load (fx) and stiffness decreased with increasing doses of fluoride. The differences were significant in the groups treated with 7.6 and 15.2 mg F^- kg^{-1} bw per day compared to the control group (0 mg F^- kg^{-1} bw per day) (Figure 12.3). This result could be explained by a lower cortical width of the midshaft in the femur of rats treated with fluoride and by a decrease in Young's modulus (a parameter of bone tissue organization).

The fracture load and stiffness of trabecular bone were also lower in the fluoride-treated groups compared with the control group. This difference may be due to a decrease in bone volume associated with a decrease in trabecular thickness and Young's modulus (Fina *et al.*, 2012a).

Histopathological analysis of the bones of rats treated with NaF showed an immature bone net, numerous active osteoblasts, osteoclasts with greater multinucleation, sinusoidal dilatation, fibrosis and edema. In contrast, the animals treated with MFP showed bone with a mature appearance and a more organized pattern (Brun *et al.*, 2012). The lack of inflammation was a distinctive feature compared with bones of rats treated with NaF (Figure 12.4). All these findings are indicative of a poor quality bone formation and would tackle the lack of effectiveness of NaF when used as therapy in postmenopausal osteoporosis.

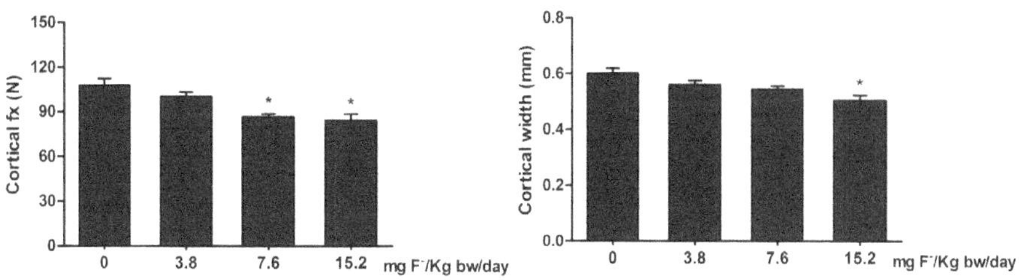

Figure 12.3 Biomechanical parameters of cortical bone in fluoride-treated rats. Cortical fracture load (fx) and width of the midshaft of femurs from fluoride-treated rats with 3.8, 7.6 and 15.2 mg F^- kg^{-1} bw per day compared to controls (0 mg F^- kg^{-1} bw per day). * indicates significant differences compared to 0 mg F^- kg^{-1} bw per day. Kruskall–Wallis test, $p < 0.05$.

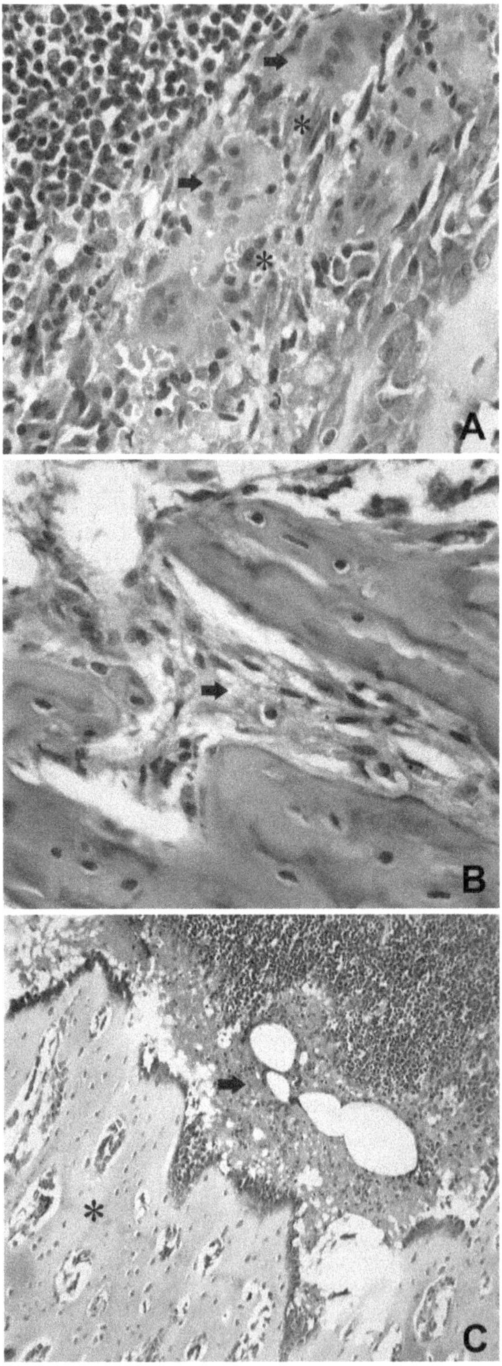

Figure 12.4 Histology of bone tissue from rats treated with sodium fluoride. Bone tissue after treatment with NaF. (A) Presence of numerous vacuolated cytoplasm and multinucleated osteoclasts (arrows), interspersed with abundant spindle cells (asterisks). On the right side of the photo active osteoblasts are seen arranged in three rows; H&E, 450×. (B) Peritrabecular fibrosis bands (arrow); H&E, 450×. (C) Area of inflammation (arrow) located between the hematopoietic tissue and the ivory-like bone (asterisk); H&E, 250×. Reproduced with Permission from Brun *et al.* (2012).

12.3 Relationship between Bone, Inflammation, Reactive Oxygen Species and Fluoride

It is known that the chronic inflammatory diseases such as rheumatoid arthritis, celiac disease and ulcerative colitis, among others, result in a loss of bone mass. Moreover, RANKL expression is increased by proinflammatory cytokines, glucocorticoids, estrogen deficiency and increased PTH. It has been observed that several cytokines have a stimulatory effect on the expression of RANKL: TNF-α, IL-1β, IL-6, IL-11 and IL-17, whereas there are others such as IFN-γ, IL-4 and TGF-β that mostly have an inhibitory effect on osteoclastogenesis (Hardy and Cooper, 2009). It has been shown that fluoride is cytotoxic and causes inflammatory responses. NaF increased the activities of fibroblasts caspases 9, 8, and 3 and consequently the level of apoptosis. Simultaneously there is a decrease in the levels of Bcl-2, involved in the antiapoptotic pathway (Lee *et al.*, 2008). Incubation of macrophages in fluoride solutions decreased significantly the synthesis of ATP and increased apoptosis and the formation of reactive oxygen species (ROS) (Gutowska *et al.*, 2010). The same result has been obtained in osteoblast cultures where an increase in the mRNA levels of caspases 3 and 9 in response to treatment with NaF 5 ppm for 72 h was observed (Yan *et al.*, 2009).

Inflammation could be the consequence of an increase in ROS as a result of a modification in the respiratory chain. The generation of free radicals, lipid peroxidation and altered antioxidant defence systems are considered to play an important role in the toxic effects of fluoride on bone.

Nowadays, the information available on the effects of fluoride on the production of ROS and the antioxidant enzymes is abundant and contradictory. The activities of catalase (CAT) and glutathione peroxidase (GPx), and sulfhydryl groups and glutathione concentrations of rats that ingested water with fluoride 25 ppm decreased, meanwhile the concentration of TBARS (thiobarbituric acid reactive substances) increased, indicating an increase in lipid peroxidation (Inkielewicz *et al.*, 2006). In a previous study, elevated TBARS and decreased CAT and glutathione-S-transferase activities were found in men suffering from fluorosis (Kalyanalakshmi *et al.*, 2007). Also, the activities of superoxide dismutase and GPx decreased markedly with a consequent increase in the levels of TBARS in chickens fed with high fluoride content diets (Chen *et al.*, 2011a). Fluoride exerts its action in a dose-dependent manner. Fluoride concentrations of 0.5–2 ppm stimulated mitochondrial activity of osteoblasts and increased antioxidant enzymes activities. However, a fluoride concentration above 11 ppm inhibits cell proliferation and enzyme activity (Xu *et al.*, 2008). Moreover, it has been shown that chronic administration of MFP increases the levels of alpha-macroglobulin, a protein involved in inhibiting proteinases. The stimulatory action of alpha-macroglobulin and its anti-inflammatory effect was confirmed by a decrease in mortality in rats as a result of experimental pancreatitis

(Di Loreto *et al.*, 2008). The same anti-inflammatory effect may be taking place in bone formation. Although it has not been shown that alpha-macroglobulin plays this role in bone formation, increased levels of proteins that react with antialpha-macroglobulin antibody has been demonstrated in bone and proven lack of inflammation in rats treated with MFP (Brun *et al.*, 2012). The local increase of the antiproteinase may be the cause of the lack of inflammatory effects in rats treated with MFP. Histological differences found between treatments with NaF or MFP confirmed they have different mechanisms of action.

It has been demonstrated that fluoride can also decrease the proliferation of osteoblasts and induce apoptosis through increased ROS production and arrested cells cycle on S-phase (Wang *et al.*, 2011). Therefore, although damaging effects of fluoride and ROS production are well documented, the cellular mechanisms by which fluoride induces ROS formation in bone tissues or in bone-like cell lines, especially in a situation resembling the *in vivo* exposure, are still unknown (Gazzano *et al.*, 2010). One possible explanation is that fluoride could trigger oxidative stress *via* inhibition of the pentose phosphate oxidative pathway (Bergandi *et al.*, 2010). In addition, fluoride induced apoptosis by oxidative stress-induced lipid peroxidation, causing the release of cytochrome-c through HL-60 cells mitochondria (Anuradha *et al.*, 2001). Presently, no mechanism for mitochondrial ROS generation by fluoride in osteoblasts has been proposed. We have previously demonstrated that in rats, *in vivo*, fluoride increases oxidative stress and decreases oxygen consumption (VO_2); and *in vitro* it decreases liver and mitochondria oxygen consumption (Fina *et al.*, 2012b). Therefore, experiments carried out with osteoblast-like cultured cells demonstrated that as soon as fluoride reaches osteoblastic cells, it significantly inhibited their respiration measured as VO_2. Fluoride decreased VO_2 and increased superoxide production by mitochondria immediately after its addition in osteoblastic cells (data not published). Furthermore, chronic treatment with fluoride decreased VO_2 and increased oxidative stress levels in osteoblastic cells. Mitochondrial VO_2 and respiratory complexes activities were inhibited by chronic treatment with fluoride (Fina *et al.*, 2013a). These results indicate that after an oral dose, fluoride inhibits the respiratory chain, increasing the release of superoxide anions. Antioxidant enzyme activities could not prevent increased free radical formation. Therefore, there is an increase in ROS that finally generates hydroxyl radical oxidations in membranes that damage the cell macromolecules (as seen by the increase in lipid peroxidation) and may be the cause of the inflammatory foci observed in the bone (Figure 12.5).

It has already been demonstrated that ROS production induced inflammatory gene expression in alveolar macrophages (Eberlein *et al.*, 2008), fibroblasts (Chen *et al.*, 2011b) and kidney (Wang *et al.*, 2009). Therefore, bone inflammatory foci could be enhanced *via* ROS-dependent activation of proinflammatory genes.

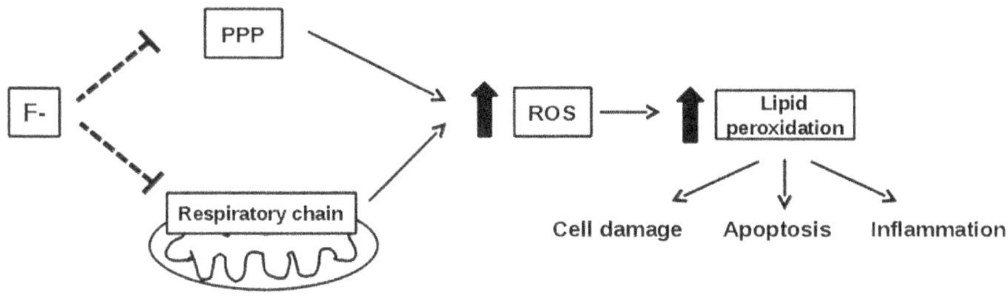

Figure 12.5 Mechanism of production of reactive oxygen species induced by fluoride. A possible mechanism of fluoride-induced oxygen radical production. F^-: fluoride; ROS: reactive oxygen species; PPP: pentose phosphate oxidative pathway. Arrows indicate processes and dashed-lines with a hyphen indicate inhibition. Thick black arrows indicate increase of a concentration.

12.4 Other Alternatives for the Administration of Fluoride with Minimal Adverse Effects

Clinically, the results of therapeutic administration of fluoride have not been successful and new drugs with better effect on bone mineral density and fracture resistance have shifted it from the pharmaceutical market (Häuselmann and Rizzoli, 2003).

Numerous clinical trials with various doses of fluoride have been conducted and yet the results are inconclusive. Low doses did not demonstrate really positive effects on bone mineral density, although changes in bone-formation markers were observed without changes in resorption markers (Grey *et al.*, 2013). Observations were made after a year of starting treatment, and perhaps a more prolonged treatment might show changes. The effects of fluoride on bone appear to be more evident when fluoride is administered as sustained release tablets. Comparison of this therapy with ibandronate, an antiresorptive drug with known efficacy on bone, showed that the increased BMD in the fluoride treatment was not significantly different with respect to treatment with intravenous ibandronate in Chron's disease patients (Klaus *et al.*, 2011). Also, the combination of MFP with estrogen for one year showed a significant increase in BMD of the spine and femur and markers of bone formation (Reid *et al.*, 2007).

It has been observed that rats treated with the same doses of NaF but administered in two different ways: by orogastric tube or drinking water, had different effects on bone properties. Administration in the drinking water may be similar to the administration of sustained release tablets of fluoride. The fluoride administered in the drinking water produced no adverse effects on bone biomechanics and morphology that were observed when fluoride was administrated by orogastric tube (Fina, 2013b). While fluoride administered by orogastric tube significantly decreased the fracture load of trabecular bone, administered *via* the drinking water did not produce this side effect

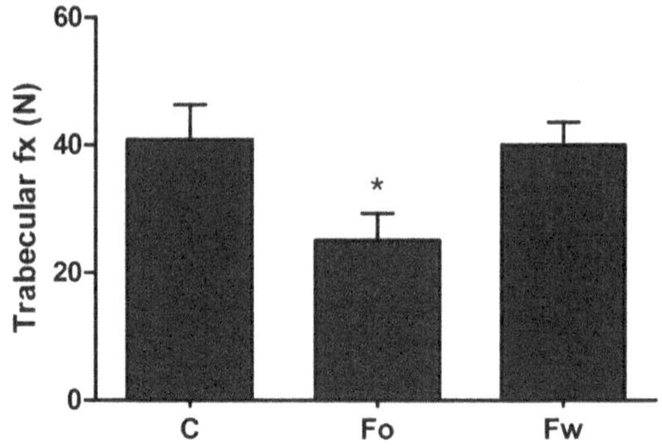

Figure 12.6 Biomechanical parameters of trabecular bone in fluoride-treated rats. Trabecular fracture load (fx) of femurs in rats treated with 7.6 mg of F^{-} kg^{-1} bw per day by orogastric tube (Fo) or by drinking water (Fw); C: control group. * Indicates significant differences from control group. One-way ANOVA, $p < 0.05$.

(Figure 12.6). However, no positive effect on bone resistance was obtained with respect to the control group.

In turn, in the groups treated with fluoride in the drinking water, the damage caused by oxidative stress was much lower than in those treated by orogastric tube. While the levels of TBARS in the groups treated *via* drinking water increased by 60%, in the groups treated by orogastric tube increased up to 240% (results not published). In conclusion, fluoride not only acts differently due to the dose but also by the method of administration, indicating the cause of such different results in papers with fluoride and possibly the lack of efficacy in the treatment of osteoporosis.

Summary Points

- Biological fluoride effects are dependent on fluoride concentration.
- Low doses of fluoride stimulate bone formation.
- High doses of fluoride produce apoptosis, oxidative stress, inflammation and bones with low resistance to fracture.
- Fluoride bioavailability from monofluorophosphate is higher than bioavailability from sodium fluoride.
- The effect of monofluorophosphate on bone mass increase is more evident when the drug is administered after a short period of high bone remodeling rate.
- Sustained release tablets of fluoride are more effective than immediate release tablets.
- The effect of fluoride on the increase of bone mass is more evident when fluoride is administered in the drinking water than when fluoride is administered as tablets.

Key Facts of Fluoride on Bone Repair

- Osteoblasts are the cells that form bone tissue. The differentiation and proliferation of osteoblasts are stimulated by fluoride.
- High concentration of fluoride produces apoptosis, oxidative stress and inflammation that affect the resistance of bones.
- Fluorosis is an illness produced by the ingestion of high amounts of fluoride. Fluorosis is characterized by alterations in teeth and in the function of different organs and systems.
- Osteoporosis is an illness produced by a decrease in bone mineral content and density with an increase in the risk of bone fracture. Osteoporosis affects one out of four women older than fifty years.
- A noncritical bone defect is a defect on bone that can be repaired without treatment. Monofluorophosphate is a drug that stimulates bone formation and reduces the time needed for the reparation of a noncritical bone defect.

Definitions of Words and Terms

Bisphosphonate: A family of drugs that prevent bone loss, used for the treatment of osteoporosis and other illnesses that produce bone loss, such as cancer, chronic renal failure, Paget' s disease, *etc.*
Biomechanical parameter: Variables that are measured to assess the resistance of a material. Bone biomechanical parameters, such as fracture load and Youngś modulus, are measured to assess bone resistance.
Bone remodeling: The process by which the bone repairs from microfracture, provide calcium to blood and reorganize the microarchitecture of bones. It comprises two processes that occur in the same place in different moments. First, bone resorption by osteoclasts, and secondly bone formation by osteoblasts.
Bone mineral density: Amount of calcium in bone per square centimetre. It is a predictor of osteoporosis and osteopenia and it can be measured by dual energy X-ray absorptiometry (DEXA).
Deoxypyridinoline: Bone marker of resorption. It is measured in urine and the higher the value the higher the bone resorption process.
Fluorosis: Chronic endemic illness produced by the ingestion of high amounts of fluoride. Fluorosis can affect teeth but also it can affect the function of different organs and systems.
Fracture load: Biomechanical parameter that indicates the force needed to produce the fracture of something. Fracture load of bone is the force needed to produce the fracture of a bone.
Histomorphometric parameter: Variable measured to assess the microarchitecture of a biological tissue. Bone histomomorphometric parameters, such as bone volume, trabecular thickness and trabecular number, are measured to assess bone distribution and architecture of trabecular bone.

Hydroxyapatite: The principal mineral of bone tissue with the formula $Ca_{10}(PO_4)_6(OH)_2$.
Osteoblast: Bone-forming cell. It synthesizes type I collagen and other substances that form bone matrix. Osteoblast also acts in bone mineralization.
Osteoclast: The bone cell that produces bone resorption. It produces demineralization and degradation of bone matrix.
Osteopenia: A decrease in bone mineral content and density.
Osteoporosis: An illness associated with a decrease in bone mineral content and density, changes in bone microarchitecture and increase in the risk of bone fracture.
Ovariectomized rats: Animal model used for the study of treatment of osteoporosis. Changes at the level of bone in ovariectomized rats are similar to that observed in human beings with osteoporosis.
Parts per million (ppm): Concentration of a substance in a sample that express the amount of milligrams of the substance in a kilogram of sample. For example, drinking water with 5 ppm fluoride contains 5 milligrams of fluoride in one kilogram (or liter) of water.
Type I collagen: Collagen that is found in bone tissue.
Young's modulus: Biomechanical parameter that indicates the stiffness of a material.
Zoledronic acid: Bisphophonate inhibitor of bone resorption used in the treatment of illness with high rate of bone remodeling such as osteoporosis, Paget's disease and cancer.

List of Abbreviations

BMD	Bone mineral density
Bw	Body weight
CAT	Catalase
CI	Confidence interval
F^-	Fluoride
fx	Fracture load
IFN-γ	Interferon gamma
IL-11	Interleukin-11
IL-17	Interleukin-17
IL-1β	Interleukin-1 beta
IL-4	Interleukin-4
IL-6	Interleukin-6
MFP	Sodium monofluorophosphate
NaF	Sodium fluoride
GPx	Glutathione peroxidase
OR	Odds ratio
ppm	Parts per million
PTH	Parathyroid hormone
RANKL	Receptor activator of nuclear factor kappa beta ligand

ROS	Reactive oxygen species
TBARS	Thiobarbituric acid reactive substances
TGF-β	Transforming growth factor beta
TNF-α	Tumor necrosis factor alpha
VO_2	Oxygen consumption
Z	Zoledronic acid

References

Anuradha, C. D., Kanno, S. and Hirano, S., 2001. Oxidative damage to mitochondria is a preliminary step to caspase-3 activation in fluoride-induced apoptosis in HL-60 cells. *Free Radical Biology & Medicine.* **31**: 367–373.

Baron, R., 2006. Anatomy and ultrastructure of bone. In: Favus, M. J., (ed.) Primer on the Metabolic Bone Diseases and Disorders of Mineral Metabolism. Washington, USA.

Bergandi, L., Aina, V., Garetto, S., Malavasi, G., Aldieri, E., Laurenti, E., Matera, L., Morterra, C. and Ghigo, D., 2010. Fluoride-containing bioactive glasses inhibit pentose phosphate oxidative pathway and glucose 6-phosphate dehydrogenase activity in human osteoblasts. *Chemico-Biological Interactions.* **183**: 405–415.

Bernstein, D. S., Sadowsky, N., Hegsted, D. M., Guri, C. D. and Stare, F. J., 1966. *The Journal of the American Medical Association.* **198**: 499–504.

Brance, M. L., Brun, L. R., Di Loreto, V. E., Lupo, M. and Rigalli, A., 2014. Sequential treatment with monofluorophosphate and zoledronic acid in osteoporotic rats. *Climacteric.* (in press).

Brun, L. R., Roma, S. M., Pérez, F. and Rigalli, A., 2012. Presencia de inflamación en el tejido óseo de ratas inducida por fluoruro de sodio. *Actualizaciones en Osteología.* **8**: 19–28.

Caverzasio, J., Palmer, G. and Bonjour, J. P., 1998. Fluoride: mode of action. *Bone.* **22**: 585–589.

Chen, T., Cui, H., Cui, Y., Bai, C. and Gong, T., 2011a. Decreased antioxidase activities and oxidative stress in the spleen of chickens fed on high-fluorine diets. *Human & Experimental Toxicology.* **30**: 1282–1286.

Chen, A. C., Arany, P. R., Huang, Y. Y., Tomkinson, E. M., Sharma, S. K., Kharkwal, G. B., Saleem, T., Mooney, D., Yull, F. E., Blackwell, T. S. and Hamblin, M. R., 2011b. Low-level laser therapy activates NF-kB via generation of reactive oxygen species in mouse embryonic fibroblasts. *PloS One.* **6**: e22453.

Di Loreto, V. E., Roma, S. M., Menoyo, I. and Rigalli, A., 2008. Effect of the administration of monofluorophosphate on alpha-macroglobulin levels and the clinical course of pancreatitis in rats. *Arzneimittel-Forschung.* **58**: 136–140.

Eberlein, M., Scheibner, K. A., Black, K. E., Collins, S. L., Chan-Li, Y., Powell, J. D. and Horton, M. R., 2008. Anti-oxidant inhibition of hyaluronan fragment-induced inflammatory gene expression. *Journal of Inflammation (London, England).* **5**: 20.

Everett, E. T., 2011. Fluoride's effects on the formation of teeth and bones, and the influence of genetics. *Journal of Dental Research.* **90**: 552–560.

Farley, J. R., Tarbaux, N., Hall, S. and Baylink, D. J., 1990. Mitogenic action(s) of fluoride on osteoblast line cells; determinants of the response in vitro. *Journal of Bone and Mineral Research.* **5**(Supp 1): S107–S113.

Fina, B. L., Lupo, M. L., Da Ros, E. R., Moreno, H., Roma, S. M. and Rigalli, A., 2012a. Effect of sodium fluoride on biomechanical and histomorphometric bone parameters. Identification of variables that determine the fracture load on NaF-treated rats. *Fluoride.* **45**: 163.

Fina, B. L., Brance, M. L., Brun, L. R. and Rigalli, A., 2012b. Fluoride inhibition of oxygen consumption and increased oxidative stress in rats. *Fluoride.* **45**: 343–348.

Fina, B. L., Rigalli, J. P. and Rigalli, A., 2013a. La inhibicion de la cadena respiratoria por fluoruro aumenta la liberación de superóxido en células osteoblásticas ros 17/2.8. In: Proceedings of Sociedad de Biología de Rosario annual conference. November 27-28. Rosario, Argentina.

Fina, B. L., 2013b. Diferentes formas de administración de fluoruro y sus efectos sobre la calidad ósea. *Actualizaciones en Osteología.* **9**: 27.

Gazzano, E., Bergandi, L., Riganti, C., Aldieri, E., Doublier, S., Costamagna, C., Bosia, A. and Ghigo, D., 2010. Fluoride effects: the two faces of janus. *Current Medicinal Chemistry.* **17**: 2431–2441.

Grey, A., Garg, S., Dray, M., Purvis, L., Horne, A., Callon, K., Gamble, G., Bolland, M., Reid, I. R. and Cundy, T., 2013. Low-dose fluoride in postmenopausal women: a randomized controlled trial. *The Journal of Clinical Endocrinology and Metabolism.* **98**: 2301–2307.

Guo, X. E., 2008. What is nanomechanics of bone and why is it important? *Journal of Musculoskeletal & Neuronal Interactions.* **8**: 327–328.

Gutowska, I., Baranowska-Bosiacka, I., Baśkiewicz, M., Milo, B., Siennicka, A., Marchlewicz, M., Wiszniewska, B., Machaliński, B. and Stachowska, E., 2010. Fluoride as a pro-inflammatory factor and inhibitor of ATP bioavailability in differentiated human THP1 monocytic cells. *Toxicology Letters.* **196**: 74–79.

Hardy, R. and Cooper, M. S., 2009. Bone loss in inflammatory disorders. *The Journal of Endocrinology.* **201**: 309–320.

Häuselmann, H. J. and Rizzoli, R., 2003. A comprehensive review of treatments for postmenopausal osteoporosis. *Osteoporosis International.* **14**: 2–12.

Inkielewicz, I., Rogowska, M. and Krechniak, J., 2006. Lipid peroxidation and antioxidant enzyme activity in rats exposed to fluoride and ethanol. *Fluoride.* **39**: 53–59.

Kalyanalakshmi, P., Vijayabhaskar, M. and Dhananjaya Naidu, M., 2007. Lipid peroxidation and antioxidant enzyme status of adult males with skeletal fluorosis in Andhra Pradesh, India. *Fluoride.* **40**: 42–45.

Klaus, J., Reinshagen, M., Herdt, K., Adler, G., von Boyen, G. B. and von Tirpitz, C., 2011. Intravenous ibandronate or sodium-fluoride – a 3.5 years study on bone density and fractures in Crohn's disease patients with osteoporosis. *Journal of Gastrointestinal and Liver Diseases.* **20**: 141–148.

Lee, J. H., Jung, J. Y., Jeong, Y. J., Park, J. H., Yang, K. H., Choi, N. K., Kim, S. H. and Kim, W. J., 2008. Involvement of both mitochondrial- and death receptor-dependent apoptotic pathways regulated by Bcl-2 family in sodium fluoride-induced apoptosis of the human gingival fibroblasts. *Toxicology.* **243**: 340–347.

Mejia Delgado, S. P. and Rigalli, A., 2013. Comparación de los efectos del tratamiento con monofluorofosfato de sodio y fluoruro de sodio sobre el proceso de reparación ósea en ratas. *Actualizaciones en Osteología.* **9**: 28–35.

Reid, I. R., Cundy, T., Grey, A. B., Horne, A., Clearwater, J., Ames, R., Orr-Walker, B. J., Wu, F., Evans, M. C., Gamble, G. D. and King, A., 2007. Addition of monofluorophosphate to estrogen therapy in postmenopausal osteoporosis: a randomized controlled trial. *The Journal of Clinical Endocrinology and Metabolism.* **92**: 2446–2452.

Reginster, J. Y., Meurmans, L., Zegels, B., Rovati, L. C., Minne, H. W., Giacovelli, G., Taquet, A. N., Setnikar, I., Collette, J. and Gosset, C., 1998. The effect of sodium monofluorophosphate plus calcium on vertebral fracture rate in post-menopausal women with moderate osteoporosis. A randomized, controlled trial. *Annals of Internal Medicine.* **129**: 1–8.

Rigalli, A., Ballina, J. C., Beinlich, A. D., Alloatti, R. and Puche. R. C., 1994a. Pharmacokinetics differences between sodium fluoride and sodium monofluorophosphate and comparative mass increasing activity of both compounds in the rat. *Arzneimittel-Forschung.* **44**: 762–766.

Rigalli, A., Cabrerizo, M. A., Beinlich, A. D. and Puche, R. C., 1994b. Gastric and intestinal absorption of monofluorophosphate and fluoride in the rat. *Arzneimittel-Forschung.* **44**: 651–655.

Rigalli, A., Morosano, M. and Puche, R. C., 1996. Bioavailability of fluoride administered as sodium fluoride or sodium monofluorophosphate to human volunteers. *Arzneimittel-Forschung.* **46**: 531–533.

Ringe, J. D. and Setnikar, I., 2002. Monofluorophosphate combined with hormone replacement therapy in postmenopausal osteoporosis. An open-label pilot efficacy and safety study. *Rheumatology International.* **22**: 27–32.

Ross, M. H. and Paulina, W., (ed.), 2006. Histology. A Text and Atlas, Lippincott Williams & Wilkins, Baltimore, USA, 974 pp.

Rubin, C. D., Pak, C. Y., Adams-Huet, B., Genant, H. K., Li, J. and Rao, D. S., 2001. Sustained-release sodium fluoride in the treatment of the elderly with es-tablished osteoporosis. *Archives of Internal Medicine.* **161**: 2325–2333.

Seeman, E. and Delmas, P. D., 2006. Bone quality. The material and structural basis of bone strength and fragility. *New England Journal of Medicine.* **354**: 2250–2261.

Simonen, O. and Laitinen, O., 1985. Does fluoridation of drinking-water prevent bone fragility and osteoporosis? *Lancet.* **2**: 432: 434.

de la Sota, M., Puche, R., Rigalli, A., Fernández, L. M., Benassati, S. and Boland, R., 1997. *Medicina.* **57**: 417–420.

Vestergaard, P., Jorgensen, N. R., Schwarz, P. and Mosekilde, L., 2008. Effects of treatment with fluoride on bone mineral density and fracture risk: a meta-analysis. *Osteoporosis International.* **19**: 257–268.

Wang, Y., John, R., Chen, J., Richardson, J. A., Shelton, J. M., Bennett, M., Zhou, X. J., Nagami, G. T., Zhang, Y., Wu, Q. Q. and Lu, C. Y., 2009. IRF-1 promotes inflammation early after ischemic acute kidney injury. *Journal of the American Society of Nephrology.* **20**: 1544–1555.

Wang, Z., Yang, X., Yang, S., Ren, G., Ferreri, M., Su, Y., Chen, L. and Han, B., 2011. Sodium fluoride suppress proliferation and induce apoptosis through decreased insulin-like growth factor-I expression and oxidative stress in primary cultured mouse osteoblasts. *Archives of Toxicology.* **85**: 1407–1417.

Waterhouse, C., Taves, D. and Munzer, A., 1980. Serum inorganic fluoride: changes related to previous fluoride intake, renal function and bone resorption. *Clinical Science.* **58**: 145–152.

Willems, H. M., van den Heuvel, E. G., Castelein, S., Buisman, J. K., Bronckers, A. L., Bakker, A. D. and Klein-Nulend, J., 2011. Fluoride inhibits the response of bone cells to mechanical loading. *Odontology.* **99**: 112–118.

Xu, H., Wang, C. H., Zhao, Z. T., Zhang, W. B. and Li, G. S., 2008. Role of oxidative stress in osteoblasts exposed to sodium fluoride. *Biological Trace Element Research.* **123**: 109–115.

Yan, X., Feng, C., Chen, Q., Li, W., Wang, H., Lv, L., Smith, G. W. and Wang, J., 2009. Effects of sodium fluoride treatment in vitro on cell proliferation, apoptosis and caspase-3 and caspase-9 mRNA expression by neonatal rat osteoblasts. *Archives of Toxicology.* **83**: 451–458.

CHAPTER 13

Fluoride and Dietary Calcium on Bone

DIANJUN SUN*[a], CHENG WANG[a], WEI ZHANG[a], AND LIJUN ZHAO[a]

[a]Center for Endemic Disease Control, Chinese Center for Disease Control and Prevention, Harbin Medical University, 157 Baojian Road, Nangang District, Harbin, Heilongjiang Province 150081, China
*E-mail: hrbmusdj@163.com

Fluorine is widely found in nature and organisms, and is the lightest and most reactive halogen in the Periodic Table of Elements. In 1990, the Joint FAO/IAEA/WHO Expert Committee regarded it as the third category element that is potentially toxic, but may be necessary for certain functions of the human body at low doses. Its main physiological function is to prevent dental caries and participate in bone mineralization. The adequate intake of fluoride can prevent dental caries. However, excessive accumulation of fluoride in the body can cause chronic fluorosis, which can lead to disorders of bone metabolism, result in dental fluorosis and skeletal fluorosis. In 2010, the WHO estimated that excessive intake of fluoride from drinking water resulted in tens of millions of patients with dental fluorosis and skeletal fluorosis worldwide (WHO, 2010). Of all the fluorosis regions, China and India are the countries having more serious conditions of fluorosis than others. In China, drinking-water-type fluorosis (Figure 13.1) threatens over 80 million population (Wang *et al.*, 2012), Meanwhile, coal-burning fluorosis (Figure 13.2) and brick-tea fluorosis (Figure 13.3) as Chinese unique fluorosis types each threaten more than 30 million residents (Sun *et al.*, 2008; Sun, 2010).

Food and Nutritional Components in Focus No. 6
Fluorine: Chemistry, Analysis, Function and Effects
Edited by Victor R Preedy

Published by the Royal Society of Chemistry, www.rsc.org

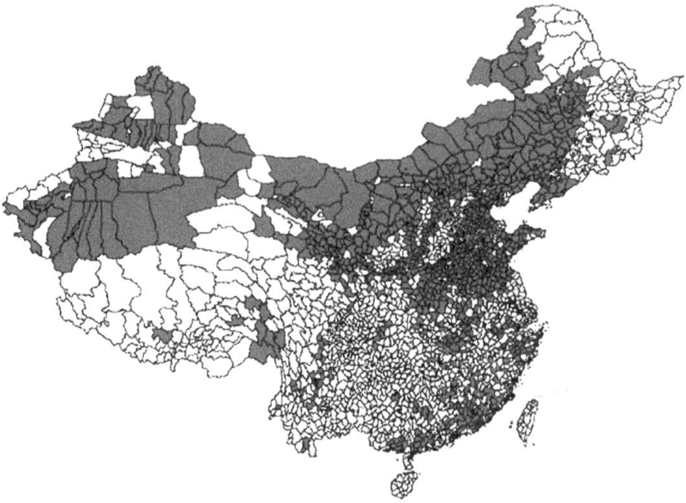

Figure 13.1 Distribution of drinking-water type of fluorosis in China.

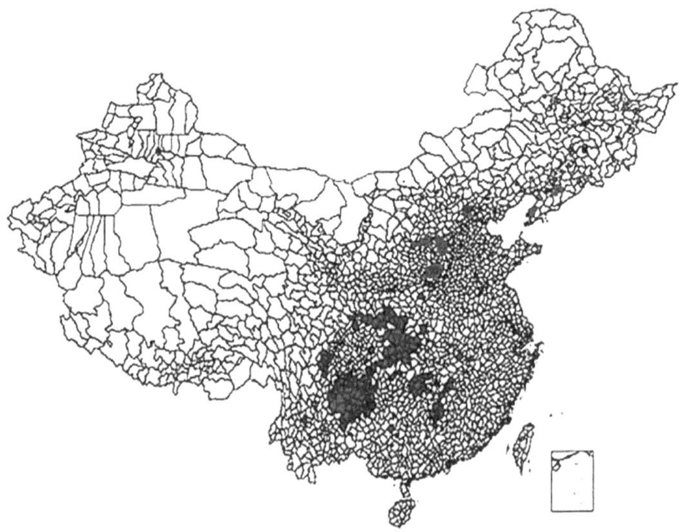

Figure 13.2 Distribution of corn-burning type of fluorosis in China.

Calcium is an essential nutrient element in the body, and closely related to the activities of bone metabolism in humans. Currently, there are different levels of calcium deficiency among different populations around the world. Calcium deficiency has become a global nutritional problem, which is more serious in children, women, and the elderly. When fluoride was taken

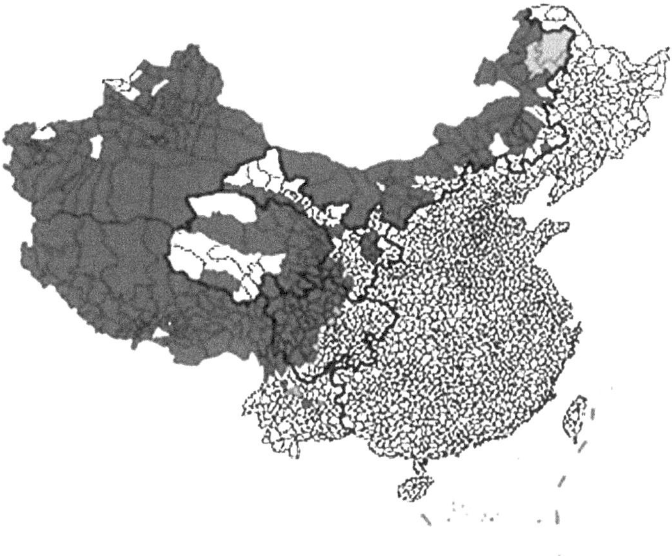

Figure 13.3 Distribution of drinking-tea type of fluorosis in China.

excessively, calcium deficiency can contribute to the occurrence and development of fluorosis (Shankar *et al.*, 2013), and fluoride and dietary calcium can act on the bone separately or jointly. In order to better understand the relationships among them, the effects of fluoride and dietary calcium on the bone and the mechanism are reviewed in this chapter.

13.1 Dietary Calcium, Fluoride and Fluorosis

Excessive accumulation of fluoride in the body can lead to fluorosis, and calcium malnutrition may promote the occurrence and development of fluorosis, which can aggravate the toxicity of fluoride to bone. This had been confirmed by experimental studies and epidemiological investigations. Ren and Li's study (Ren and Li, 1999) indicated that rats with calcium-rich diet (7900 mg calcium/kg forage) showed strong tolerance to fluorosis (100 mg L^{-1} fluoride in drinking water) and a low-calcium diet (630 mg calcium/kg forage) greatly increased the bone toxicity caused by fluoride, and in the same fluorine concentration (100 mg L^{-1}) of drinking water, the fluoride accumulation in the bone of rats with low-calcium diet was 2.3 times higher compared with calcium-rich diet rats. Griffiths *et al.* (1975) investigated the effects of fluoride and calcium on osteoporosis through rhesus monkeys fed with diet of different fluoride and calcium amount for five years. The results showed that osteoporosis occurred in the rhesus monkeys with low-calcium (0.15% calcium) diet and normal fluoride-contained water. Even though feeding with low calcium and 50 mg L^{-1} fluoride in monkeys can prevent the occurrence of osteoporosis, osteomalacia was induced because of interfered osteoid mineralization. Calcium-rich diet (calcium 1%) combined with 50 mg L^{-1} fluoride in drinking water could not induce the skeletal fluorosis, which indicated that the calcium-rich diet had a protective effect on bone toxicity caused by fluoride. However, Yu *et al.*'s study (Yu *et al.*, 2013) showed that excessive calcium supplementation (2% calcium diet) could not attenuate, but on the other hand did not aggravate, the bone toxicity of fluoride in rats (The content of calcium in normal diet is 0.5%).

As early as 1955, there were correlative epidemiological investigations into fluorosis. In India, Siddiqui (1955) reported the occurrence of skeletal fluorosis in the immigrants after they moved to Kamaguda village for 1–4 years, which may be associated with drinking fluoride-contained water (9.2–11.8 mg L^{-1}) and low-calcium diet. To elucidate the effect of fluoride on human bone diseases and the relationship with calcium diet, Teotia *et al.* (1998) summarized the surveys for more than 30 years from 1963 to 1997. This study collected 45 725 children with skeletal fluorosis. All of them lived in the fluorosis areas after birth, and fluoride contents in drinking water ranged from 1.5 mg L^{-1} to 25 mg L^{-1}. They found that the children with calcium-deficient diet (<300 mg day^{-1}) showed more serious and complicated fluorosis symptoms compared to the children with calcium-adequate diet (>800 mg day^{-1}), under the same fluoride intake (9.5 ± 1.9 mg day^{-1}). The incidence rate of metabolic bone diseases, such as rickets, osteoporosis, parathyroid hormone-related bone disease and leg deformities in the children with a calcium-deficient diet was over 90%. The incidence rate of metabolic bone diseases in the children with calcium-adequate diet was less than 25%, and most of them were sclerotic skeletal fluorosis accompanied by a slight secondary hyperparathyroidism. This study also found that the toxic effects of fluorosis appeared in the children with calcium deficiency, even though the exposure dose was a marginal value (> 2.5 mg fluoride/day).

In contrast to inadequate intake of dietary calcium, excessive fluoride can also interfere with the normal metabolism of calcium in the body and lead to the increasing demand for calcium, resulting in a relative calcium deficiency. Calcium has a strong affinity with fluorine, and calcium fluoride can be formed immediately when fluorine is absorbed into the blood, which in turn leads to a temporary decrease of serum Ca^{2+} (Mertz, 1994). After fluorine is incorporated into the bone mineral crystals, crystallinity can be improved and the solubility of bone mineral can be decreased (Grynpas and Cheng, 1988), which can cause the slow release of calcium from the bone, and thereby reduce the humoral Ca^{2+} concentration. In addition, excessive fluoride can activate osteoblast and increase bone formation, and therefore increase the demands for Ca^{2+} of bone forefront (Krishnamachari, 1986), which lead to a relative deficiency of calcium in the body. Adequate dietary calcium can antagonize the toxicity of fluoride and reduce the absorption of fluoride in intestinal. When the calcium level in the body is low, the absorption of fluorine in the intestinal and the accumulation of fluorine in the bone are both enhanced. Moreover, the low level of calcium itself can cause an increased bone turnover by increasing the activity of bone formation and bone absorption (Ren and Li, 1999). Furthermore, the increased bone turnover is also an important feature of bone lesions of skeletal fluorosis (Ren *et al.*, 1998). Therefore, the roles of low calcium and high fluoride are isotropic to bone metabolism, and low calcium is an important pathogenesis basis for the aggravation of fluorosis.

13.2 Cellular Ca^{2+}, Fluoride and Bone Metabolism

Excessive intake of fluoride can lead to fluorosis, dietary calcium can trigger and aggravate the occurrence of the disease, which contributes to the occurrence of skeletal fluorosis. Fundamental changes of skeletal fluorosis include the increase of bone mass (osteosclerosis), the increase of bone-like tissue (osteomalacia), osteoporosis, and boned soft tissue surrounding bone. In teams of bone metabolism, fundamental changes of skeletal fluorosis consist of the enhanced activity of osteoblast, activation of osteoclasts, and acceleration of bone turnover. Among them, the enhanced activity of osteoblasts is an early and dominant link in skeletal fluorosis. The relationships among cellular Ca^{2+}, fluoride and bone metabolism will be discussed in this section (Figure 13.4).

13.2.1 Cellular Ca^{2+} and Bone Metabolism

As a second messenger, Ca^{2+} plays a key role of signal transduction in the regulation of cell function. Some hormones, such as PTH, which can promote bone resorption, stimulate calcium ions into osteoblast (Liu and Zhang, 2008). Voltage-gated calcium channels in osteoblast, mainly L-type, include three major isomers: α1-C, α1-D and α1-S. α1-C is the main channel of Ca^{2+} influx in the proliferated osteoblasts, and plays an important role in controlling the intracellular and extracellular calcium balance. In addition, there is a T-type calcium channel. In osteoblasts, an L-type calcium channel involves in cell proliferation, while a T-type calcium channel involves differentiation of osteoblasts after bone remodeling promoted by vitamin D3. Many bone-regulating factors can activate calcium channels of osteoblasts, and contribute in the reconstruction of bone by changing Ca^{2+} levels (Barry,

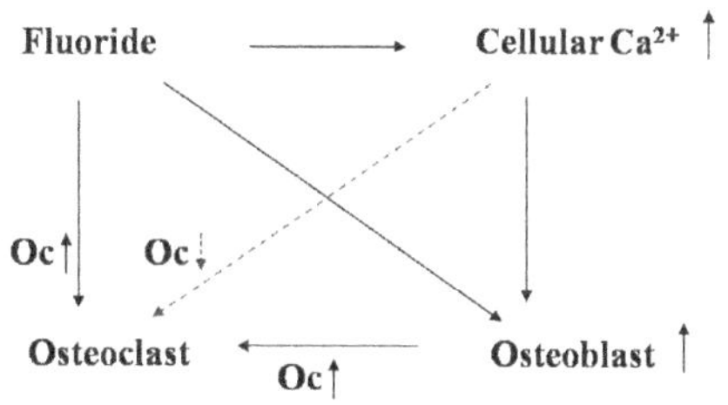

Figure 13.4 The relationships among cellular Ca^{2+}, fluoride and bone metabolism (Oc, Osteoclast).

2000). PTH may cause Ca^{2+} influx through opening voltage-gated calcium channels, increase intracellular calcium, and cause depolarization of osteoblasts (Edelman *et al.*, 1986). Ca^{2+} can affect osteoclasts directly and affect osteoclasts indirectly by acting on osteoblasts. For example, the increasing concentration of extracellular calcium could cause the changes in the skeleton of osteoclasts, activate phospholipase C (PLC) and increase intracellular calcium, and thereby inhibit osteoclast resorption (Bennett *et al.*, 2001). However, in the presence of osteoblasts, extracellular high calcium (3–5 mmol L^{-1}) can stimulate the formation of osteoclast-like cells as well as the resorption activity of mature osteoclasts through acting on osteoblasts (Kaji *et al.*, 1996).

13.2.2 Cellular Ca^{2+}, Fluoride and Bone Metabolism

In 1983, Farley *et al.*' s study (Farley *et al.*, 1983) confirmed for the first time that fluoride can directly stimulate proliferation of osteoblasts, and enhance the activity of alkaline phosphatase, and thereby strengthen osteogenesis through osteoblasts culture of chick embryo. Alkaline phosphatase (ALP) might reflect directly the activity or functional status of osteoblasts, which is the best biomarker to evaluate the bone mineralization disorders of body. In view of the ALP activity that was higher in differentiated osteoblasts than in preosteoblasts, they speculated that fluoride not only stimulated proliferation of osteoblasts, but also stimulated their differentiation. In 2000, the results of Yu and Cui's study (Yu and Cui, 2000) showed that fluoride could affect the differentiation process of osteoblasts, which can cause the increase in ALP activity and osteocalcin levels of osteoblasts, and stimulate human marrow stromal cells to produce bone gla-protein. Fluoride could increase calcium influx, reducing the concentrations of extracellular Ca^{2+}. The results of Zerwekh *et al.*'s study (Zerwekh *et al.*, 1990) showed that fluoride (25 ng ml^{-1}) could increase calcium of the bone cells. There was a dose-dependent relationship between fluoride (25–500 ng ml^{-1}) and calcium of the bone cells, but calcium channel blockers could inhibit this effect. The levels of blood calcium might also be reduced by fluoride through bone formation, which increased PTH secretion by feedback and activated osteoclasts.

Khokher *et al.*'s report (Khokher and Dandona, 1990) indicated that NaF (250 μmol L^{-1}) in osteoblasts could increase uptake of H3-labeled thymidine and the release of ALP, which could also be inhibited by calcium channel blockers (verapamil). This suggests that fluoride could stimulate osteoblasts proliferation and release of ALP by increasing the Ca^{2+} influx, and then affect bone metabolism. By checking the electrophysiological responses of osteoblast to fluoride using the patch clamp technique, Gofa and Davidson found that fluoride might increase amplitude and open of K^{+} selective ion channels, which depended on extracellular Ca^{2+}, and could be blocked by calcium channel blockers. The results illustrate that a grade chain reaction of second messenger Ca^{2+} and the activity of selected potassium channel could regulate the early response of osteoblasts to fluoride (Gofa and Davidson, 1996).

In summary, fluoride could enhance the activity of osteoblasts and activate osteoclasts to affect bone metabolism by affecting Ca^{2+} influx directly or indirectly. In addition, the effects of fluoride on bone metabolism were regulated by calcitropic hormones, including parathyroid hormone, calcitonin, 1, $25(OH)_2D_3$, corticosteroids, and sex hormones.

13.3 Calcitropic Hormone and Fluoride and Bone Metabolism

High fluoride exposure can affect hormone secretion and cause the changes of various inorganic ions. Thereby, the surrounding environment in the bone tissue including cytokines, transcription factors, *etc.* could be changed, and some abnormality occurred in bone tissue cells, including intracellular and intercellular signal transduction, cell proliferation, differentiation, apoptosis, function implementation, *etc.* In addition, bone metabolism is characterized by high bone turnover. The equilibrium between bone formation and resorption is destroyed, and eventually leads to the skeletal fluorosis. Fluoride can affect some calcitropic hormones, which were closely related to bone metabolism, including parathyroid hormone, 1, 25 $(OH)_2D_3$, calcitonin, glucocorticoids, sex hormone, *etc.* (Figure 13.5). These hormones play the important roles in the occurrence and development of skeletal fluorosis.

13.3.1 Parathyroid Hormone

Parathyroid hormone (PTH) is secreted by parathyroid glands, which regulate bone turnover, especially promoting osteoclast resorption. PTH can act directly on osteoblast-line and indirectly on osteoclasts. PTH could regulate

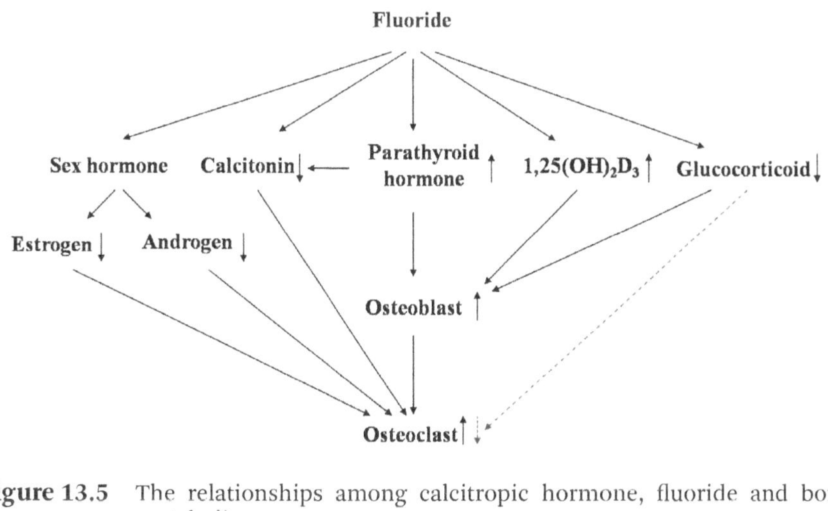

Figure 13.5 The relationships among calcitropic hormone, fluoride and bone metabolism.

the differentiation and activation of bone marrow stromal cells to osteoblasts, promote bone lining cells into osteoblasts and inhibit the apoptosis of mature osteoblasts (Jilka *et al.*, 1999), thus promoting bone formation. Parathyroid hormone receptor (PTHR) can be expressed in osteoblasts-line including osteoblasts, osteoblast progenitor cells, bone lining cells and chondrocytes. However, PTHR cannot be expressed in osteoclasts. Therefore, PTH first acts on the osteoblasts-line, which can secret osteoclasts differentiation factor (ODF), interleukin-1 (IL-1), interleukin-6 (IL-6), *etc.* These cytokines can induce the differentiation or activation of osteoclasts, then start to control the activation of osteoclasts, finally play its role in bone resorption (Suda *et al.*, 1999).

Parathyroid cell membrane contains a calcium-sensing receptor. PTH secretion is mainly regulated by serum calcium concentration. Fluoride could increase the level of intracellular calcium in parathyroid cell, which can promote the secretion of PTH (Ridefelt *et al.*, 1992). Fluoride can also decrease the level of cAMP, and thus inhibit the secretion of PTH (Sugimoto *et al.*, 1990). So, fluoride can indirectly affect bone metabolism by affecting the serum PTH level. In addition, fluoride can directly stimulate the proliferation and activation of osteoblasts and their precursor cells, simultaneously increasing the sensitivity of osteoblasts to local cytokines (Reed *et al.*, 1993). Hence, fluoride can directly activate the osteoblasts, as well as facilitate bone formation by changing PTH level. Moreover, it can also indirectly activate the osteoclasts as well as bone turnover status, which causes the diverse pathological manifestations of skeletal fluorosis.

13.3.2 Calcitonin

The main function of calcitonin (CT) is to inhibit the bone resorption. CT can act directly on the CT receptor of osteoclasts, then rapidly inhibit the activity of osteoclasts and the proliferation, differentiation and aggregation of osteoclast precursor cells, and also reverse the formation of osteoclasts. So, the number of osteoclasts and the release of bone calcium are reduced, then the bone resorption can also be significantly reduced. Therefore, it has been widely used in the treatment of osteoporosis (Carter and Schipani, 2006).

In the occurrence and development of fluorosis, both the decrease of the serum calcium caused by calcium malnutrition or fluoride and the secondary increase of PTH secretion can inhibit the CT secretion. Then, the reduced CT attenuates its effect on inhibiting bone resorption. The results of Yu *et al.* (2013) showed fluoride could strengthen the formation of osteoclasts, and increase bone resorption, then bone turnover was further boosted, and bone damages were aggravated.

13.3.3 1, 25-Dihydroxy-vitamin D_3

1, 25-dihydroxy-vitamin D_3 (1, 25$(OH)_2D_3$) is the most active product of vitamin D metabolized by the liver and kidney. It participates in the bone metabolism and affects the development and stability of mineralized bone. The biological function of 1, 25$(OH)_2D_3$ is related to the distribution of its receptor (VDR) in bone tissue cells. VDR can be expressed in osteoblasts and its precursor cells, but cannot be expressed in osteoclasts. 1, 25$(OH)_2D_3$ can act on osteoblasts, enhance the synthesis of the extracellular matrix, and promote the calcium deposition of bone. In addition, it can promote the osteoblasts synthesizing osteoclast activating factor, which can indirectly activate the osteoclasts and boost the osteoclast resorption. 1, 25$(OH)_2D_3$ plays the bidirectional roles in bone turnover. After long-term low-dose administration of 1, 25$(OH)_2D_3$, osteoblasts and their precursor cells can be activated, then bone formation is promoted and the bone resorption is inhibited. High-dose administration of 1, 25$(OH)_2D_3$ can indirectly promote the activation and recruitment of osteoclasts and their precursor cells in a short time, at the same time inhibit the apoptosis of osteoclasts and promote bone resorption, then convert to promote bone formation after several days (Brendan and Weisbrode, 1985).

Study on the rat model of osteomalacia-type skeletal fluorosis with low-calcium diet showed a decreased calcium in serum could enhance serum 1, 25$(OH)_2D_3$ and up-regulate the calcium-sensing receptor and vitamin D receptor (Shankar *et al.*, 2013). PTH synergism with increased 1, 25$(OH)_2D_3$ induced by low calcium can promote the osteoclast resorption, form high bone turnover status and aggravate the pathological lesions caused by fluorosis.

13.3.4 Glucocorticoid

Glucocorticoid is a kind of steroid hormone synthesized and secreted by the adrenal cortex cells, which involved a variety of metabolic processes in the body, including bone metabolism. Glucocorticoid has played an important role in the bone-conversion process including the normal bone turnover and abnormal bone turnover in disease. Adequate glucocorticoid is needed by cell growth and differentiation in all kinds of bone, excessive glucocorticoids can damage the proliferation, differentiation and function of osteoblast as well as inhibition of bone formation by inducing apoptosis of osteoblasts and osteocyte. Furthermore, it can also stimulate the formation of osteoclasts, enhance the osteoclast function and osteoclast resorption with PTH synergistically. The administration of excessive glucocorticoid could lead to the occurrence of bone osteoporosis. Fractures would be found in 30–50% of patients with osteoporosis who received long-term glucocorticoid therapy (Mazziotti *et al.*, 2007).

Plasma glucocorticoid levels in patients with chronic fluoride toxicity was significantly low (Das and Susheela, 1991), which can attenuate the osteoclast resorption mediated by glucocorticoid, and reduce the inhibition on bone formation. Thereby, such changes contribute to the enhancement of the osteogenic activity. In addition, a relatively low dose of endogenous glucocorticoid was necessary for bone marrow stromal stem cells differentiated into osteoblast (Ahdjoudj *et al.*, 2001). So, a low dose of glucocorticoid in individuals with fluorosis can contribute to the proliferation of osteoblast precursor cells and differentiation into osteoblast, and lead to enhancement of osteogenesis activity.

13.3.5 Sex Hormone

Sex hormone mainly includes estrogen and androgen, which have important roles in both growth and relative stability of bone. Some of the receptors of estrogen and androgen are distributed in bone tissue, including marrow stromal cells, osteoblasts, osteocytes and osteoclasts, which can regulate the bone metabolism mediated by corresponding receptors. The androgen mainly includes testosterone and dihydrotestosterone. They are directly involved in bone metabolism to stimulate the proliferation of osteoblasts, regulate calcium homeostasis, and regulate calcium absorption temporarily (Hope *et al.*, 1992). In addition, androgen can also be involved in bone metabolism *via* its aromatized metabolites (estrogen). Estrogen, mainly including estradiol, can inhibit the proliferation and differentiation of osteoclast precursor, the activity of osteoclasts, and induce apoptosis, thereby inhibiting the resorption of bone. Estrogen may also act on marrow stromal cells, osteoblasts and osteocytes regulating them to secrete matrix proteins, enzymes and cytokines to promote formation of bone. In addition, it can inhibit simultaneously the secretion of some cytokines that can activate osteoclasts, such as IL6 and TNF, *etc.*, thereby indirectly inhibit resorption of bone (Kassem *et al.*, 1996).

It was reported that fluoride could reduce the level of serum androgen. No matter whether skeletal fluorosis occurred or not, the levels of serum testosterone decline anyway among individuals, but this phenomenon was more significant in individuals with skeletal fluorosis (Chen *et al.*, 1997). Skeletal fluorosis itself is in the state of high bone turnover. The decrease of androgen could lead to further enhancement of bone resorption, and increased pathological damage of skeletal fluorosis, which can make the disease develop into osteoporosis and osteomalacia. The results of animal experiments showed that if fluoride intake is normal (1.5 mg kg^{-1} day), the fluorine could enhance estrogen to maintain bone mass (Zheng, 1995). But, if it is excessive (4.5 mg kg^{-1} day), fluorine could produce toxic effects on the gonads, which could decrease estrogen secretion, and thereby cause bone loss (Wei *et al.*, 2003).

Fluoride enters into body to affect bone metabolism by affecting a variety of hormones. Wherein the increased secretion and frequent fluctuation in PTH can both activate osteoblasts and promote resorption of osteoclasts. This plays an important role in the occurrences and development of fluorosis. In the mature organism, the homeostasis of bone resorption and bone formation can be regulated by a complex network including calcitropic hormone and several cytokines. Therefore, the study on the relationships among calcitropic hormone, fluoride and bone metabolism is more conductive to identify the effects of fluoride and dietary calcium on bone.

13.4 Effects of Fluoride on the Signal Transduction Pathways of Bone Metabolism

Recently, researchers have carried out studies on the roles of fluoride in the metabolic regulation of bone cells from different aspects, and have made great progresses. Among them, osteoprotegerin/ligand of receptor activator of NF-κB/receptor activator of NF-κB system (OPG/RANKL/RANK system), phosphatidylinositol 3 kinase/Akt pathway (PI–3K/Akt pathway), mitogen-activated protein kinase (MAP kinase) pathway, and other signal transduction pathways are the important parts of regulating differentiation from bone marrow stromal cells to bone cells. The roles of fluoride on the various factors of transduction pathways of bone-cell metabolism mentioned above are to be reviewed in detail in this chapter.

13.4.1 OPG/RANKL/RANK System is a Great Breakthrough in the Field of Bone in Recent Years

OPG is a negative regulator of osteoclasts, which inhibits the activation of RANKL/RANK on osteoclasts signal transduction pathways by binding with RANKL (Brendan and Xing, 2007). RANKL is also known as the osteoclast differentiation factor (ODF), the receptor of which is RANK that locates on the membrane of osteoclasts. The combination of RANKL with RANK located in osteoclasts is a critical step in the gene activation of osteoclasts. When the ratio of RANKL/OPG decreases, osteogenesis is the dominant process. Conversely, when the ratio of RANKL/OPG increases, the bone resorption becomes the dominant process gradually. It was reported that the mRNA and protein expression levels of OPG and RANKL are significantly increased after exposure to fluoride (Wang *et al.*, 2005), which accelerated the turnover of bone. In the human osteosarcoma cell line OS732, the expression levels of OPG and RANKL were significantly increased in the early stimulation of fluoride. However, in the late stage, the level of OPG is upregulated, the RANKL level is significantly reduced, which suggested that the leading direction of bone formation or resorption is closely related to fluoride exposure time and concentration in fluoride-induced enhancement of bone turnover (Zhang, 2012).

13.4.2 PI3K/Akt Signaling Pathway Plays Important Roles in Both Cell Proliferation and Differentiation of Osteoblasts or Osteoclasts

It has been reported (Chen *et al.*, 2004) that the PI3K/Akt signaling pathway could change the cytoskeleton of osteoclasts, regulate apoptosis and differentiation, migration and survival of osteoclasts. Furthermore, it has an indispensable role in the differentiation of osteoblasts. PI3K can be activated by fluoride, and then the mRNA and protein expression levels of RANKL and macrophage colony-stimulating factor (M-CSF) were both increased. Therefore, it is crucial in the differentiation, proliferation and activation of osteoclasts. Bone morphogenetic protein −2 (BMP-2) secreted by osteoblasts is an important local growth factor with bone-inducing activity, which differentiate mesenchymal cells into osteogenic cell or chondrocyte. Sugimori *et al.* (2005) found that antiapoptotic effects of BMP-2 were from the activation of PI3K, then stimulated the differentiation of osteoblasts and increased activity of alkaline phosphatase (ALP), which was a differentiation marker of osteoblasts. Akt belongs to the serine/threonine protein kinases family. Akt activated by PI3K plays a wide range of biological roles including antiapoptosis, promoting cell proliferation *etc.* through activation or inhibition of the downstream target proteins by the phosphorylation. In addition, Akt may also participate in the pathogenesis of fluorosis accompanied by a MAP kinase pathway (Gustin *et al.*, 2004).

13.4.3 Activated MAP Kinase Pathway Plays an Important Role in the Regulation of Cell Cycle in Osteoblasts

The MAP kinase pathway includes the extracellular signal-regulated kinases (ERKs), c-Jun N -terminal kinase (JNKs), p38 and so on. The ERKs pathway is also a common pathway in many signal transduction pathways of growth factors, such as BMP- 2, TGF-β, bFGF, *etc.* MAP kinase pathway involves in differentiation process from bone marrow mesenchymal stem cells to osteoblasts as well as the synthesis of collagen matrix. It plays an important role in the regulation of the cell cycle in osteoblasts by mediating differentiation and maturation of osteoblasts, and inhibiting differentiation of osteoclasts and bone resorption. It was reported that the activation of MAP kinase by 100 μM (1.9 mg l^{-1}) fluoride has been attributed to the inhibition of a phosphotyrosine phosphatase in human TE85 osteosarcoma cells (Wu *et al.*, 1997). Lau *et al.* (Lau and Baylink, 1998) reported that fluoride could inhibit fluorine-sensitive phosphotyrosine phosphatase particularly in osteoblasts, which could phosphorylate the tyrosine included in all signaling proteins in the MAP kinase signal transduction pathway. Therefore, the MAPKs accesses into a nucleus, and acts in a catalytic role on many nuclear transcription factors, accordingly regulating gene transcription. In addition, with excessive fluoride stimulation, transcription factor c-fos has been highly expressed in both osteoblasts of bone tissue in rat and *in vitro*, and then may enhance proliferation of osteoblasts (Zhang *et al.*, 2003).

Besides, there are other signaling pathways affecting the roles of fluoride on the bone metabolism pathway, including the Runx2/Cbfal pathway, NF-κB pathway, notch signaling pathway and so on. Such signal transduction pathways are functionally interconnected with each other by sharing important products and molecules, which establishes a complex network effects. In view of the important roles of these signaling pathways in the effects of fluoride on the proliferation and differentiation in osteoblasts and osteoclasts, understanding their variations and functions in endemic skeletal fluorosis will help to elucidate the pathogenesis of skeletal fluorosis.

To summarize, the effect of dietary calcium and fluoride on bone and their mechanisms are reviewed in this chapter. Excessive accumulation of fluoride in the body can lead to dental fluorosis and skeletal fluorosis, the occurrence and development of which are promoted by the inadequate intake of calcium. Fluoride can enhance the osteogenesis and activate osteoclasts, and thus affect the metabolism of bone directly or affect the influx of Ca^{2+}. In addition, the effects of fluoride and dietary calcium on the bone are generally regulated by calcitropic hormones such as parathyroid hormone, calcitonin, and 1, $25(OH)_2D_3$ as well as signal transduction pathways involving 3-phosphoinositide kinase/(PI3K/Akt) and mitogen-activated protein kinase. The improvement of calcium nutrition and control of fluoride exposure are complementary. Therefore, we need to control the intake of fluoride and improve residents' calcium nutrition at the same time to prevent and control fluorosis eventually.

Summary Points

- Both calcium and fluoride are essential nutrients for the body, and are closely related to the vital movement of human, especially bone metabolism.
- The appropriate amount of fluoride can prevent dental caries, but excessive accumulation of fluoride in the body will lead to chronic fluorosis, which cause disorders of bone metabolism, and further the occurrence of dental fluorosis and skeletal fluorosis.
- The inadequate intake of dietary calcium can cause the occurrence and development of fluorosis.
- Fluoride can affect bone metabolism by directly or indirectly acting Ca^{2+} influx to enhance the osteoblast activity and osteoclast activation.
- The effects of fluoride on bone metabolism were also regulated by calcitropic hormones including parathyroid hormone, calcitonin, *etc.* and PI3K/Akt, mitogen-activated protein kinase, *etc.* that are involved in the regulation of fluoride-related signal transduction pathways.

Key Facts of Endemic Fluorosis

Endemic fluorosis is a geochemistry disease occurred in particular geographical environment. It is a chronic, systemic, cumulative toxicosis caused by excessive intake of fluoride mainly through drinking water, air or food, and these people suffering from such fluorosis mainly live in high fluoride environment under natural conditions. Dental fluorosis and skeletal fluorosis were the major clinical manifestations. Skeletal fluorosis is a severe disease status since it can lead to the numbness and pain in waist, legs and all joints, hunched, functional disorders, and finally paralysis (Figure 13.6). In 2010, the WHO reported that excessive intake of fluoride from drinking water resulted in tens of millions of patients with dental fluorosis and

Figure 13.6 A man with skeletal fluorosis induced by drinking high-fluorine water.

skeletal fluorosis worldwide, of which China and India are the countries with severe prevalence of fluorosis. In addition, in China, there are tens of millions of rural residents threatened by endemic fluorosis from coal burning activity and drinking brick tea habit. Malnutrition, especially inadequate calcium intake, may aggravate the occurrence and development of fluorosis. In areas with high water fluoride, the occurrence and development of fluorosis can be effectively prevented and controlled by finding a low-fluoride water source for drinking or supplying calcium properly.

Definitions of Words and Terms

Bone metabolism: Bone metabolism involves the renovation and rebuilding of bone by the interactions of bone cells, erythropoietic cell of bone marrow and stromal cell. Wherein the osteoclasts and osteoblasts are responsible for the osteoclastic absorption and synthesis of bone matrix, respectively, and play the important roles in bone metabolism. They distribute in periosteum, bone trabeculae and cortical bone. The part of two-cell interactions is called as basic multicellular units. In every basic multicellular units, bone can disappear because of osteoclastic absorption, and can be replaced by resynthetic osteoblast.

Skeletal fluorosis: It is a kind of chronic metabolic bone disease caused by excessive intake of fluoride, and with major clinical manifestations of pain in the neck, waist and large joints in limbs, motor dysfunction and abnormal signs of bone X-ray.

Calcitonin (CT): CT is a single-chain polypeptide hormone synthesized and secreted in thyroid parafollicular cells, and its main biological role is to reduce blood calcium, the main target organ is bone tissue.

1, 25-dihydroxy-vitamin [D3(1, 25(OH)$_2$D$_3$): 1, 25(OH)$_2$D$_3$ is the most active metabolite of vitamin D metabolized by the liver and renal, and its main biological role is to maintain homeostasis through regulating calcium and phosphorus metabolism, its synthesis and secretion is regulated by PTH, calcium, and phosphate.

Bone alkaline phosphatase (BALP): BALP is synthesized by the osteoblast. When calcium malnutrition occurs in the body, serum calcium is decreased, parathyroid hormone is risen, thereby the kidney to synthesize 1, 25 (OH)$_2$D$_3$ was prompted, which can make the still osteoblasts into active osteoblasts. In addition, the lack of calcium can also prevent the osteoblasts becoming osteocytes, which results in the feedback hyperplasia of osteoblasts, and thus large amounts of bone alkaline phosphatase synthesized. Therefore, it can reflect directly the bone cell activity or functional status, and it is a sensitive and reliable indicator for evaluating bone mineralization disorders of body.

Osteocalcin (OCN): Osteocalcin is a noncollagenous protein synthesized and secreted by osteoblasts. The osteocalcin value changes with age and bone turnover rate. The faster bone turnover rate is often with the higher osteocalcin value. Now, it is widely regarded as a specific marker that reflects the process of the bone turnover.

Osteoprotegerin (OPG): OPG is also known as the osteoclast inhibitory factor, which works primarily by binding osteoprotegerin ligand. OPG not only inhibits osteoclast formation, but also inhibits osteoclasts resorption.

Runt-related gene 2/Core binding factor al (Runx2/Cbfal): Runx2 is an important transcription factor in osteoblasts. This pathway participates in the differentiation of osteoblasts by regulating cell-specific gene expression of extracellular matrix and cell-cycle-related genes. Runx2/Cbfalis downstream transcription factor of BMPs is involved in osteoblasts differentiation. Defects in this pathway would result in serious defects in bone.

NF-κB pathway: Nuclear factor –κB (NF-κB) is a nuclear transcription factor with multi-directional regulation, they are widely existed in a variety of tissues, involving in immune response, stress, inflammation, cell proliferation and apoptosis once activated.

Notch signaling pathway: The notch signal transduction pathway is widely present in vertebrates and invertebrates, and plays an important role in cellular signal transduction pathways involved in cell proliferation, differentiation and apoptosis, is the intersection of many important cell signaling pathway, a number of studies have shown that the notch signaling pathway is involved in osteoblasts differentiation and bone formation.

List of Abbreviations

1, 25$(OH)_2D_3$	1, 25-dihydroxy-vitamin D_3
VDR	1, 25$(OH)_2$ D_3 receptor
BAP	Bone alkaline phosphatase
BMP-2	Bonemorphogeneticprotein-2
CT	Calcitonin
JNKs	C-jun-N-terminal kinases
cAMP	Cyclic adenosine monophosphate
ERKs	Extracellular signal-regulated kinases
IL-1 I	nterleukin-1
IL-6	Interleukin-6
M-CSF	Macrophage colony-stimulating factor
MAP	Mitogen-activated protein
ODF	Osteoclast differentiation factor
OPG	Osteoprotegerin
OPG/RANKL/ RANK	Osteoprotegerin/ligand of receptor activator of NF-κB/ receptor activator of NF-κB
PTH	Parathyroid hormone
PTHR	Parathyroid hormone receptor
Runx2/Cbfal	Runt-related gene 2/Core combining factors al
PI-3K/Akt	Phosphatidylinositol 3 kinase

References

Ahdjoudj, S., Lasmoles, F., Oyajobi, B. O., Lomri, A., Delannoy, P. and Marie, P. J., 2001. Reciprocal control of osteoblast/chondroblast and osteoblast/adipocyte differentiation of multipotential clonal human marrow stromal F/STRO-1(+) cells. *Journal of Cell Biochemistry.* **81**: 23–38.

Barry, E. L., 2000. Expression of mRNAs for the alpha 1 subunit of voltage-gated calcium channels in human osteoblast-like cell lines and in normal human osteoblasts. *Calcified Tissue International.* **66**: 145–150.

Bennett, B. D., Alvarez, U. and Hruska, K. A., 2001. Receptor-operated osteoclast calcium sensing. *Endocrinology.* **142**: 1968–1974.

Brendan, R. W. and Weisbrode, S. E., 1985. Histogenesis of hyperosteoidosis in 1,25(OH)2D3-treated rats fed high levels of dietary calcium. *Bone.* **6**: 105–112.

Brendan, B. F. and Xing, L. P., 2007. Biology of RANK, RANKL, and osteoprotegerin. *Arthritis Research & Therapy.* **9** Suppl 1: S1.

Carter, P. H. and Schipani, E., 2006. The roles of parathyroid hormone and calcitonin in bone remodeling: prospects for novel therapeutics. *Endocrine, Metabolic & Immune Disorders Drug Targets.* **6**: 59–76.

Chen, P. Z., Meng, X. C. and Hao, J. T., 1997. The effects of fluoride on endocrine function in male adult. *Endemic Bulletin.* **12**: 57–58. (in Chinese).

Chen, X. S., Yu, X., Wang, S. Y., Zhang, M. W. and Mao, N., 2004. Blockage of PI3K signal pathway significantly inhibits osteoblast differentiation inpre-osteoblast MC3T3-El. *Bulletin of the Academy of Military Medical Sciences.* **8**: 513–515.

Das, T. K. and Susheela, A. K., 1991. Effect of chronic fluoride toxicity on glucocorticoid levels in plasma and urine. *Fluoride.* **24**: 23–28.

Edelman, A., Fritsch, J. and Balsan, S., 1986. Short term effects of PTH on cultured rat osteoblasts: changes in membrane potential. *American Journal of Physiology.* **251**: C483–C490.

Farley, J. R., Wergedal, J. E. and Baylink, D. J., 1983. Fluoride directly stimulates proliferation and alkaline phosphatase activity of bone-forming cells. *Science.* **222**: 330–332.

Gofa, A. and Davidson, R. M., 1996. NaF potentiates a K(+)-selective ion channel in G292 osteoblastic cells. *Journal of Membrane Biology.* **149**: 211–219.

Griffiths, H. J., Hunt, R. D., Zimmerman, R. E., Finberg, H. and Cuttino, J., 1975. The role of calcium and fluoride in osteoporosis in rhesus monkeys. *Investigative Radiology.* **10**: 263–268.

Grynpas, M. D. and Cheng P. T., 1988. Fluride reduced the rate of dissolution of bone. *Bone and Mineral.* **5**: 1–9.

Gustin, J. A., Ozes, O. N., Akca, H., Pincheira, R., Mayo, L. D., Li, Q., Guzman, J. R., Korgaonkar, C. K. and Donner, D. B., 2004. Cell type-specific expression of the kappa B kinases determines the significance of phosphatidylinositol 3-kinase/Akt signaling to NF-kappa B activation. *Journal of Biological Chemistry.* **279**: 1615–1620.

Hope, W. G., Ibarra, M. J. and Thomas, M. L., 1992. Testosterone alters duodenal calcium transport and longitudinal bone growth rate in parallel in the male rat. *Proceedings of the Society for Experimental Biology and Medicine.* **200**: 536–541.

Jilka, R. L., Weinstein, R. S., Bellido, T., Roberson, P., Parfitt, A. M. and Manolagas, S. C., 1999. Increased bone formation by prevention of osteoblast apoptosis with parathyroid hormone. *Journal of Clinical Investigation.* **104**: 439–446.

Kaji, H., Sugimoto, T., Kanatani, M. and Chihara K., 1996. High extracellular calcium stimulates osteoclast-like cell formation and bone-resorbing activity in the presence of osteoblastic cells. *Journal of Bone & Mineral Research.* **11**: 912–920.

Kassem, M., Harris, S. A., Spelsberg, T. C. and Riggs, B. L., 1996. Estrogen inhibits interleukin-6 production and gene expression in a human osteoblastic cell line with high levels of estrogen receptors. *Journal of Bone & Mineral Research.* **11**: 193–199.

Khokher, M. A. and Dandona, P., 1990. Fluoride stimulates [3H] thymidine incorporation and alkaline phosphatase production by human osteoblasts. *Metabolism.* **39**: 1118–1121.

Krishnamachari, K. A., 1986. Skeletal fluorosis in humans: a review of recent progress in the understanding of the disease. *Progress in Food & Nutrition Science.* **10**: 279–314.

Lau, K. H. and Baylink, D. J., 1998. Molecularmechanism of action of fluoride on bone cells. *Bone & Mineral Research.* **13**: 1660–1667.

Liu, Q. Y. and Zhang, W. Q., 2008. The effect of parathyroid hormone (PTH 1-34) on mechanical response of osteoblasts. *Chinese Journal of Endemiology.* **8**(10): 1874–1876. (in Chinese).

Mazziotti, G., Giustina, A., Canalis, E. and Bilezikian, J. P., 2007. Glucocorticoid-induced osteoporosis: clinical and therapeutic aspects. *Arquivos Brasileiros de Endocrinologia e Metabologia.* **51**: 1404–1412.

Mertz, W., 1994. *Trace elements in human and animal nutrition.* Translated by Zhu, L. Z. Qing Dao Publishers, Qing Dao, pp. 297–339.

Reed, B. Y., Zerwekh, J. E., Antich, P. P. and Pak, C. Y., 1993. Fluoride-stimulated [3H] thymidine uptake in a human osteoblastic osteosarcoma cell line is dependent on transforming growth factor beta. *Bone & Mineral Research.* **8**: 19–25.

Ren, L. Q. and Li G. S., 1999. Effect of calcium nutrition on skeletal toxicity of fluoride. *Chinese Journal of Endemiology.* **14**: 3–5. (in Chinese).

Ren, L. Q., Li, G. S. and Sun, B., 1998. The effects and mechanism of moderate and long period of chronic fluorosis on of bone turnover in rats. *Chinese Journal of Endemiology.* **17**: 75–78. (in Chinese).

Ridefelt, P., Hellman, P., Rastad, J., Larsson, R., Akerström, G. and Gylfe, E., 1992. Fluoride interactions with stimulus-secretion coupling of normal and pathological parathyroid cells. *Acta Physiologica Scandinavica.* **145**: 275–285.

Shankar, P., Ghosh, S., Bhaskarachary, K., Venkaiah, K. and Khandare, A. L., 2013. Amelioration of chronic fluoride toxicity by calcium and fluoride-free water in rats. *British Journal of Nutrition.* **110**: 95–104.

Siddiqui, A. H., 1955. Fluorosis in Nalgonda District, Hyderabad-Deccan. *British Medical Journal.* **2**: 1408–1413.

Suda, T., Takahashi, N., Udagawa, N., Jimi, E., Gillespie, M. T. and Martin, T. J., 1999. Modulation of osteoclast differentiation and function by the new members of the tumor necrosis factor receptor and ligand families. *Endocrine Reviews.* **20**: 345–357.

Sugimori, K., Matsui, K., Motomura, H., Tokoro, T., Wang, J., Higa, S., Kimura, T. and Kitajima, I., 2005. BMP-2 prevents apoptosis of the N1511 chondrocytic cell line through PI3K/Akt-mediated NF-kappa B activation. *Journal of Bone and Mineral Metabolism.* **23**: 411–419.

Sugimoto, T., Ritter, C., Slatopolsky, E. and Morrissey, J., 1990. Role of guanine nucleotide binding protein, cytosolic calcium and cAMP in fluoride-induced suppression of PTH secretion. *Mineral and Electrolyte Metabolism.* **16**: 224–231.

Sun, D. J., 2010. Strategy for the prevention and control of endemic fluorosis in China. *Chinese Journal of Endemiology.* **29**: 119–120.

Sun, D. J., Gao, Y. H., zhao, L. J., Yu, G. Q., Wu, L. Y. and Li, Q. L., 2008. A cross-sectional survey on drinking brick-tea type fluorosis in China. *Chinese Journal of Endemiology.* **27**: 513–517.

Teotia, M., Teotia, S. P. and Singh, K. P., 1998. Endemic chronic fluoride toxicity and dietary calcium deficiency interaction syndromes of metabolic bone disease and deformities in India: year 2000. *Indian Journal of Pediatrics.* **65**: 371–381.

Wang, C., Gao, Y. H., Wang, W., Zhao, L. J., Zhang, W., Han, H. P., Shi, Y. X., Yu, G. Q. and Sun, D. J., 2012. A national cross-sectional study on effects of fluoride-safe water supply on the prevalence of fluorosis in China. *BMJ Open.* doi: 10.1136/bmjopen-2012–001564.

Wang, C. H., Jing, L. and Li, G. S., 2005. Influence of fluoride on proliferation and expression of RANKL mRNA and OPG mRNA of bone marrow stromal cells. *Chinese Journal of Endemiology.* **24**: 55–58. (in Chinese).

Wei, Y., Xie, C., Zhang, H. and Zhou, L. Y., 2003. The effects of estrogen and age on fluorosis. *Chinese Journal of Endemiology.* **22**: 396–398. (in Chinese).

World Health Organization (WHO), 2010. *Exposure to inadequate or excess fluoride: A major public health concern.* Available at: http://www.who.int/entity/ipcs/features/fluoride.pdf. Accessed 27 September 2013.

Wu, L. W., Yoon, H. K., Baylink, D. J., Graves, L. M. and Lau, K. H., 1997. Fluoride at mitogenic doses induces a sustained activation of p44mapk, but not p42 mapk, in human TE85 osteosacrcoma cells. *Journal of Clinical Endocrinology and Metabolism.* **82**: 1126–1135.

Yu, D. L. and Cui, S., 2000. *Proliferation and differentiation of sodium fluoride on osteoblasts.* Acta of Tongji Medical University. **29**: pp. 471–473.

Yu, J., Gao, Y. H. and Sun, D. J., 2013. Effect of fluoride and low versus high levels of dietary calcium on mRNA expression of osteoprotegerin and osteoprotegerin ligand in the bone of rats. *Biological Trace Element Research.* **152**: 387–395.

Zerwekh, J. E., Morris, A. C., Padalino, P. K., Gottschalk, F. and Pak, C. Y., 1990. Fluoride rapidly and transiently raises intracellular calcium in human osteoblasts. *Journal of Bone & Mineral Research.* **5** Suppl 1: S131–S136.

Zhang, J. M., 2012. The research of endoplasmic reticulum stress research on the pathogenesis of skeletal fluorosis. *Master's Thesis*, Jilin University.

Zhang, W. L., Cui, Y. N., Gao, S., Zhang, X. Y. and Li, G. S., 2003. Expression of proto-oncogenes c-fos and c-jun in osteoblasts activated by excess fluoride. *Chinese Journal of Preventive Medicine.* **37**: 246–250. (in Chinese).

Zheng, F. Y., 1995. The effects of application with estrogen, calcium and sodium fluoride alone or in combination on preventing the bone loss in the ovariectomized rats. *Chinese Journal of Orthopaedics.* **15**: 226–270. (in Chinese).

CHAPTER 14

Sodium Fluoride and PET Bone Scans

GHOLAM R. BERENJI*[a], YUXIN LI[a], ROBERTO GONZALEZ-ODRIOZOLA[a], AND ANURADA THENKONDAR[a]

[a]Department of Nuclear Medicine, VA Greater Los Angeles Healthcare System, 11301 Wilshire Blvd, Los Angeles, California 90073, USA
*E-mail: Gholam.Berenji@va.gov

14.1 Diagnostic Imaging Using Radionuclides

18 F sodium fluoride has been used as a bone-imaging agent as early as 1962 (Blau *et al.*, 1962). Its ability to detect osteogenic activity (bone growth) has proven beneficial in diagnosing primary bone tumors and metastatic bone disease. After the introduction of positron emission tomography (PET) to the nuclear medicine community, 18 F sodium fluoride was utilized as a PET radiotracer and was eventually approved by the U.S. Food and Drug Administration (FDA) in 1972 (Segall *et al.*, 2010). Since 1975, technetium 99m (Tc-99m) has become the most widely used bone-imaging agent, and it has almost completely replaced 18 F sodium fluoride. This was mainly due to the availability of the newly designed Anger gamma camera (Li *et al.*, 2012b). In the last decade, with the wider availability of PET scanners and the possibility of a Tc-99m shortage, there has been a renewed interest in using 18 F sodium fluoride as a bone tracer.

14.1.1 Radionuclide and Radiotracer

Scientific discoveries during the late 19th and early 20th century by scientists such as Curie, Becquerel, de Hevesy, and many others led to the discovery of radionuclides and radiotracers (Brucer, 1978; Myers, 1979; Lever, 2002).

Food and Nutritional Components in Focus No. 6
Fluorine: Chemistry, Analysis, Function and Effects
Edited by Victor R Preedy

Published by the Royal Society of Chemistry, www.rsc.org

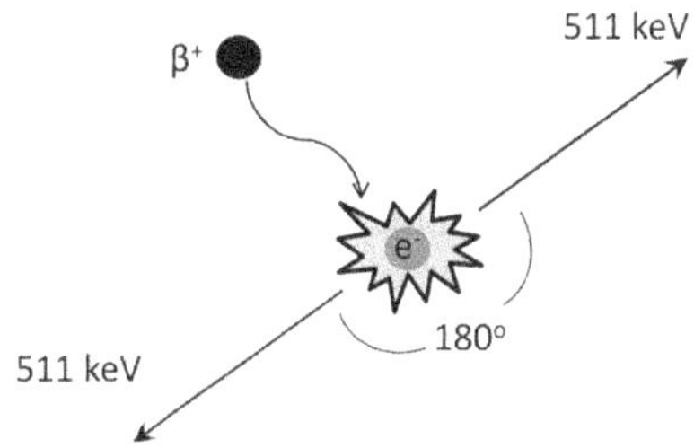

Figure 14.1 Schematic representation of positron annihilation. The positron travels a short distance in tissue before electron positron annihilation. A pair of 511 keV photons are emitted at 180 deg.

Radionuclides are unstable atoms that undergo radioactive decay. This results in their releasing gamma rays or high-energy photons, which permit external detection by imaging devices. Because most radionuclides keep the chemical and physiological properties of their stable states, they can be easily embedded in, or linked to chemical or biological compounds to create radioactive tracers (radiotracers).

The two main classes of radiotracers used in nuclear medicine are positron-emitting radiotracers and nonpositron-emitting radiotracers. 18 F sodium fluoride contains fluorine 18, which emits positrons. Although there are many different modes of radioactive decay, positron emission decay is briefly discussed further below.

In positron emission, proton-rich radionuclides have a tendency to change into more stable nuclides with fewer numbers of protons and increased numbers of neutrons by releasing a positron (positively charged beta particle). This process produces a new nuclide with one less proton and one more neutron than the parent radionuclide. The released positron has the same mass and charge of an electron, except it is positively charged. Once released, the positron travels a short distance, interacts with the atoms around it, loses its kinetic energy, and comes in close contact with an electron. Together, they annihilate or convert into two high-energy photons simultaneously, with each photon measuring 511 keV (Figure 14.1). The decay of 18 F is 97% by positron decay, which changes to 18-O as the stable, nonradioactive product (Cherry *et al.*, 2003).

14.1.2 Radionuclide Production

Three major production methods for radioisotopes are described in Table 14.1. In this chapter, we briefly discuss the cyclotron production of radioisotopes, which is the main method of producing 18 F sodium fluoride.

14.1.2.1 Cyclotron

To produce proton-rich, positron-emitting radionuclides, a high-energy beam composed of positively charged particles is required to bombard stable nuclei. The charged bombarding particles must be accelerated to reach

Table 14.1 Common methods of radionuclide production. Most radionuclides are produced in three different methods. Reactor-produced such as 131-I, which is used in imaging and treatment of thyroid cancer. Cyclotron-produced radioisotopes such as 18 F, which can be used in imaging. Generator-produced radionuclide such as 99m-Tc, which is used in diagnostic imaging.

Method	Example	Common Use
Reactor produced	131-I	Diagnostic imaging and therapy
Cyclotron produced	18 F	Positron emission tomography
Generator produced	99m-Tc	Diagnostic imaging

Table 14.2 Commonly used cyclotron-produced radionuclide. The most frequently used cyclotron-produced radionuclide is 18 F due to its longer half-life. Researcher use 13 N, 11 C and 15-O with shorter half-lives in research studies.

Radionuclide	Half-life (min)
11 C	20.4
13 N	10.0
18 F	109.7
15 O	2.04

high kinetic energies to overcome the repulsive coulombic barrier of the target nuclei. This is achieved by particle accelerators, *e.g.*, linear accelerators or cyclotrons. Most PET tracers are produced by a cyclotron due to its versatile and economic solution. A cyclotron is much smaller and less expensive than a linear accelerator. A cyclotron is basically a cyclic accelerator in which the particles are accelerated in a spiral path by the application of radiofrequency and magnetic fields. The cyclotron was invented by Ernest Lawrence at the University of California, Berkeley in 1931. It is widely used to produce radionuclides for medical and research application (Strijckmans, 2001; Morris and Perkins, 2012). A typical cyclotron uses electromagnets to create a uniform magnetic field. Using this magnetic field, very high energy particles, such as protons or alpha particles, are directed to a target to create an unstable nucleus (radionuclide). Table 14.2 demonstrates some examples of cyclotron-produced radionuclides and their half-lives (Cherry *et al.*, 2003).

14.1.3 Imaging Methods

14.1.3.1 Rectilinear Scanner

The clinical utility of 18 F sodium fluoride as a bone-seeking radiopharmaceutical in many aspects reflects the history of nuclear medicine imaging technology. The discovery of sodium fluoride as an excellent bone-seeking agent initiated the idea of the 18 F sodium fluoride bone scan using the general-purpose rectilinear scanner and a dedicated positron detector in the early 1960s. The rectilinear scanner was composed of a detector and a photomultiplier tube (PMT). The detector was made of a NaI (sodium iodine) crystal scintillator, which detected the ionizing radiation emitted by a radiotracer and transformed it into visible light. The light signal was amplified by a PMT. The detector was connected to a light bulb, which was moved over the photographic film in the same pattern as the detector, and produced a dotted image on the film outlining the spatial distribution of the counts. The rectilinear scanner was the earliest imaging device used in clinical nuclear medicine. It was limited, however, by slow acquisition and poor image quality.

14.1.3.2 Gamma Camera

The rectilinear scanner was soon replaced by the gamma camera, also known as the Anger camera or the scintillation camera. The gamma camera was developed in the late 1950s by Hal Anger. It contains a large detector head composed of NaI crystal, and it is coupled to an array of PMTs (Figure 14.2). The gamma camera converts the light energy collected from PMTs into three electronic signals, two of which define the *X* and *Y* coordinates of the detected gamma ray, while the third defines the energy level. These signals are then used to create digital images. Gamma cameras are faster than their predecessors and are able to cover a larger area of the body. In addition to the crystal scintillator and the PMT, the collimator is another essential component of the gamma camera (Figure 14.2). The purpose of the collimator is to filter gamma rays to let those parallel to a specific direction pass. The gamma ray cannot be focused or polarized by a lens, like visible light. The only way to filter the direction of a gamma ray is by absorptive collimation. Collimators are made from heavy metals (almost always lead due to cost) to effectively block the gamma rays that do not travel along the holes of the collimator. For example, in most commonly used parallel-hole collimators, only gamma rays that travel through the parallel holes of the collimator will reach the NaI crystal (Figure 14.2). Due to the detector's and collimator's limitations, gamma cameras were not able to acquire high-quality imaging of positron-emitting radiotracers.

The development of the 99-Mo/99m-Tc generator in 1958 and the subsequent development of 99m-Tc-labeled phosphate compounds in 1971 as well as the limited availability of positron emission tomography (PET) scanners led to the replacement of 18 F sodium fluoride as a bone radiotracer (Tucker *et al.*, 1958). Currently, 99m-Tc labeled diphosphonates, particularly 99m-Tc methylene diphosphonate (MDP), is the standard radiotracer for bone scans.

Most 99m-Tc MDP bone scans were acquired using planar (two-dimensional) imaging. Frequently, tomographic imaging or three-dimensional imaging can be acquired with single photon emission computed tomography (SPECT). SPECT significantly increases image sensitivity compared to planar imaging. In a SPECT camera, the detector heads of the gamma camera rotate around the patient and acquire signals at each preset angle. The data is further analyzed by a computer through a tomographic reconstruction algorithm to generate the 3D dataset. Because SPECT requires multiple image acquisitions, it takes more time to acquire when compared to planar imaging.

14.1.3.3 Positron Emission Tomography

In contrast to the gamma camera, which uses a collimator to determine the direction and location of each photon, imaging positron emission requires a different technique known as coincidence detection. In coincidence detection, two detector elements on opposite sides of the patient must detect the paired annihilation photons at the same time (or "in coincidence") and then localize the event along the line of the two detector elements (Figure 14.3). By localizing the events from different angles and applying mathematical algorithms, tomography images can be reconstructed. The concept was introduced by Kuhl and Edwards in the late 1950s (Kuhl and Edwards, 1968) and was the basis for positron emission tomography (PET). However, the development of PET was augmented after computed tomography (CT) was introduced in the early 1970s by Hounsfield, in which the filtered back projection reconstruction algorithms developed for CT were applied to PET. PET cameras consist of a whole ring of detectors acquiring counts from all different angles surrounding the region of interest (Figure 14.3). These signals are then processed by a computer to yield sectional or three-dimensional

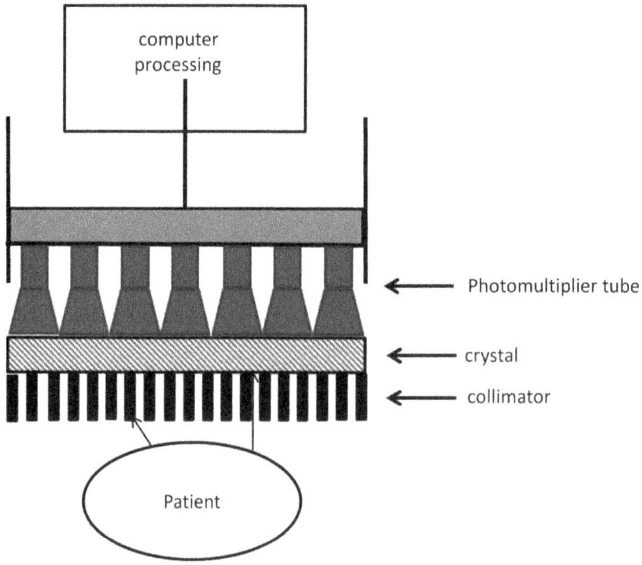

Figure 14.2 Schematic representation of a gamma camera. The gamma camera head consists of a collimator, a NaI crystal, and photomultiplier tubes. Only the gamma rays that have been filtered through the collimator can be detected.

images. Today's PET cameras have multiple ring detectors, which significantly reduce the image acquisition time. The major advantages of the PET camera compared to the gamma camera are significantly improved sensitivity, increased image resolution, and decreased acquisition time. The gamma camera is particularly known for its low geometric efficiency because the collimator blocks most of the gamma rays. Only ~0.01% of emitted photons are acquired for imaging. In contrast, a modern PET camera does not need a collimator, which results in the order of ~1% of photons being detected. This leads to a much higher sensitivity (by approximately two to three orders of magnitude) for PET over gamma camera imaging. The PET camera also has a much better spatial resolution compared to the gamma camera. The imaging resolution of the gamma camera mainly depends on the collimator and type of radiotracer used for imaging. For 99m-Tc imaging, the spatial resolution is 5–10 mm. The resolution for 18 F can be improved to 3–5 mm in the modern PET camera. There has been continuous improvement in image quality in PET due to the advancement of crystal scintillators, hardware, and image processing. However, the intrinsic characteristics of positron emission have limited the ultimate improvement of imaging resolution. What PET detects are not positrons; rather, PET detects the annihilation photons that occur a short distance away from the positron emission (Figure 14.1). The distance that the positron travels before annihilation depends on its energy as

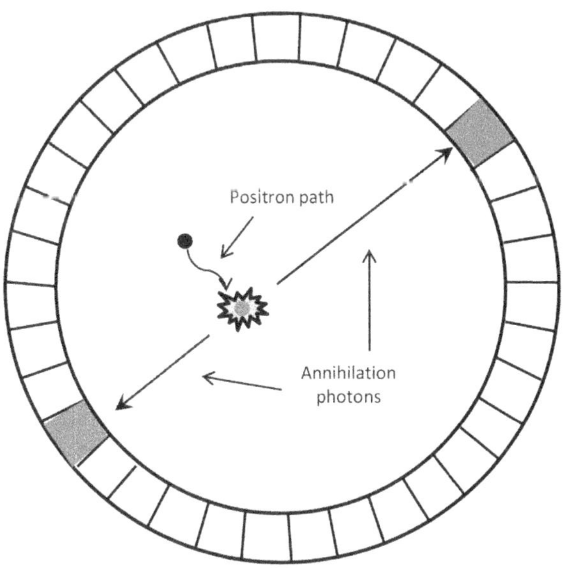

Figure 14.3 Coincidence detection of annihilation photons by PET. A pair of annihilation photons are detected by a pair of detectors at the opposite direction among the ring detectors in PET camera. The detection of the paired annihilation photons occurs at the same time, which is called coincidence detection.

defined by Einstein's famous equation $E = MC^2$. Among all clinically used positron-emitting radionuclides, 18 F has the lowest positron energy; thus, it travels the shortest distance before annihilation and thereby has the highest imaging resolution. Luckily, 18 F also has a long enough physical half-life (110 min) to allow for the commercial distribution of 18 F-labeled radiotracers to remote PET centers (Table 14.2).

14.1.4 Commonly Used Radionuclide in PET Imaging

The most commonly used radionuclide worldwide is 18 F fluorodeoxyglocuse or FDG. Its major application is in the detection of solid tumors, and currently, is the standard of care for most malignant tumors (Figure 14.4). Other commonly used radiotracers include 13 N ammonium for imaging of the heart and 11 C-labeled radiotracers, which are mainly used for research purposes. Unlike F-18, the short half-lives of 13 N and 11 C (Table 14.2) make them undesirable for commercial distribution and limit their use in centers with an on-site cyclotron.

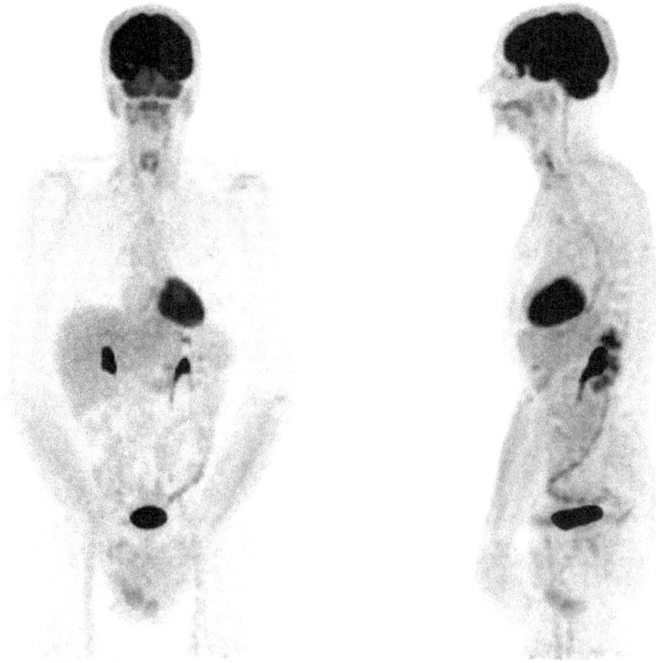

Figure 14.4 FDG PET study in a normal subject. Two views of a normal FDG PET scan in a male subject. No significant bone uptake is usually seen in FDG PET scans. Brain and liver activity is normal. Activity in the kidneys and urinary tract is normally seen and is due to excretion of FDG in the urine.

14.2 Fluoride as an Imaging Radionuclide

14.2.1 Production and Kinetics

Historically, nonradioactive sodium fluoride has been used to treat metabolic bone diseases. The ability of fluoride to accumulate in the bone has been well documented since 1937.

For the PET bone scan, 18 F sodium fluoride is administered intravenously. After injection, 18 F sodium fluoride leaves the plasma and extracellular space immediately to bind with bone through passive diffusion. Once it reaches the bone surface, the fluoride moves into the bone crystal matrix, which is made of hydroxyapatites. The fluoride will exchange places with the hydroxyl ions to become fluoroapatites. This process allows the fluoride to become part of the bone crystal matrix. Fluoride is considered inside the bone after it enters the bone surface. 18 F sodium fluoride is excreted through the kidneys. The clearance of 18 F sodium fluoride by kidneys, or renal clearance, is dependent on urine flow rate. Some of the 18 F sodium fluoride is taken up by red blood cells (mature and immature) and bone marrow. This does not affect the absorption of 18 F sodium fluoride by the bone, as 18 F sodium fluoride can easily move from blood cells to bone surface. The movement of the radioisotope to the bone is dependent on the molecular size of the agent. 18 F sodium fluoride molecules are much smaller than molecules of 99m-Tc MDP. The smaller size of the 18 F sodium fluoride molecules allows them to easily enter the bone surface. The first pass extraction fraction for 18 F sodium fluoride is near 100%.

The immediate renal clearance and bone accumulation of 18 F sodium fluoride produces a high bone-to-soft tissue contrast within 1 h after injection (Figure 14.5). For 99m-Tc MDP, the highest bone-to-soft tissue contrast is within 3 to 6 h of injection. Additionally, 18 F sodium fluoride has a faster clearance compared to 99m Tc MDP. One hour after the 18 F sodium fluoride injection, only 10% of it remains in plasma. The accumulation of 18 F sodium fluoride in bone tissue is proportional to blood flow and bone tissue turnover rate (Li *et al.*, 2012b). 18 F sodium fluoride has a high affinity for bone tissue, particularly osteoblastic activity or sites of new bone formation (Grant *et al.*, 2008; Jacobson *et al.*, 2012).

14.2.2 Toxicity and Radiation Safety

Most radiopharmaceuticals do not have any pharmacodynamic effects due to the extremely low quantity being used. "The dose makes the poison" is a principle of toxicology that was expressed by Paracelsus, and it has precluded the chemical toxicity of most of the radiopharmaceuticals. The diagnostic radiopharmaceuticals used in nuclear medicine studies today rarely exceed 1110 MBq, which equals 1.1 x 10^9 decays per second. Hypothetically, If one were to label one of the most toxic man-made compounds TCDD (2, 3, 7, 8-Tetrachlorodibenzo-*p*-dioxin), 1110 MBq of the radiotracer is equivalent to

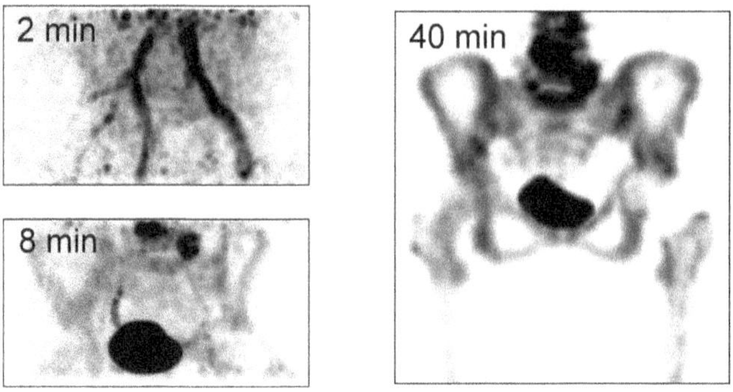

Figure 14.5 Rapid bone uptake of sodium fluoride. Maximum intensity projection (MIP) images of the pelvis acquired at 2 min, 8 min, and 40 min after 18 F sodium fluoride administration. Blood flow and iliac vessels were demonstrated 2 min after tracer injection. After 8 min, rapid clearance of tracer from the plasma shows low activity in the vessels, and early skeletal uptake of fluoride. Delayed image at 40 min demonstrates excellent bone–soft tissue contrast, with most of tracer activity located in the bone or bladder.

5.9×10^{-10} mg of molecules, which is less than 1/100 000 of the lethal dose 50 (LD_{50}) for a hamster (0.045 mg kg^{-1}) and will not cause any toxic effects. The recommended dose of F 18 sodium fluoride is 185–370 MBq. This is less than equivalent to 5.2×10^{-11} mg of sodium fluoride and should have close to zero pharmacodynamic effects in humans.

The main concern about radiopharmaceuticals' adverse events comes from radiation exposure. In the past few years, there has been an increase in public concern regarding medical radiation exposure. Currently, medical equipment represents the largest source of radiation exposure for the United States population, exceeding the natural background radiation from the environment. Increased radiation exposure has been associated with increased cancer risk, usually at high doses of radiation exposure. Because most medical imaging procedures in radiology and nuclear medicine expose people to low levels of radiation, the increased risk of cancer from exposure to any single imaging test is likely to be negligible. Further, because there is no threshold to tell which dose of radiation is totally safe, minimizing medical radiation exposure is a prudent approach. Therefore, the "as low as reasonably achievable" (ALARA) principle has been applied to the practice of imaging for many years in an attempt to minimize the adverse effects of radiation exposure. Many imaging societies have introduced guidelines to lower the radiation exposure during imaging procedures, particularly in children. The industry has responded to these concerns and introduced more efficient imaging devices with the ability to reduce patients' radiation exposure.

In nuclear medicine studies, the radiation exposure to the whole body is nonuniform and is generally expressed as the effective dose (ED) calculated by summing the average organ equivalent dose of several organs using tissue weighting factors (Cherry *et al.*, 2003). The effective dose is used to estimate the risk to a patient, and it reflects the equivalent whole-body radiation exposure in milliSieverts (mSV). The PET/CT bone scan using 185 MBq of F-18 sodium fluoride has similar radiation exposure as a chest CT, and it is not significantly higher than the conventional MDP bone scan.

14.2.3 Differences between MDP Bone Scan and 18 F Fluoride PET Bone Scan

The main advantage of the 18 F sodium fluoride PET over the 99m-Tc MDP bone scan is increased sensitivity and image quality (Figure 14.6). This is due to the fact that only ~0.01% of emitted photons are acquired with 99m-Tc MDP bone imaging, as mentioned before. Additionally, the spatial resolution is much better in the PET scan compared to the 99m-Tc MDP bone imaging. The bone-to-soft tissue contrast is better in 18 F sodium fluoride PET imaging due to the higher affinity of 18 F sodium fluoride for bone tissue and lower plasma protein binding (Li *et al.*, 2012b). Imaging can be done 30 to 45 min after injection of 18 F sodium fluoride, while in the 99m-Tc MDP bone scan images are acquired two to three hours after injection. The shorter imaging time reduces the chance of patient motion and, therefore, results in fewer artifacts seen in imaging (Grant *et al.*, 2008). Attenuation happens when photons from deeper tissue are not well detected, while photons from superficial tissues are better detected. This may lead to attenuation artifacts in imaging and degrade image quality. Attenuation correction to reduce artifacts can be done by different methods, namely using computed tomography data (Ziessman *et al.*, 2013). The guidelines for 18 F sodium fluoride PET/CT by the Society of Nuclear Medicine recommended computed tomography (CT) for attenuation correction and anatomical localization (Segall *et al.*, 2010). Without attenuation correction, small lesions can be missed, especially spinal lesions (Li *et al.*, 2012b). This is an added benefit of performing PET bone imaging using a PET/CT scanner.

14.2.4 Application and Indications

14.2.4.1 Benign Conditions and Bone Metabolism

Any bone lesion that can cause increased osteoblastic activity or increased blood flow will demonstrate increased sodium fluoride uptake in F-18 NaF PET bone scan. Besides malignant bone disease, a variety of benign bone disorders can result in an abnormal PET bone scan. These include fractures, Paget's disease, degenerative bone disease, and benign spine lesions, among others. However, evaluating benign bone disease with a sodium fluoride PET bone study has not been examined as thoroughly as metastatic bone disease. Promising results have been reported in the evaluation of fractures in child abuse, which requires highly sensitive imaging modalities (Drubach *et al.*, 2010). Drubach *et al.* found that the 18 F NaF PET bone scan had a higher sensitivity than a skeletal survey for the detection of fractures from child

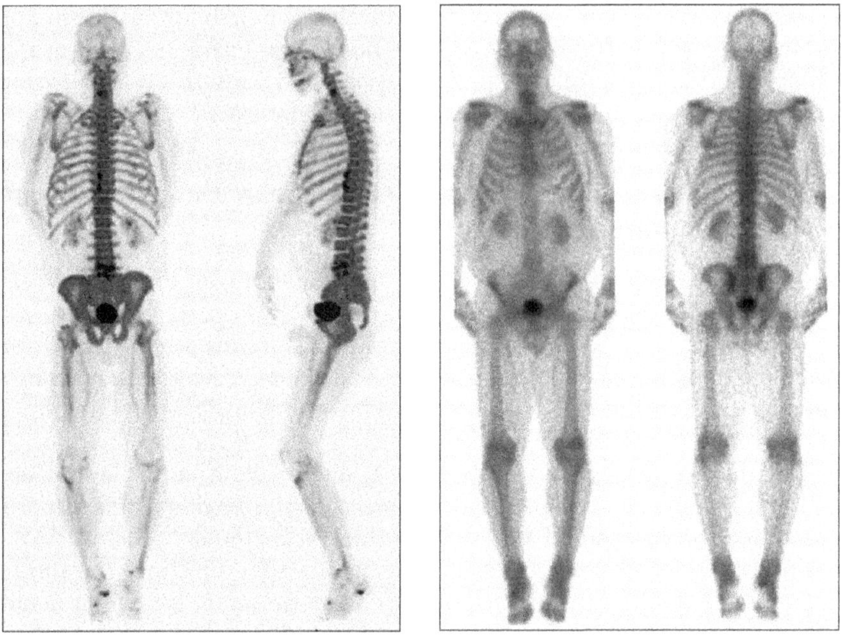

Figure 14.6 Image quality between 18 F sodium fluoride PET and MDP bone scintigraphy. Improved image quality of 18 F NaF PET/CT bone scan (left) compared to the conventional MDP bone scintigraphy (right) in the same patient. Degenerative changes with increased 18 F sodium fluoride uptake is noted in the lower thoracic and lower lumbar spine, but not shown on the planar images of the MDP bone scintigraphy.

abuse, in particular for rib fractures. The results suggest that 18 F NaF PET may serve as the sole global skeletal assessment tool for the evaluation of child abuse in children older than 12 months (Drubach *et al.*, 2010).

Another promising technique that is gaining interest, is the use of dynamic PET imaging with 18 F NaF for an accurate quantitative evaluation of bone metabolism. The more favorable kinetics of 18 F NaF, as well as the advantage of superior imaging characteristics of PET over planar scintigraphy or SPECT, has made the 18 F NaF PET bone scan the technique of choice for an accurate quantitative evaluation of bone metabolism. Consecutive images are acquired for 1–2 h after injection of 18 F NaF to evaluate the entire process of how it enters the blood stream, diffuses through the body, and eventually reaches and accumulates in the target tissues (bones). It is based on a compartmental kinetic modeling that was introduced by Hawkins *et al.* in 1992 and is the most accurate approach to evaluate regional blood flow and turnover by using 18 F NaF and PET (Hawkins *et al.*, 1992). A quantitative assessment of regional bone metabolism by 18 F NaF PET provides valuable clinical information that may potentially affect the management of metabolic bone disease. Promising results have been reported in a variety of clinical situations, including in osteoporosis, Paget's disease, renal osteodystrophy, bone graft evaluation, trauma, and metastatic bone disease (Messa *et al.*, 1993; Berding *et al.*, 1995; Schiepers *et al.*, 1997; Schiepers *et al.*, 1998; Cook *et al.*, 2000; Cook *et al.*, 2002; Brenner *et al.*, 2004; Installe *et al.*, 2005; Frost *et al.*, 2009; Uchida *et al.*, 2009; Doot *et al.*, 2010).

14.2.4.2 Malignancies

Similar to 99m-Tc MDP bone scan, the most common clinical application for 18 F NaF PET/CT study is in oncology. Although F-18 NaF can accumulate in primary bone malignancies, it is mainly used to detect metastatic bone disease, especially from prostate, breast, and lung cancers.

Metastatic bone lesions demonstrate a higher accumulation of 18 F NaF than normal bone parenchyma due to locally increased osteoblastic activity as well as increased blood flow. It is important to note that the increased F-18 sodium fluoride uptake represents the reaction of normal bone tissue to the metastatic tumor cell growth, rather than the metabolic activity of the malignant lesions. Bone metastases can be classified as sclerotic, lytic, or mixed. Sclerotic bone metastases usually have much greater oseoblastic activity than lytic bone metastases. Therefore, a bone scan is most useful to evaluate bone metastasis from prostate cancer, which predominantly has sclerotic bone lesions. In contrast, lung cancer mainly has lytic bone metastasis and favors osteoclastic activity, promoting the local resorption of the bone matrix (osteolysis). Because of its increased resolution and the sensitivity of PET, it has been suggested that an 18 F NaF PET bone scan is more sensitive than a 99m-Tc MDP bone scan in detecting osteolytic bone metastases. The minimal osteoblastic activity associated with lytic bone metastases may not be identified by 99m-Tc MDP bone scintigraphy, but it may be seen with an 18 F NaF PET scan (Petren-Mallmin *et al.*, 1998; Schirrmeister *et al.*, 1999; Even-Sapir *et al.*, 2004). On the other hand, the accumulation of 18 F NaF is not tumor specific and can also be observed in a variety of benign bone disorders, most commonly in degenerative bone disease, which frequently occurs in older people. It can be very difficult to distinguish between benign and metastatic bone lesions. In clinical practice, most of the nuclear medicine physicians or radiologists use their clinical experience to differentiate between benign and malignant conditions, which sometimes results in inaccurate diagnoses. This pitfall was greatly improved with the introduction of hybrid imaging, such as PET/CT, where the CT scan portion provides more detailed anatomical information, which is very helpful in distinguishing between benign and metastatic bone lesions, as shown in Figure 14.7. Thus, the F-18 sodium fluoride PET/CT study not only has increased sensitivity, but also has increased specificity compared to conventional bone scans. This has been consistently demonstrated by numerous investigators in evaluating bone metastasis in various cancers, including prostate, breast, lung, and thyroid (Schirrmeister *et al.*, 1999; Even-Sapir *et al.*, 2004; Kawaguchi *et al.*, 2010; Withofs *et al.*, 2011). A recent *meta*-analysis including 11 studies and 425 patients has shown that the sensitivity and specificity for 18 F NaF PET

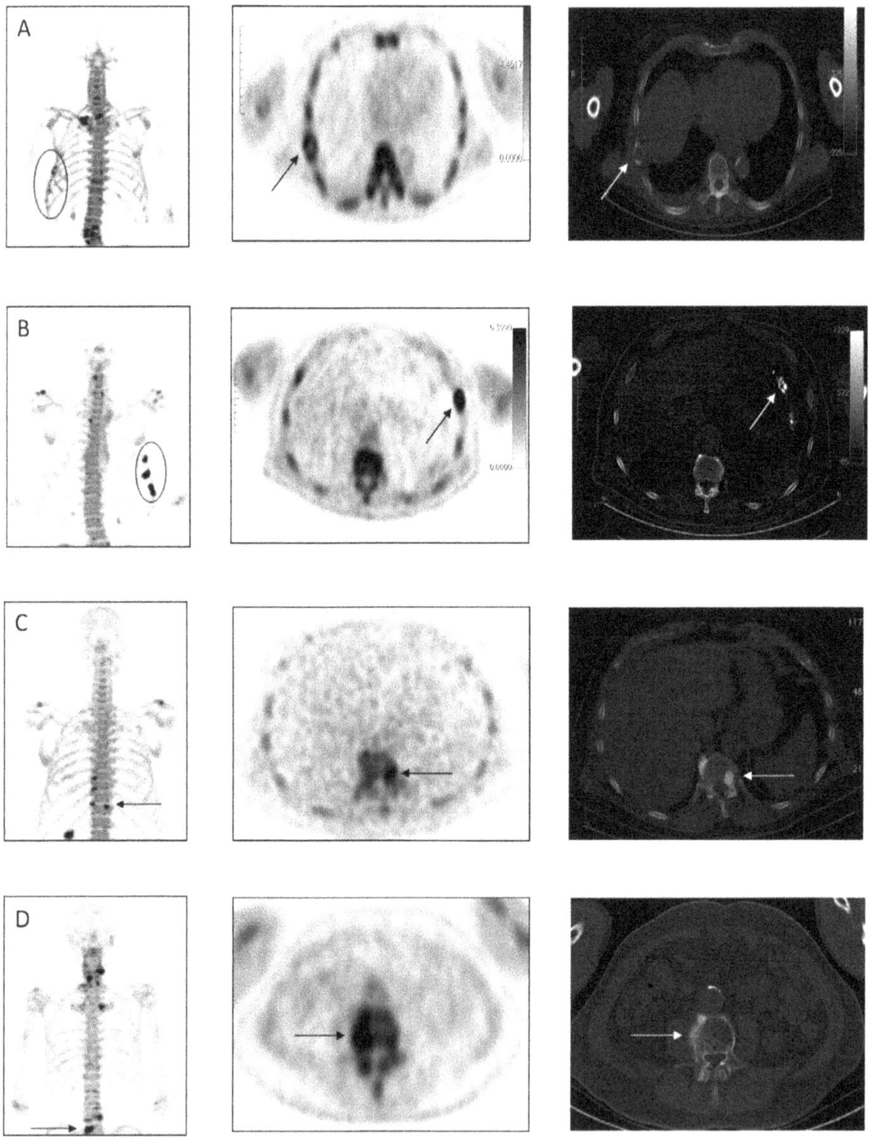

Figure 14.7 CT improves accuracy of image interpretation in evaluating bone metastases. The 18 F NaF PET images of patient A and Patient B demonstrate similar rib tracer uptake (arrow). CT images demonstrate bone metastasis in patient A and rib fractures in patient B. The PET images of patient C and D also demonstrate similar spine uptake. CT images demonstrate sclerotic bone metastasis in patient C and typical degenerative changes in patient D.

or PET/CT were 96.2% and 98.5%, respectively, on a patient basis and 96.9% and 98.0%, respectively, on a lesion basis (Tateishi *et al.*, 2010). These results are higher than the conventional MDP bone scan, which has poor specificity.

14.2.4.3 Cardiovascular

Coronary atherosclerotic plaque rupture is the leading cause of acute myocardial infarction and sudden cardiac death. High-risk and ruptured atherosclerotic plaques are challenging to predict because most plaques are nonobstructive and, therefore, cannot be diagnosed with nuclear stress testing or coronary angiography. High-risk atherosclerotic lesions at risk of rupture demonstrate certain histopathological characteristics, including positive remodeling, microcalcifications, and a large necrotic core that will lead to coronary calcifications. The F-18 sodium fluoride PET bone scan has been found to be a promising test to detect active vascular calcification (Derlin *et al.*, 2010). Coronary calcifications detected by a sodium fluoride PET bone scan are significantly correlated with increased cardiovascular events (Figure 14.8) (Li *et al.*, 2012a). A recent prospective study has found

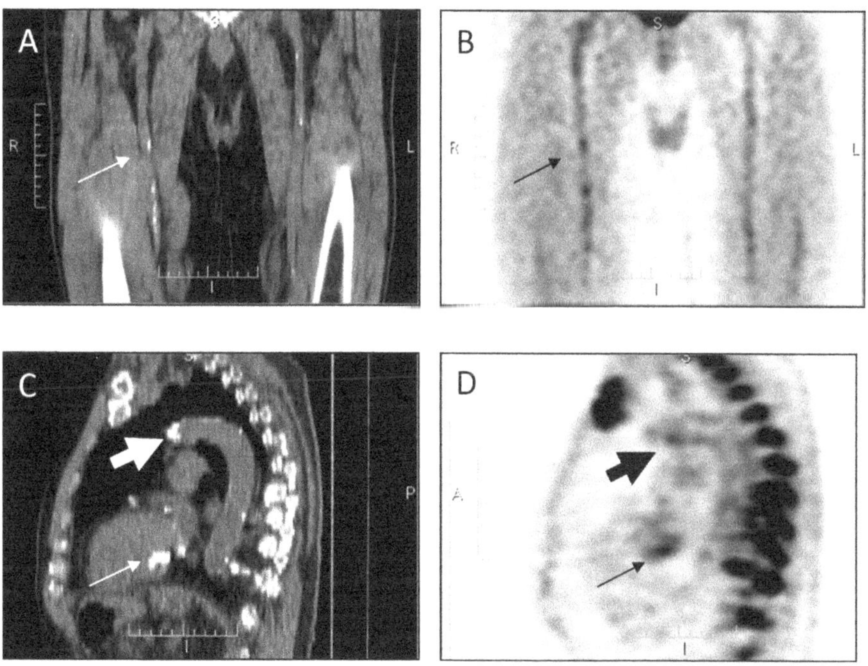

Figure 14.8 Vascular calcification can be visualized by 18 F sodium fluoride PET/CT bone scan. Coronal view of CT (A) and PET (B) images demonstrate prominent calcifications and 18 F sodium fluoride uptake in the femoral arteries. Sagittal view of CT (C) and PET (D) demonstrate calcifications and 18 F sodium fluoride uptake in the aorta (big arrow) and left circumflex coronary artery (small arrow).

that a F-18 sodium fluoride PET bone scan can identify and localize ruptured and high-risk coronary plaque as well as perform better than FDG PET (Joshi *et al.*, 2013). This is a very exciting finding, and it has the potential to change how clinicians identify and manage patients with coronary artery disease. Future studies are needed to further evaluate this clinical application.

Summary Points

- This chapter has discussed application of 18 F sodium fluoride in PET bone scans.
- 18 F sodium fluoride was the first radiotracer used for bone imaging.
- The majority of bone scans are currently performed using 99m-TC MDP as a radiotracer due to wide availability of gamma cameras and 99m-TC.
- There has been a growing interest in 18 F sodium fluoride as a bone scan radiotracer in the past decade due to availability of PET scanners and higher quality of bone images.
- 18 F sodium fluoride is a cyclotron-produced radiotracer with physical half-life of 110 min.
- The majority of 18 F sodium fluoride decays by positron emission.
- Positron emission creates two high-energy photons, which travel in opposite directions.
- PET scanners detect the annihiliation photons using a coincidence detection method.
- Hybrid PET/CT scanners use CT images to correct attenuation of photons and accurately localize abnormal PET findings.
- A 18 F sodium fluoride PET bone scan is commonly used for detection of benign or malignant bone diseases.
- New research has revealed a potential application of 18 F sodium fluoride in cardiovascular imaging and evaluation of atherosclerosis.

Key Facts of PET Bone Scans

- PET bone scans are performed using 18 F sodium Fluoride.
- The physical half-life of 18 F sodium fluoride is about 110 min.
- The spatial resolution of PET bone scan is 3–6 mm.
- The time to imaging after intravenous injection of radiotracer is 30 to 60 min.
- The image quality of PET bone scan is excellent.
- Most PET bone scans are performed with a concurrent CT scan (PET/CT) to increase the sensitivity and to better locate the abnormal areas.
- PET bone scans are three-dimensional.
- The bladder receives the highest amount radiation during PET bone scans.

Definitions

Annihiliation: When a positron interacts with an electron, the collision of these two particles results in two gamma ray photons at high energy.

As low as reasonably achievable (ALARA) principles: Methods to reduce radiation exposure to patients and professionals. An example of ALARA is using the proper distance from patients during procedures.

Collimator: A part of a gamma camera detector made of heavy metals to filter gamma rays.

Computed tomography: A technology that uses a computer to process acquired radiation signals to produce tomographic images. This can be used in X-ray computed tomography (X-ray CT), positron emission tomography (PET), or single-photon emission computed tomography (SPECT).

Filter back projection: A method for reconstructing an image out of multiple projections.

Gamma rays: A form of ionizing electromagnetic radiation released from an unstable atomic nucleus with high frequency and high energy.

Metastasis: Dissemination of cancer to other organs.

Osteogenic activity: This usually refers to bone growth, mostly done by osteoblasts in the bone tissue.

Photon: The quantum of electromagnetic energy.

Osteoporosis: A condition that bone density decreases and may increase risk of fractures.

Scintigraphy: The process of acquiring a two-dimensional image using a gamma camera after administration of a radiotracer.

Abbreviations

ALARA	as low as reasonably achievable
CT	computed tomography
ED	effective dose
FDA	Food and Drug Administration
keV	kilo electron volt
MBq	mega Becquerel
mSV	milliSieverts
PET	positron emission tomography
PMT	photomultiplier
SPECT	single-photon emission tomography

References

Berding, G., Burchert, W., van den Hoff, J., Pytlik, C., Neukam, F. W., Meyer, G. J., Gratz, K. F. and Hundeshagen, H., 1995. Evaluation of the incorporation of bone grafts used in maxillofacial surgery with [18F]fluoride ion and dynamic positron emission tomography. *European Journal of Nuclear Medicine.* 22(10): 1133–1140.

Blau, M., Nagler, W. and Bender, M. A., 1962. Fluorine-18: a new isotope for bone scanning. *Journal of Nuclear Medicine.* **3**: 332–334.
Brenner, W., Vernon, C., Conrad, E. U. and Eary, J. F., 2004. Assessment of the metabolic activity of bone grafts with (18)F-fluoride PET. *European Journal of Nuclear Medicine and Molecular Imaging.* **31**(9): 1291–1298.
Brucer, M., 1978. Nuclear medicine begins with a boa constrictor. *Journal of Nuclear Medicine.* **19**(6): 581–598.
Cherry, S. R., Sorenson, J. A. and Phelps, M. E., 2003. *Physics in nuclear medicine*. Saunders, Philadelphia, PA.
Cook, G. J., Blake, G. M., Marsden, P. K., Cronin, B. and Fogelman, I., 2002. Quantification of skeletal kinetic indices in Paget's disease using dynamic 18F-fluoride positron emission tomography. *Journal of Bone and Mineral Research.* **17**(5): 854–859.
Cook, G. J., Lodge, M. A., Blake, G. M., Marsden, P. K. and Fogelman, I., 2000. Differences in skeletal kinetics between vertebral and humeral bone measured by 18F-fluoride positron emission tomography in postmenopausal women. *Journal of Bone and Mineral Research.* **15**(4): 763–769.
Derlin, T., Richter, U., Bannas, P., Begemann, P., Buchert, R., Mester, J. and Klutmann, S., 2010. Feasibility of 18F-sodium fluoride PET/CT for imaging of atherosclerotic plaque. *Journal of Nuclear Medicine.* **51**(6): 862–865.
Doot, R. K., Muzi, M., Peterson, L. M., Schubert, E. K., Gralow, J. R., Specht, J. M. and Mankoff, D. A., 2010. Kinetic analysis of 18F-fluoride PET images of breast cancer bone metastases. *Journal of Nuclear Medicine.* **51**(4): 521–527.
Drubach, L. A., Johnston, P. R., Newton, A. W., Perez-Rossello, J. M., Grant, F. D. and Kleinman, P. K., 2010. Skeletal trauma in child abuse: detection with 18F-NaF PET. *Radiology.* **255**(1): 173–181.
Even-Sapir, E., Metser, U., Flusser, G., Zuriel, L., Kollender, Y., Lerman, H., Lievshitz, G., Ron, I. and Mishani, E., 2004. Assessment of malignant skeletal disease: initial experience with 18F-fluoride PET/CT and comparison between 18F-fluoride PET and 18F-fluoride PET/CT. *Journal of Nuclear Medicine.* **45**(2): 272–278.
Frost, M. L., Blake, G. M., Cook, G. J., Marsden, P. K. and Fogelman, I., 2009. Differences in regional bone perfusion and turnover between lumbar spine and distal humerus: (18)F-fluoride PET study of treatment-naive and treated postmenopausal women. *Bone.* **45**(5): 942–948.
Grant, F. D., Fahey, F. H., Packard, A. B., Davis, R. T., Alavi, A. and Treves, S. T., 2008. Skeletal PET with 18F-fluoride: applying new technology to an old tracer. *Journal of Nuclear Medicine.* **49**(1): 68–78.
Hawkins, R. A., Choi, Y., Huang, S. C., Hoh, C. K., Dahlbom, M., Schiepers, C., Satyamurthy, N., Barrio, J. R. and Phelps, M. E., 1992. Evaluation of the skeletal kinetics of fluorine-18-fluoride ion with PET. *Journal of Nuclear Medicine.* **33**(5): 633–642.
Installe, J., Nzeusseu, A., Bol, A., Depresseux, G., Devogelaer, J. P. and Lonneux, M., 2005. (18)F-fluoride PET for monitoring therapeutic response in Paget's disease of bone. *Journal of Nuclear Medicine.* **46**(10): 1650–1658.

Jacobson, M. S., Steichen, R. A. and Peller, P. J., 2012. PET radiochemistry and radiopharmacy. In: Peller, P., Subramaniam, R. and Guermazi, A. (ed.) *PET-CT and PET-MRI in Oncology, Medical Radiology*. Springer Berlin Heidelberg, pp. 19–30.

Joshi, N. V., Vesey, A. T., Williams, M. C., Shah, A. S., Calvert, P. A., Craighead, F. H., Yeoh, S. E., Wallace, W., Salter, D., Fletcher, A. M., van Beek, E. J., Flapan, A. D., Uren, N. G., Behan, M. W., Cruden, N. L., Mills, N. L., Fox, K. A., Rudd, J. H., Dweck, M. R. and Newby, D. E., 2013. F-fluoride positron emission tomography for identification of ruptured and high-risk coronary atherosclerotic plaques: a prospective clinical trial. *Lancet.* **383**(9918): 705–713.

Kawaguchi, M., Tateishi, U., Shizukuishi, K., Suzuki, A. and Inoue, T., 2010. 18F-fluoride uptake in bone metastasis: morphologic and metabolic analysis on integrated PET/CT. *Annals of Nuclear Medicine.* **24**(4): 241–247.

Kuhl, D. E. and Edwards, R. Q., 1968. Reorganizing data from transverse section scans of the brain using digital processing. *Radiology.* **91**(5): 975–983.

Lever, S. Z., 2002. Evolution of radiopharmaceuticals for diagnosis and therapy. *Journal of Cellular Biochemistry. Supplement.* **39**: 60–64.

Li, Y., Berenji, G. R., Shaba, W. F., Tafti, B., Yevdayev, E. and Dadparvar, S., 2012a. Association of vascular fluoride uptake with vascular calcification and coronary artery disease. *Nuclear Medicine Communications.* **33**(1): 14–20.

Li, Y., Schiepers, C., Lake, R., Dadparvar, S. and Berenji, G. R., 2012b. Clinical utility of (18)F-fluoride PET/CT in benign and malignant bone diseases. *Bone.* **50**(1): 128–139.

Messa, C., Goodman, W. G., Hoh, C. K., Choi, Y., Nissenson, A. R., Salusky, I. B., Phelps, M. E. and Hawkins, R. A., 1993. Bone metabolic activity measured with positron emission tomography and [18F]fluoride ion in renal osteodystrophy: correlation with bone histomorphometry. *Journal of Clinical Endocrinology and Metabolism.* 77(4): 949–955.

Morris, P. and Perkins, A., 2012. Diagnostic imaging. *Lancet.* **379**(9825): 1525–1533.

Myers, W. G., 1979. Georg Charles de Hevesy: the father of nuclear medicine. *Journal of Nuclear Medicine.* **20**(6): 590–594.

Petren-Mallmin, M., Andreasson, I., Ljunggren, O., Ahlstrom, H., Bergh, J., Antoni, G., Langstrom, B. and Bergstrom, M., 1998. Skeletal metastases from breast cancer: uptake of 18F-fluoride measured with positron emission tomography in correlation with CT. *Skeletal Radiology.* **27**(2): 72–76.

Schiepers, C., Broos, P., Miserez, M., Bormans, G. and De Roo, M., 1998. Measurement of skeletal flow with positron emission tomography and 18F-fluoride in femoral head osteonecrosis. *Archives of Orthopaedic and Trauma Surgery.* **118**(3): 131–135.

Schiepers, C., Nuyts, J., Bormans, G., Dequeker, J., Bouillon, R., Mortelmans, L., Verbruggen, A. and De Roo, M., 1997. Fluoride kinetics of the axial skeleton measured in vivo with fluorine-18-fluoride PET. *Journal of Nuclear Medicine.* **38**(12): 1970–1976.

Schirrmeister, H., Guhlmann, A., Kotzerke, J., Santjohanser, C., Kuhn, T., Kreienberg, R., Messer, P., Nussle, K., Elsner, K., Glatting, G., Trager, H., Neumaier, B., Diederichs, C. and Reske, S. N., 1999. Early detection and accurate description of extent of metastatic bone disease in breast cancer with fluoride ion and positron emission tomography. *Journal of Clinical Oncology.* **17**(8): 2381–2389.
Segall, G., Delbeke, D., Stabin, M. G., Even-Sapir, E., Fair, J., Sajdak, R., Smith, G. T. and Snm, 2010. SNM practice guideline for sodium 18F-fluoride PET/CT bone scans 1.0. *Journal of Nuclear Medicine.* **51**(11): 1813–1820.
Strijckmans, K., 2001. The isochronous cyclotron: principles and recent developments. *Computerized Medical Imaging and Graphics.* **25**(2): 69–78.
Tateishi, U., Morita, S., Taguri, M., Shizukuishi, K., Minamimoto, R., Kawaguchi, M., Murano, T., Terauchi, T., Inoue, T. and Kim, E. E., 2010. A meta-analysis of (18)F-Fluoride positron emission tomography for assessment of metastatic bone tumor. *Annals of Nuclear Medicine.* **24**(7): 523–531.
Tucker, W. D., Greene, M. W., Weiss, A. J. and Murrenhoff, A., 1958. Methods of preparation of some carrier-free radioisotopes involving sorption on alumina. *Transactions American Nuclear Society.* 1.
Uchida, K., Nakajima, H., Miyazaki, T., Yayama, T., Kawahara, H., Kobayashi, S., Tsuchida, T., Okazawa, H., Fujibayashi, Y. and Baba, H., 2009. Effects of alendronate on bone metabolism in glucocorticoid-induced osteoporosis measured by 18F-fluoride PET: a prospective study. *Journal of Nuclear Medicine.* **50**(11): 1808–1814.
Withofs, N., Grayet, B., Tancredi, T., Rorive, A., Mella, C., Giacomelli, F., Mievis, F., Aerts, J., Waltregny, D., Jerusalem, G. and Hustinx, R., 2011. (1)(8)F-fluoride PET/CT for assessing bone involvement in prostate and breast cancers. *Nuclear Medicine Communications.* **32**(3): 168–176.
Ziessman, H., O'Mally, J. and Thrall, J., 2013. *Nuclear Medicine: The Requisites:* Elsevier.

CHAPTER 15

Fluoride-Induced Oxidative Damage in Hippocampal Cells

IWONA INKIELEWICZ-STEPNIAK*[a] AND NARCYZ KNAP[a]

[a]Department of Medical Chemistry, Medical University of Gdansk, 1 Debinki St., 80-211 Gdansk, Poland
*E-mail: iinkiel@gumed.edu.pl

15.1 A Few Words about Fluoride

Fluoride has a very narrow safety margin. Drinking water is the major source of fluoride in the general population, except for people occupationally exposed. It is well documented that high concentration of fluoride (>1.5 mg L^{-1}) in tap water, as found in many regions of the world, results in fluorosis (Choi *et al.*, 2012). However, WHO data (2003) indicate that other components of the diet, such as minerals, vitamins, fats, proteins as well as pharmacotherapeutics and various supplements may contribute to the development of fluorosis. In order to prevent fluorosis, adequate intake (AI) of fluoride from all possible sources (ground water, food, air, drugs, cosmetics, tooth paste and dental applications) has been set at 0.05 mg kg^{-1} body weight per day (WHO, 2003). Excessive fluoride intake leads to structural, functional and metabolic dysfunction in various cells and tissues through multiple mechanisms including oxidative stress.

15.2 Oxidative Stress

Oxygen is essential for the survival of aerobic organisms. Because of its high redox potential, it serves as the terminal electron acceptor in the process of metabolic energy production. However, its high redox potential can also be

Food and Nutritional Components in Focus No. 6
Fluorine: Chemistry, Analysis, Function and Effects
Edited by Victor R Preedy

Published by the Royal Society of Chemistry, www.rsc.org

1.
$$\text{R-OOH} + 2\,\text{GSH} \xrightarrow{\textbf{GPx}} \text{R-OH} + \text{GSSG} + H_2O$$

2.
$$2H_2O_2 \xrightarrow{\textbf{CAT}} 2H_2O + O_2$$

3.
$$O_2^{\bullet-} + O_2^{\bullet-} + 2H^+ \xrightarrow{\textbf{SOD}} H_2O_2 + O_2$$

Figure 15.1 Reactions catalyzed by antioxidant enzymes. Reactions catalyzed by antioxidant enzymes: glutathione peroxidase (1); catalase (2); superoxide dismutase (3). GSH is reduced glutathione, GPx is glutathione peroxidase, R–OOH is an organic peroxide, CAT is catalase, $O_2^{\bullet-}$ is superoxide anion, SOD is superoxide dismutase.

harmful to cells. An incomplete reduction of molecular oxygen may lead to the generation of highly reactive oxygen and nitrogen species (ROS/RNS), such as singlet oxygen, superoxide anion, hydrogen peroxide, hydroxyl and peroxyl radicals, nitric oxide, nitrogen dioxide, dinitrogen trioxide and peroxynitrite. Both ROS and RNS may play either physiological or (when generated in excess), potentially pathological roles in biological systems. At high concentrations, ROS and RNS can mediate damage to cell structures, including membrane lipids, carbohydrates, proteins and nucleic acids.

Over millions of years of evolution, cells have developed highly complex antioxidant systems (enzymatic and nonenzymatic), which work synergistically, and in combination with each other to protect against oxidative damage. The antioxidants can be endogenous or obtained exogenously, as part of a diet or dietary supplements. The most efficient enzymatic antioxidative systems involve glutathione peroxidase, catalase and superoxide dismutase. Reactions catalyzed by the antioxidative enzymes presented in Figure 15.1.

Nonenzymatic antioxidants include Vitamins A, E and C, thiol antioxidants (glutathione, thioredoxin and lipoic acid), melatonin, bilirubin, coenzyme Q10, carotenoids, natural flavonoids, and other compounds. Glutathione is the most abundant and important endogenous antioxidant. It is synthesized from 3 amino acids: L-cysteine, L-glutamic acid and L-glycine, see Figure 15.2.

An imbalance of the equilibrium between free-radical generation and antioxidant defense systems leading to the accumulation of free radicals was described in 1991 by Sies as *oxidative stress*, see Figure 15.3.

Oxidative stress has been implicated in various pathologies (thrombosis, cancer, diabetes mellitus, neurological disorders, *etc.*) also caused by xenobiotics (drugs, dietary supplements, alcohol, cigarette smoking, *etc.*). The assessment of oxidative damage requires both the detection of ROS/RNS-mediated structural changes in molecules and their functional impairment. Table 15.1 illustrates the most commonly used markers of oxidative stress (damage).

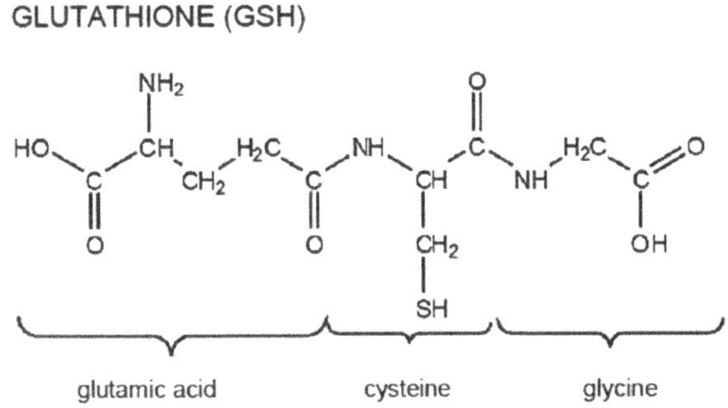

Figure 15.2 Chemical structure of glutathione (GSH). Glutathione is a tripeptide consisting of glutamic acid, cysteine and glycine.

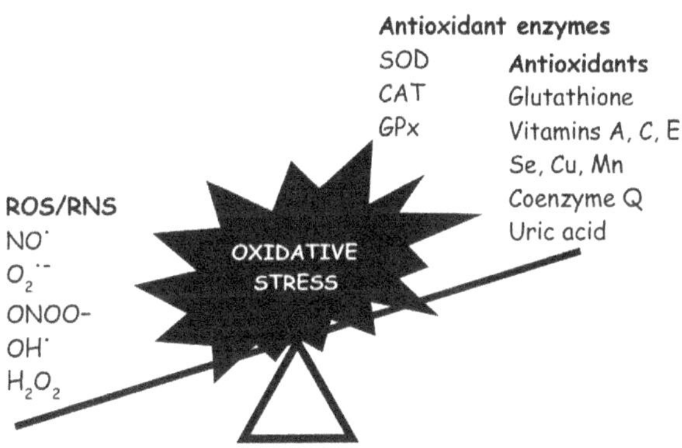

Figure 15.3 Oxidative stress. The imbalance between ROS/RNS production and antioxidant defence (enzymatic and nonenzymatic) leads to oxidative stress.

15.2.1 Why is the Central Nervous System so Highly Susceptibility to Oxidative Damage?

The brain, which accounts for only 2% of body weight, consumes 20% of the total oxygen inspired. The brain is particularly vulnerable to oxidative injury because of its high rate of oxidative metabolic activity, intense production of reactive oxygen metabolites, high content of polyunsaturated fatty acids, the presence of catecholamines prone to oxidation, low antioxidant capacity and poor repair mechanisms, as well as nonreplicating nature of neurons. Oxidative stress has been indicated as a key mechanism involved in fluoride-induced toxicity in hippocampal cells.

Table 15.1 Markers of Oxidative Stress. The Table Lists the Most Commonly Determined Markers of Oxidative Stress: Products of Lipid, Protein, DNA Peroxidation and Element of Antioxidant Defense System: Enzymatic and Nonenzymatic.

Lipid peroxidation markers:
Malondialdehyde (MDA)
4-hydroxynonenal (4-HNE)
8-iso prostaglandin F2α (8-iso-PGF2α)
Oxidative DNA damage:
8OH-Deoxyguanosine (8OH-dG)
Protein oxidation products:
Advanced oxidation protein product (AOPPs)
Carbonyl group
Antioxidant defense system:
Antioxidant enzymes: Glutathione peroxidase (GPx) catalase (CAT), superoxide dismutase (SOD)
Antioxidants: Reduced glutathione/oxidized glutathione ratio (GSH/GSSG), uric acid, vitamins: A, C, E antioxidative minerals: Se, Zn, Mn

15.3 Fluoride and the Central Nervous System (CNS)

In 1937, KajRoholm published his findings on skeletal fluorosis with neurological symptoms: excessive tiredness, sleepiness, indisposition, headache, and giddiness observed in 22.1% of cryolite workers. Thereafter, many researchers reported neurological manifestations such as tingling, loss of the perception of pain, temperature and touch, altered reflexes as diagnosed in patients suffering from fluorosis. In 1995, Mullenix *et al.* published a study on the neurotoxicity of sodium fluoride in rats. The authors stated: "This is the first laboratory study to demonstrate that CNS functional output is vulnerable to fluoride, that the effects on behavior depend on the age at exposure and that fluoride accumulates in brain tissue." The study indicated a potentially negative effect of fluoride on the brain manifesting clinically as motor dysfunction, IQ deficits and learning disabilities. The authors demonstrated that plasma fluoride levels in their animal model (0.059 to 0.640 ppm) are similar to those reported in humans exposed to high levels of fluoride. Guan *et al.*, (1986) reported that the high concentration of fluoride affects the behavioral development of the offspring visible primarily as slight delay in response times, particularly with regard to motor and coordination function and well as muscle strength. Next, in 1998, Varner *et al.* documented that administration of 1 ppm fluoride for 52 weeks resulted in distinct morphological alterations in the brain of rats as well as in the reduction of the neuronal density, chromatin clumping, enhanced protein staining, pyknosis and vacuolation. Furthermore, fluoride can affect the nervous system directly without initial induction of fluorosis-dependent skeletal deformations. In our previous study we demonstrated that daily administration of fluoride (25 mg) with drinking water for 12 weeks resulted in an increase in the brain fluoride level in rats as compared to the control animals (Inkielewicz and Krechniak, 2003). Numerous studies have documented that the placental barrier does not prevent the passage of fluoride from a pregnant mother's bloodstream to the fetus. As a result, the fetus can be harmed by fluoride ingested during pregnancy (Han *et al.*, 1989; Madhusudhan *et al.*, 2010; Yu *et al.*, 2008). It was found that maternal exposure to 100 ppm fluoride in drinking water, caused fluoride accumulation up to 2.14 $\mu g\ g^{-1}$ wet tissue in the brain of young rats as compared to controls: 0.27–0.64 $\mu g\ g^{-1}$ tissue (Madhusudhan *et al.*, 2010). The passage of fluoride thorough the placenta and its accumulation within the fetal brain was shown to have an adverse impact on the brain development. Han *et al.* (1989) indicated that fluoride can retard the growth and division of cells in the cerebral cortex, influencing the intellectual development after birth. Yu *et al.* (2008) investigated the fluoride effect on the fetal brain and found evidence of significant neurological damage, including neuronal degeneration and decrease in neurotransmitters such as norepinephrine. Du *et al.* (2008) found that Purkinje cells of fetuses from an endemic fluorosis area were abnormally disorganized and had a thicker granulated layer in the cerebellum. Other dysmorphologies, including higher nucleus–cytoplasm ratio of brain cones, and hippocampus cones, indicate that fluoride has an adverse effect on brain development. Microscopic analyses also showed reduced number of the cortical neurons in the brain, decreased numerical density, volume density, and surface density. Bera *et al.* (2007) presented that perinatal exposure to sodium fluoride at dosage levels below those associated with gross malformations and/or overt neurotoxic effects, produced both short- and long-term sex and dose-specific neurobehavioral alterations in rat offspring. Basha *et al.* (2011) carried out a study to assess the effect of fluoride: 100, 200 ppm in drinking water on the oxidative stress in the brain, for three generations of rats. Interestingly, they observed that fluoride increased lipid peroxidation and decreased the cellular antioxidant potential to a larger extent in the second and third as compared to the first generation of the rats.

Over the last 23 years numerous studies conducted in China, India, Iran, and Mexico have proved that fluoride exposure is associated with IQ deficits in children. In 2012, a team of Harvard researchers published a "meta-analysis" of 27 studies that have investigated the relationship between fluoride and human intelligence (Choi *et al.*, 2012). They found strong indications that fluoride may adversely affect cognitive development in children. Children who lived in areas with high fluoride exposure had lower IQ scores than those who lived in low-exposure or control areas. The exposed groups had access to drinking water with fluoride concentrations up to 11.5 $mg\ L^{-1}$. Thus, in many cases the concentrations were high above the levels recommended by WHO: 0.7–1.2 $mg\ L^{-1}$ (WHO, 2003). On the other hand, Ding *et al.* (2011) suggested that low levels of water fluoride (0.24–2.84 $mg\ L^{-1}$) had a significantly negative effect on children's intelligence.

Nowadays, there is no doubt that fluoride can cross the blood–brain barrier and it can cause adverse effects on the brain-cell architecture, metabolism, the oxidant/antioxidant status deteriorating intellectual functions. Transport of fluoride through the blood–brain barrier is based on an active transport system that is similar to that of other halogens and ionic substances, and the normal cerebrospinal fluid (CSF)/blood fluoride ratio is less than 1.0.

15.4 Hippocampus

Fluoride accumulates within various parts of the brain especially in the hippocampus (Zhai *et al.*, 2003; Chirumari and Reddy, 2007). The hippocampus is a critical component of the brains of humans and other mammals. It belongs to the limbic system and plays important roles in long-term memory, emotion, learning, sexual, social behavior and spatial navigation. Like the cerebral cortex, with which it is closely associated, it is a paired structure, with mirror-image halves in the left and right sides of the brain. In humans and other primates, the hippocampus is located inside the medial temporal lobe, beneath the cortical surface. The hippocampus itself consists of two interlocking C-shaped structures: the cornu ammonis and the dentate gyrus. Histologically, the hippocampus is further divided into four sections: CA1 to CA4.

15.4.1 Fluoride and the Hippocampus – *In Vivo* Study

Numerous *in vivo* studies investigated effects of fluoride on the structure and function of the hippocampus. Mullenix *et al.* (1995) observed an accumulation of fluoride in the hippocampus (mean 0.993 ppm F at 125 ppm water fluoride during weaning), which was found to increase as the fluoride levels in the drinking water increased. Also, Zhai *et al.* (2003) found that fluoride went through the blood–brain barrier, accumulated in rat hippocampus, and inhibited the activity of cholinesterase (ChE). The cholinesterase activity in the brain is also known to be important for maintaining physiological brain processes associated with learning and memorizing abilities. A decrease in ChE activity in the hippocampus of rats presenting with fluorosis correlated significantly with higher levels of fluoride in the same region of the brain. Nasir and Asad (2013) found that in rats treated with 20 ppm of sodium fluoride for 7 weeks, the body weight and physical activity was significantly reduced in the treated group, whereas fluoride level in the hippocampus greatly increased. Microscopy analysis of the hippocampus, showed reduced neuronal density more pronounced in the CA3 region. Another study, in rats exposed to fluoride showed increased DNA damage as well as increased apoptosis in parts of the brain that are related to cognition – the hippocampus, cerebellum and pallium (Chen *et al.*, 2002).

It was also found that fluoride increased activity of protein kinase C (PKC) in the hippocampus of fluoride-treated animals and resulted in the impairment of motor behavior as well as learning and memory (Chirumari and Reddy, 2007). Exposure to sodium fluoride leads to the activation of polyphosphoinositide phospholipase C in rat cerebral cortex, causing an increase in the intracellular free calcium pool and activation of PKC through the secondary messengers inositol triphosphate and diacylglycerol.

Bhatnagar *et al.* (2002) found that fluoride-intoxicated animals performed poorly in motor coordination tests and maze tests. The inability to perform well increased with higher fluoride concentration in drinking water. Additionally, a light microscopic study of hippocampal subregions showed a significant number of degenerated nerve cell bodies in the CA3, CA4 and dentate gyrus (Dg) areas of sodium fluoride administered to adult female mice. They ultrastructural observations demonstrated neurodegenerative characteristics like involution of cell membranes, swelling of mitochondria and clumping of chromatin material in the cell bodies of CA3, CA4 and Dg.

Also, Pereira *et al.* (2011) showed that sodium fluoride impaired open-field habituation and increased noradrenaline (NA) and serotonin (5-HT) in the striatum, neocortex and hippocampus. Adult Long–Evans rats drinking water containing 1 ppm of fluoride for 52 weeks were demonstrated to present with a significant reduction in the number of cells within certain areas of the hippocampus relative to the control group. A toxic effect of high-fluoride intake (100 ppm) during the early developing stages of life on the growth, differentiation, and subcellular organization of brain structures including the hippocampus in rats, was demonstrated by Shivarajashankara *et al.* (2002) Interestingly, Zhu *et al.* (2011) found that postsynaptic density-95 (PSD-95) in CA3 region of the hippocampus is one of the target molecules for fluoride. PSD-95 is a 95 kDa scaffolding protein that assembles signaling complexes at synapses. They observed in weaning male Sprague–Dawley rats treated with 15, 30, and 60 mg L^{-1} sodium fluoride solution for 9 months a decrease in synaptic membrane fluidity and postsynaptic density-95 expression.

It is known that glutamate, a major excitatory neurotransmitter in the central nervous system, plays an important role in the process of learning and memorization. Niu *et al.* (2009) assessed the glutamate level and the activities of glutamate-metabolism-related enzymes including asparate aminotransferase (AST), alanine aminotransferase (ALT) and glutamic acid decarboxylase (GAD) in the hippocampus, as well as the learning abilities of rat offspring exposed to fluoride at postnatal week 6, 8, 10 and 12. During lactation, the rat pups ingested fluoride *via* the maternal milk from their mothers that had been exposed to sodium fluoride (150 mg L^{-1} in drinking water) from the day of delivery. After weaning at postnatal day 21, the rat pups were exposed to the same treatment as their mother. It was demonstrated that the learning abilities and hippocampus glutamate levels were significantly decreased by fluoride. The activities of AST and ALT in treatment groups were significantly inhibited, while the activities of GAD were increased. The obtained result suggested that alteration of hippocampus glutamate by fluoride might in part reduce learning ability in rats.

Intracellular Ca^{2+} overload is recognized as a common pathway of neuronal damage induced by various stimuli. One of the main pathways for such Ca^{2+} overload is associated with the action of CaMKIIα, c-fos, Bax, and Bcl-2. Due to an increase in the intracellular [Ca^{2+}], Ca^{2+}-sensitive protein kinases such as CaMKIIα are activated. Zhang and Zhang (2013) indicated that the signaling pathway of CaMKIIα, c-fos, Bax, and Bcl-2 may therefore be one of the related molecular mechanisms of fluorosis affecting the central nervous system. They found that the expression of CaMKIIα, c-fos, Bax, and Bcl-2 statistically significantly changed upon fluoride exposure in the CA3 area of the rat hippocampus. Interestingly, they noticed that the expression of CaMKIIα and c-fos in the high-F (60 mg NaF/L) group of rats was slightly less than that in the medium (30 mg NaF/L) group. Another study showed that acute exposure of rats to 150 mg NaF/L in drinking water caused lower expression patterns of CaMKIIα in the hippocampus (Luo *et al.*, 2011). These results suggest that toxicity induced by higher concentrations of fluoride or chronic exposure to fluoride are mediated by oxidative stress or regulated by Bcl-2. The Bcl-2 family comprises a group of structurally related proteins that play a fundamental role in the regulation of the intrinsic apoptotic pathway by controlling mitochondrial membrane permeability and release of the proapoptotic factor – cytochrome c (Zhang and Zhang, 2013). Chirumari and Reddy (2007) found that NaF administered *via* intraperitoneal injection to female Wistar rats for 14 days at increasing concentrations (1, 5, 10, and 20 mg NaF/kg bw/day) caused dose-dependent neurotoxicity, triggering lipid peroxidation and enhanced free-radical generation. They also revealed alterations in the levels of antioxidative enzyme activities in the hippocampus.

Thus, *in vivo* investigations prove that complex mechanism, including the generation of free radicals due to enhanced oxidative stress is responsible for fluoride-induced structural and functional damage to the hippocampus.

15.4.2 Fluoride and Oxidative Injury in Hippocampal Cells – *In Vitro* Study

A harmful effect of fluoride on hippocampal cells was also demonstrated *in vitro*. In our previous study we observed that exposition to fluoride (50, 100, 250 μM) for 3 days decreased viability, see Figure 15.4, and enhanced apoptosis in the mouse hippocampal cell line (HT22) (Inkielewicz-Stepniak *et al.*, 2012).

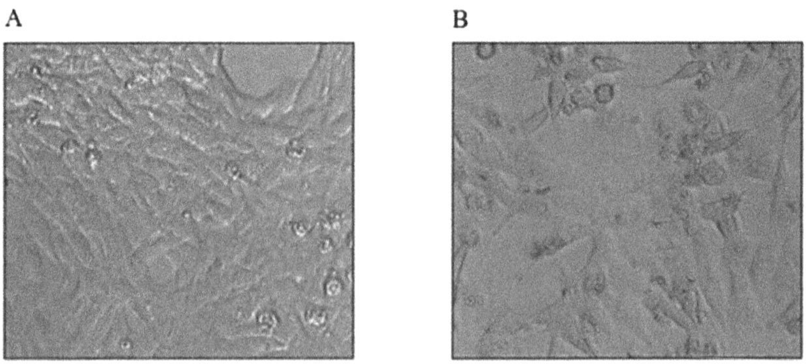

Figure 15.4 Hippocampal cells (HT22). Hippocampal cells (HT22): A. control cells; B. cells treated with 250 μg mL^{-1} fluoride for 72 h.

Murphy *et al.* (2011) found that fluoride affected cell death, and specifically apoptosis by means of the increased ROS/RNS generation. Many agents, which are oxidants or stimulators of cellular oxidative metabolism play a major role in various forms of cell death, including apoptosis. Indeed, we noticed that fluoride induced the generation of ROS/RNS, mainly hydrogen peroxide (H_2O_2) and nitrogen oxide (NO) in HT22 cells. Both NO and H_2O_2 play important role in the regulation of hippocampal metabolism; however, their excessive production caused cytotoxicity (Khalid 2007; Zhang *et al.* 2008; Inkielewicz-Stepniak *et al.*, 2012). Much of NO-mediated pathogenicity depends on the formation of secondary intermediates such as peroxynitrite anion ($ONOO^-$) and nitrogen dioxide (NO_2) that are yet more reactive, and thus potentially toxic than NO *per se*. Our previous, *in vivo* study provided evidence that exposition to fluoride led to overproduction of NO in the brain (Inkielewicz-*Stepniak* and Czarnowski, 2010). During *in vitro* experiments we noticed that pre-incubation with catalase and L-NMMA together (prior to fluoride treatment) completely inhibited apoptosis in HT22 cells. These results clearly indicate that NO and H_2O_2 cause fluoride-induced apoptosis in hippocampal cells and discount the involvement of other apoptotic stimuli (Inkielewicz-Stepniak *et al.*, 2012). In addition to that, we observed that fluoride impaired the enzymatic and nonenzymatic antioxidant defense systems in HT22 cell line. We found a significant decrease in the activity, mRNA expression and protein level of SOD2 (Mn-SOD), see Figure 15.5.

SOD2 exists as a tetramer and is a key mitochondrial antioxidative enzyme. Its perturbation leads to oxidative cell death, which results in various pathologies. In our study we noticed that fluoride decreased the level of GSH in a concentration-dependent manner, see Figure 15.6.

GSH participates not only in the antioxidant defense systems, but also in many metabolic processes for example the detoxification of endogenous compounds and xenobiotics. These findings indicate that impaired antioxidant defence system may be an important mechanism of fluoride-induced damage as observed in the hippocampal cells.

Furthermore, Zhang *et al.* (2008) reported that fluoride (20, 40, and 80 mg L^{-1} NaF) induced apoptosis in rat primary hippocampal neurons possibly mediated by elevated oxidative stress. As compared with the control group, lipid peroxidation was significantly enhanced, while glutathione levels and glutathione peroxidase activities were decreased in all the fluoride-treated groups, accompanied by the markedly reduced superoxide dismutase activity in fluoride-treated group (80 mg L^{-1}). Moreover, they found that fluoride decreased mRNA and protein expression levels of neural cell adhesion molecules (NCAM) after 24 h incubation. NCAM is associated with neural development and some momentous brain functions, such as formation, stabilization, and regeneration of learning and memorization. It was also observed that sodium fluoride at concentration of 40 mg L^{-1} and 80 mg L^{-1} induced cell-cycle arrest at S-phase and sub-G1 phase in primary hippocampal neurons (Zhang *et al.*, 2007). Fluoride could induce expression of the nuclear factor kappa-B (NF-κB) at the transcriptional level in hippocampal

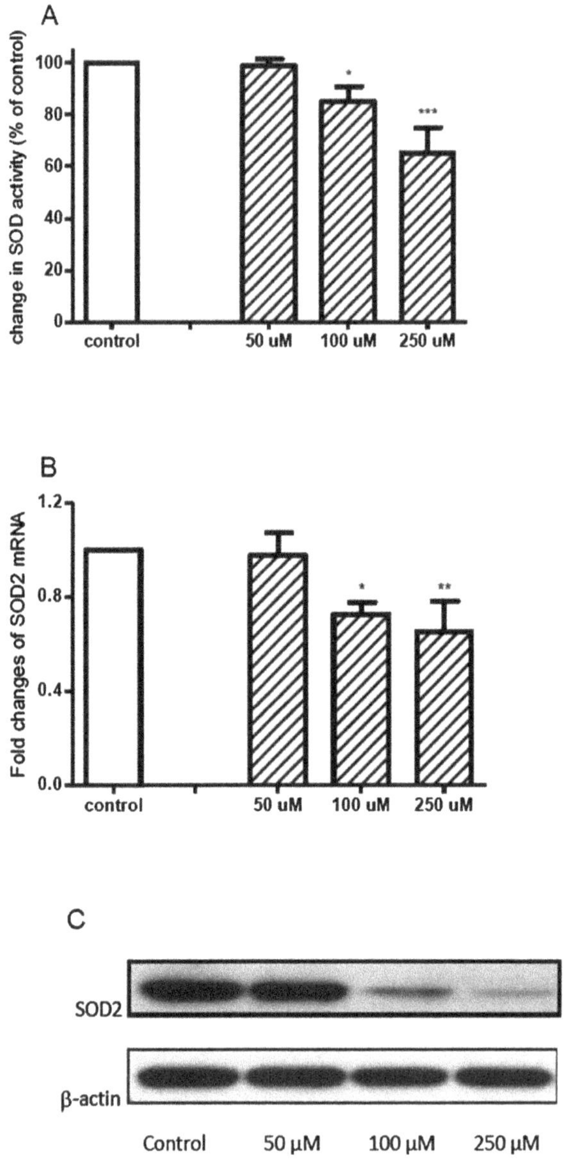

Figure 15.5 The effect of fluoride on SOD activity, expression of SOD2 mRNA and SOD2 protein level in HT22 cells. Decreased SOD activity (A), SOD2 mRNA expression (B) and SOD2 protein level (C) in HT22 cells after exposure to fluoride. Data are means ± SD of 3–4 separate determinations. *$p < 0.05$; **$p < 0.001$; ***$p < 0.001$; treatments *vs.* control.

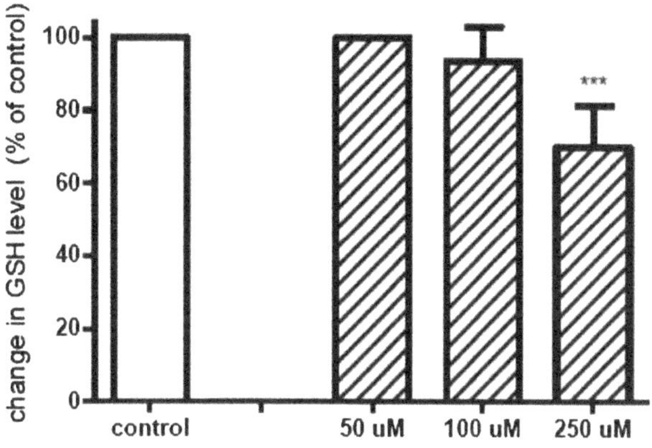

Figure 15.6 The effect of fluoride on glutathione (GSH) level in HT22 cells. Decreased GSH level in HT22 cells following exposure to fluoride. Data are presented as means ± SD of 3–4 separate determinations. $^{***}p <$ 0.001; treatments *vs.* control.

neurons, which was at least partially mediated by the elevated oxidative stress. It has been demonstrated that oxidation plays a major role in the alterations of NF-κB activity during the process of synaptic plasticity in the hippocampus. NF-kappaB is a family of seven structurally related transcription factors that play a key role in cell growth, cell phase regulation, stress response, and inflammatory process by controlling gene network expression. In the brain, NF-κB signaling is associated with survival of neurons, but in some cases of injury, such as ischemic insult, it has been linked to neuronal death.

Numerous epidemiological investigations, animal experiments and *in vitro* studies indicated that fluoride is a neurotoxin and it has a serious adverse impact on the hippocampus/hippocampal cells. Fluoride induces the generation of free radicals, increases lipid peroxidation, impairs intracellular and extracellular antioxidant potential, inhibits key enzymes of the critical metabolic pathways like energy production and protein synthesis. A deeper understanding of the mechanisms responsible for fluoride neurotoxicity might be of importance in prevention and treatment of fluoride-induced brain damage.

Summary Points

- This chapter focuses on fluoride-induced toxicity in the hippocampus.
- The general population may be exposed to fluoride through consumption of drinking water, foods, and fluoride-containing oral care products.
- Excessive intake of fluoride (>0.05 mg kg^{-1} body weight) results in fluorosis. Nowadays, it is known that fluorosis is a chronic systemic disease.
- Fluoride can cross the blood–brain barrier and causes adverse effects in the hippocampal cell architecture, metabolism and function.

- Free-radical mechanisms and oxidative stress are involved in neurotoxicity of fluoride.
- Fluoride enhances lipid, protein, DNA peroxidation, impairs antioxidant status in different regions of the brain, including the hippocampus.
- NO and H_2O_2 cause fluoride-induced apoptosis in hippocampal cells.
- Fluoride toxicity to the hippocampus has been confirmed by a plethora of epidemiological studies, as well as *in vitro* and *in vivo* experimental models.

Key Facts of the Hippocampus

- The hippocampus belongs to the limbic system.
- The hippocampus plays an important role in the mediation of memory, emotion, learning, spatial navigation, sexual, and social behavior processes.
- Hippocampal neurons are very sensitive to oxidative damage due to their large dependence on oxidative phosphorylation for energy production as compared to other cells.
- Fluoride is able to cross the blood–brain barrier and accumulate in the hippocampus.
- The hippocampus has been postulated to be one of the target sites affected by fluoride.
- Chronic exposure to fluoride impairs the hippocampal function and structure.
- Chronic exposure to fluoride induces damage to the hippocampus and subsequently impairs child's intelligence and neurobehavioral functions including learning and memory processes.
- Epidemiological and experimental studies showed that exposure to fluoride increases oxidative stress markers in the hippocampus.
- Reactive oxygen/nitrogen species (ROS/RNS) play a key role in fluoride-induced hippocampal neurodegeneration.

Definitions and Explanation of Key Terms

Fluoride: A chemical ion of the element fluorine; fluoride has one extra electron that gives it a negative charge.

Fluorosis: A disease caused by excessive intake of fluorine compounds, most commonly with drinking water, fluorosis affects the teeth, bones and soft tissues.

Dental fluorosis: A diffuse symmetric hypomineralization (irregular calcification) disorder of ameloblasts (enamel-forming cells) caused by excessive exposure to high concentrations of fluoride during tooth development.

Skeletal fluorosis: A chronic metabolic bone and joint disease caused by ingesting large amounts of fluoride, that leads to disturbances in the homeostasis of bone mineral metabolism.

Reactive oxygen species (ROS): Chemically reactive molecules containing oxygen. At low levels, these species may participate in cell-signaling processes, at higher levels, however, ROS may damage cellular molecules and induce apoptosis.

Reactive nitrogen species (RNS): A family of nitrogen molecular entities that are highly reactive and derivatives of nitric oxide ($^{\bullet}NO$) produced endogenously by inducible nitric oxide synthase 2 (NOS2) and interacting chemically with ROS.

Free radical: A reactive atom or group of atoms with one or more unpaired electrons.

Oxidative stress: A disturbance in the balance between the production of reactive oxygen/species and antioxidant defenses system in the body in favor of the former; oxidative stress is involved in numerous physiological processes and disease.

The antioxidant defense system: A system composed of both endogenous (produced by the human body) antioxidants, such as enzymes and small molecules, and exogenous ones (provided with the diet), such as vitamins, plant-derived antioxidants and cofactors (micronutrients) responsible for controlling oxidation–reduction processes.

Antioxidant: Any substance that is present at low concentrations compared to an oxidizable substrate and can significantly delay or prevent the oxidation of that substrate.

Glutathione peroxidase (GPx): A family of enzymes playing important roles in the protection of organisms from oxidative damage; it reduces the peroxide group to a relatively unreactive alcohol group, using glutathione as a reducing agent.

Superoxide dismutase (SOD): A family of antioxidative enzymes that catalyze the dismutation of superoxide ($O_2^{\bullet -}$) into oxygen and hydrogen peroxide.

Catalase (CAT): An enzyme that protects the cell from oxidative damage by reactive oxygen species; catalyzes the decomposition of hydrogen peroxide to water and oxygen.

Glutathione (GSH): An essential endogenous antioxidant; is a linear tripeptide made of L-glutamine, L-cysteine, and glycine, whose thiol group is a reducing agent.

Abbreviations

CSF	cerebrospinal fluid
ROS	reactive oxygen species
RNS	reactive nitrogen species
GSH	reduced glutathione
GPx	glutathione peroxidase
CAT	catalase
SOD	superoxide dismutase

CNS	central nervous system
IQ	intelligence quotient
PKC	protein kinase C
NA	noradrenaline
5-HT	serotonin
NF-κB	nuclear factor kappa-B
NCAM	neural cell adhesion molecules
PSD-95	postsynaptic density-95
AST	aspartate aminotransferase
ALT	alanine aminotransferase
GAD	glutamic acid decarboxylase
MDA	malondialdehyde
4-HNE	4-hydroxynonenal
8-iso-PGF2α	8-iso prostaglandin F2α
8OH-Dg	8OH-deoxyguanosine
AOPP	advanced oxidation protein products.

References

Basha, P. M., Rai, P. and Begum, S., 2011. Evaluation of fluoride-induced oxidative stress in rat brain: a multigeneration study. *Biological Trace Element Research.* **142**: 623–637.

Bera, I., Sabatini, R., Auteri, P., Flace, P., Sisto, G., Montagnani, M., Potenza, M. A., Marasciulo, F. L., Carratu, M. R., Coluccia, A., Borracci, P., Tarullo, A. and Cagiano, R., 2007. Neurofunctional effects of developmental sodium fluoride exposure in rats. *European Review for Medical and Pharmacological Sciences.* **11**: 211–224.

Bhatnagar, M., Rao, P., Sushma, J. and Bhatnagar, R., 2002. Neurotoxicity of fluoride: neurodegeneration in hippocampus of female mice. *Indian Journal of Experimental Biology.* **40**: 546–554.

Chen, J., Chen, X., Yang, K., Xia, T. and Xie, H., 2002. Studies on DNA damage and apoptosis in rat brain induced by fluoride. *Chinese Journal of Preventive Medicine.* **36**: 222–224.

Chirumari, K. and Reddy, P. K., 2007. Dose-dependent effects of fluoride on neurochemical milieu in the hippocampus and neocortex of rat brain. *Fluoride.* **40**: 101–110.

Choi, A. L., Sun, G., Zhang, Y. and Grandjean, P., 2012. Developmental fluoride neurotoxicity: a systematic review and meta-analysis. *Environmental Health Perspectives.* **120**: 1362–1368.

Ding, Y., Gao, Y., Sun, H., Han, H., Wang, W., Ji, X., Liu, X., Sun, D., 2011. The relationships between low levels of urine fluoride on children's intelligence, dental fluorosis in endemic fluorosis areas in Hulunbuir, Inner Mongolia, China. *Journal of Hazardous Materials.* **186**: 1942–1946.

Du, L., Wan, C., Cao, X. and Liu, J., 2008. The effect of fluorine on the developing human brain. *Fluoride.* **41**: 327–330.

Guan, Z., Yu, Y. and Liu, J., 1986. Changes in brain morphology in the offspring of chronically fluoride poisoned rats. *Chinese Journal of Pathology.* **15**: 297–299.

Han, H., Cheng, Z. and Liu, W., 1989. Effects of fluorine on the human fetus. *Chinese Journal of Control of Endemic Diseases.* **4**: 136–138.

Inkielewicz, I. and Krechniak, J., 2003. Fluoride content in soft tissues and urine of rats exposed to sodium fluoride in drinking water. *Fluoride.* **36**: 263–236.

Inkielewicz-Stepniak, I. and Czarnowski, W., 2010. Oxidative stress parameters in rats exposed to fluoride and caffeine. *Food and Chemical Toxicology.* **48**: 1607–1611.

Inkielewicz-Stepniak, I., Radomski, M. W. and Wozniak, M., 2012. Fisetin prevents fluoride- and dexamethasone-induced oxidative damage in osteoblast and hippocampal cells. *Food and Chemical Toxicology.* **50**: 583–589.

Khalid, R., 2007. Studies on free radicals, antioxidants, and co- factors. *Clinical Interventions in Aging.* **2**: 219–236.

Luo, G. Y., Niu, R. Y., Sun, Z. L., Zhang, J. H., Wang, J. M., Wang, C., *et al.*, 2011. Reduction of CaMKII expression in the hippocampus of rats from ingestion of fluoride and/or lead. *Fluoride.* **44**: 63–69.

Madhusudhan, N., Basha, P. M., Rai, P., Ahmed, F. and Prasad, G. R., 2010. Effect of maternal fluoride exposure on developing CNS of rats: protective role of Aloe vera, Curcuma longa and Ocimum sanctum. *Indian Journal of Experimental Biology.* **48**: 830–836.

Mullenix, P. J., Denbesten, P. K., Schunior, A. and Kernan, W. J., 1995. Neurotoxicity of sodium fluoride in rats. *Neurotoxicology and Teratology.* **17**: 169–177.

Murphy, M. P., Holmgren, A., Larsson, N. G., Halliwell, B., Chang, C. J., Kalyanaraman, B., Rhee, S. G., Thornalley, P. J., Partridge, L., Gems, D., Nyström, T., Belousov, V., Schumacker, P. T. and Winterbourn, C. C., 2011. Unraveling the biological roles of reactive oxygen species. *Cell Metabolism.* **13**: 361–366.

Nasir, N. and Asad, M. R., 2013. Effects of flouride on CA3 region of hippocampus in adult albino rats. *Journal of Asian Scientific Research.* **3**: 729–733.

Niu, R., Sun, Z., Cheng, Z., Li, Z. and Wang, J., 2009. Decreased learning ability and low hippocampus glutamate in offspring rats exposed to fluoride and lead. *Environmental Toxicology and Pharmacology.* **28**: 254–258.

Pereira, M., Dombrowski, P. A., Losso, E. M., Chioca, L. R., Da Cunha, C. and Andreatini, R., 2011. Memory impairment induced by sodium fluoride is associated with changes in brain monoamine levels. *Neurotoxicity Research.* **19**: 55–62.

Roholm K., 1937. *Fluorine intoxication. A clinical-hygienic study with a review of the literature and some experimental investigations.* H. K. Lewis & Co. Ltd, London, England.

Shivarajashankara, Y. M., Shivashankara, A. R., Gopalakrishna Bhat, P., Muddanna Rao, S., Hanumanth Rao, S., 2002. Histological changes in the brain of young fluoride intoxicated rats. *Fluoride.* **35**: 12–21.

Sies, H. *Oxidative stress: introduction.* In: Sies, H. (ed.) *Oxidative stress, oxidants and antioxidants.* Academic Press, London, England, pp. 15–22.

Varner, J. A., Jensen, K. F., Horvath, W. and Isaacson, R. L., 1998. Chronic administration of aluminum-fluoride or sodium-fluoride to rats in drinking water: alterations in neuronal and cerebrovascular integrity. *Brain Research.* **784**: 284–298.

World Health Organization (WHO), 2003. Diet, nutrition and the prevention of chronic diseases. *WHO Technical Report Series 196, Report of a Joint WHO/FAO Expert Consultation,* WHO, Geneva.

Yu, Y., Yang, W., Dong, Z., Wan, C., Zhang, J., Liu, J., Xiao, K., Huang, Y. and Lu, B., 1996. Neurotransmitter and receptor changes in the brains of fetuses from areas of endemic fluorosis. *Chinese Journal of Endemiology.* **15**: 257–259.

Yu, Y., Yang, W., Dong, Z., Wan, C., Zhang, J., Liu, J., Xiao, K., Huang, Y. and Lu, B., 2008. Neurotransmitter and receptor changes in the brains of fetuses from areas of endemic fluorosis. *Fluoride.* **41**: 134–138.

Zhai, J. X., Guo, Z. Y., Hu, C. L., Wang, Q. N. and Zhu, Q. X., 2003. Studies on fluoride concentration and cholinesterase activity in rat hippocampus. *Chinese Journal of Industrial Hygiene and Occupational Diseases.* **21**: 102–104.

Zhang, J. and Zhang, Z., 2013. Effects of chronic fluorosis on camkiiα, c-fos, bax, and bcl-2 channel signalling in the hippocampus of rats. *Fluoride.* **46**: 135–141.

Zhang, M., Wang, A., Xia, T. and He, P., 2008. Effects of fluoride on DNA damage, S-phase cell-cycle arrest and the expression of NF-kappaB in primary cultured rat hippocampal neurons. *Toxicology Letters.* **179**: 1–5.

Zhang, M., Wang, A., He, W., He, P., Xu, B., Xia, T., Chen, X. and Yang, K., 2007. Effects of fluoride on the expression of NCAM, oxidative stress, and apoptosis in primary cultured hippocampal neurons. *Toxicology.* **236**: 208–216.

Zhu, W., Zhang, J. and Zhang, Z., 2011. Effects of fluoride on synaptic membrane fluidity and PSD-95 expression level in rat hippocampus. *Biological Trace Element Research.* **139**: 197–203.

CHAPTER 16

Fluoride-Induced Oxidative Stress in the Liver

SEYED FAZEL NABAVI[a], MARIA DAGLIA[b], ANTONI SUREDA[c], AND SEYED MOHAMMAD NABAVI*[a]

[a]Applied Biotechnology Research Center, Baqiyatallah University of Medical Sciences, Tehran, Iran; [b]Department of Drug Sciences, Medicinal Chemistry and Pharmaceutical Technology Section, University of Pavia, via Taramelli 12, 27100 Pavia, Italy; [c]Department of Fundamental Biology and Health Sciences, CIBERobn, Instituto de Salud Carlos III, University of the Balearic Islands, Ctra. Valldemossa, km 7.5, CP 07122, Palma de Mallorca, Balearic Islands, Spain

*E-mail: Nabavi208@gmail.com

16.1 Introduction

Elemental fluorine (known by the symbol F, with atomic number 9) was first isolated by Henri Moissan in 1886 through electrolysis of potassium hydrogen difluoride solution (KHF_2) in liquid hydrogen fluoride (HF) (Roesky, 2010). The name of this element was derived from the Latin name of the original mineral. Fluorine is the most electronegative element, the lightest member of the halogen series in the periodic table and it has a stable isotope, the fluorine-19. It is a diatomic molecule (F_2) and a pale yellow gas in standard conditions (Roesky, 2010).

The presence of fluorine in some natural products, including alkaloids, nucleosides, carbohydrates or lignans, has been reported to have several therapeutic effects. These compounds behave as xenobiotics in biological systems (Bégué and Bonnet-Delpon, 2006). About a dozen of fluorine containing natural

Food and Nutritional Components in Focus No. 6
Fluorine: Chemistry, Analysis, Function and Effects
Edited by Victor R Preedy

Published by the Royal Society of Chemistry, www.rsc.org

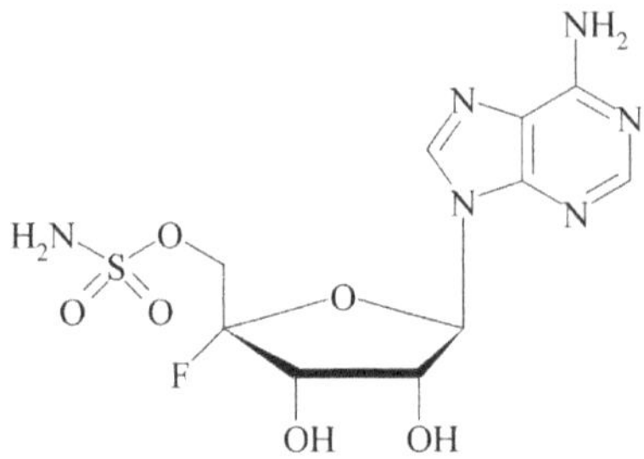

Figure 16.1 Chemical structures of nucleocidin.

products have been isolated, such as nucleocidin (Figure 16.1) which was too toxic for therapeutic uses (Bégué and Bonnet-Delpon, 2006).

Fluorine is a naturally occurring element on Earth ($1.5 \times 10^{-4}\%$); it is highly distributed and found in trace amounts in air, water, lithosphere and every living beings. It is well known that a correct exposure to fluorides (F^-), the ionic form of F, has protective effects on human health preventing dental caries and strengthening bones, while overexposure to environmental F^- causes toxic effects, including the dental and skeletal fluorosis that are the most frequent pathologies related to high fluoride intake.

The occurrence of F in the atmosphere, and therefore in water, can be ascribed to different causes depending on the geographical area. The volcanic emissions, which are one of the main causes of the presence of fluoride in air and water, contain significant amounts of HF, which can be evaluated in tens of thousands of tons/year that are taken back to Earth by the meteoric phenomena and get into the groundwater and then in the drinking water. It was estimated that drinking water collected from rainwater fallen in Ambrym Island, where there is one of the world's most active volcanoes, contained fluoride levels from 0.7 to 9.5 ppm, exceeding the recommended concentration of 1.0 ppm (Allibone *et al.*, 2012). It can be inferred that people living in the vicinity of active volcanoes (about 7% of the global population), especially where rainwater comprises the primary potable water supply for humans or animals, are subject to the risk of developing fluorosis due to a high exposure to F^-. Another cause of high F^- concentration in air and water is pollution. In developed countries the high F content of air can be ascribed to industrial production of phosphate fertilizers and coal ash deriving from coal burning. For instance, in China, in the southwestern province of Guizhou, more than 10 million people suffer from various forms of fluorosis and indoor burning of fluoride-rich coal is believed to be the prevailing cause of endemic fluorosis among the children of this Chinese Province (Qin *et al.*, 2009).

F is present in rocks, soil, coal, and clay is due in part to the minerals, such as fluorite (CaF_2), cryolite (Na_3AlF_6) and fluorapatite ($Ca_5(PO_4)_3F$) (Figure 16.2) and in part to the deposition of F by pesticides, irrigation with contaminated rainwater, and the addition of phosphate fertilizers.

Figure 16.2 Fluorite, cryolite and fluorapatite, mineral sources of fluorine.

F located in soils may be accumulated in plants. The plant uptake depends upon the type of plant and soil and the fluoride content found in the soil. It has been demonstrated that plant fluoride is closely related to the available fluoride in soil. A study with onion (*Allium cepa* L.) show that plants grown on soils contaminated by inorganic fluoride (NaF) to different extent (from 0 to 800 mg NaF kg^{-1} of soil) accumulate F in roots, shoots, and bulbs and showed visible symptoms of phytotoxicity such as tip burning and plant death (at concentrations higher than 400 mg NaF kg^{-1} of soil) (Jha *et al.*, 2009). The tea plant (*Camellia sinensis* L.) is another plant capable of accumulating large amounts of F, especially in leaves, also when grown on soils containing normal F concentrations (less than 10 ppm kg^{-1}) without showing any symptom of phytotoxicity (Ruan *et al.*, 2004). This is why drinking tea is one the three main causes responsible for endemic fluorosis in most parts of China (Zheng and Huang, 1989), together with contaminated drinking water and pollution from coal burning. Recently, many researches have been attempting to find new methods to reduce F accumulation by tea plants (Gao *et al.*, 2012). It should be noted that the contamination of leafy vegetables can also occur through contamination of air in heavily industrialized areas or in the vicinity of active volcanoes.

Animals, including humans, may accumulate large amounts of F in their bodies by both ingestion and inhalation. As reported above, F sources can be drinking water, tea beverages, vegetables, marine fish, certain tainted wines (Soriano and Manchon, 1966), fluoridated toothpaste (Roos *et al.*, 2005), and other unusual exposures (Krishnamachari, 1987; Whitford, 1992).

It is estimated that the content of F in the human body is about 2.6 g (Belitz *et al.*, 2009). The absorption takes place in the gastrointestinal tract. F introduced with foods has an absorption ranging from 30 to 60%, which depends on the inorganic form (NaF is well absorbed, while the fluoroapatite, being poorly soluble, is less absorbed). On the contrary, F^- introduced with drinking water has a rapid and almost complete absorption. Absorbed F is quickly distributed throughout the body and enters tissues affecting their metabolism. The excretion occurs mainly *via* kidneys. For this reason, urine is used to assess the exposure to F. Nevertheless, this biomarker is unreliable due to: 1) fluctuations in urinary flow and pH which influences fluoride output and 2) slow release of fluoride accumulated in the bone that may influence the urinary fluoride output. Therefore, at present, measurement of plasma fluoride is the only exposure biomarker that gives reliable information on recent fluoride intake (Scientific Committee on Health and Environmental Risks, 2011).

F is deposited mainly in bones and teeth, but also in soft tissues: in human bones the content can vary between 0.3 and 7 mg g^{-1} dry bone tissue, in relation to the total intake, and tends to increase with age, provided that the contribution is constant. By contrast, F levels in soft tissues remain stable or increase only slightly. Thus, long-term consumption of excess of F^- can cause adverse effects not only on skeletal tissue, but also on nonskeletal tissues such as brain, liver, pancreas, kidney, muscles, and testes (Shivarajashankara and Shivashankara, 2012). Moreover, it was shown that humans suffering from skeletal fluorosis suffer from an increased occurrence of gastrointestinal disorders that often disappear when fluoride intake is reduced (Dasarathy *et al.*, 1996). The most common symptoms are abdominal pain, nausea, loss of appetite, constipation, diarrhea, and gastrointestinal abnormalities (chronic atrophic gastritis, loss of microvilli, cracked-clay appearance, and the presence of surface abrasions of mucosal cells).

The liver is an indispensable organ with a wide range of functions that are vital to sustain the body's homeostasis. These functions include drug and chemical detoxification, synthesis of protein and hormones, storing of glucose as glycogen, lipid metabolism, *etc.* Given these important roles for human health, any interference with liver functions, could severely threaten life. Fluoride in ionic form is easily absorbed through intestinal mucosa and distributed into bodies *via* bloodstream. The liver plays an important role in fluoride detoxification and therefore this anion is injurious to liver normal function. This chapter will focus on fluoride-induced hepatotoxicity and role of oxidative stress.

16.2 Fluorine Applications

The use of fluorinated compounds in drug development has increased from 2% in 1970 to more than 18% in 2012. There are more than 150 fluorinated drugs and among them, atorvastatin (a hypolipidemic drug), fluoxetine (an antidepressant drug) and ciprofloxacin (an antibiotic) have the largest usage (Figure 16.3) (Müller *et al.*, 2007). In medicinal chemistry, it is well known that fluorine can improve binding efficacy and selectivity of drugs. The effect of carbon–fluorine single bonds on docking interactions of drugs has recently gained an increasing attention (Müller *et al.*, 2007). Nonetheless, further research about the therapeutic applications of organofluorine compounds is needed.

In different countries over the world, a variety of fluoridated dental products such as fluorinated salt, mouth rinses, toothpastes, *etc.*, has been used in large amounts to prevent tooth decay (Azarpazhooh and Main, 2008). In recent years, an increasing amount of literature demonstrated the beneficial effects of low concentrations of fluoride in saliva against dental plaque

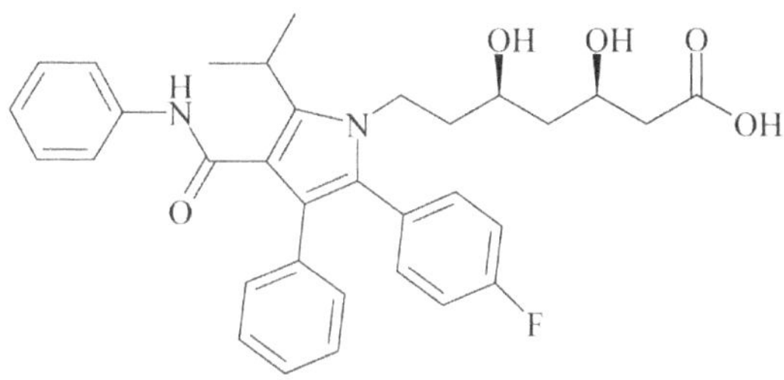

Atrovastatin

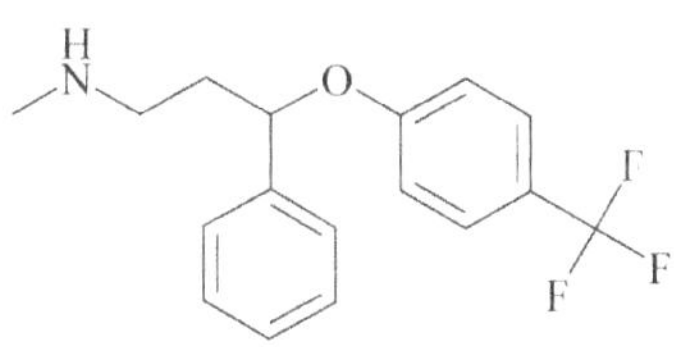

Fluoxetine

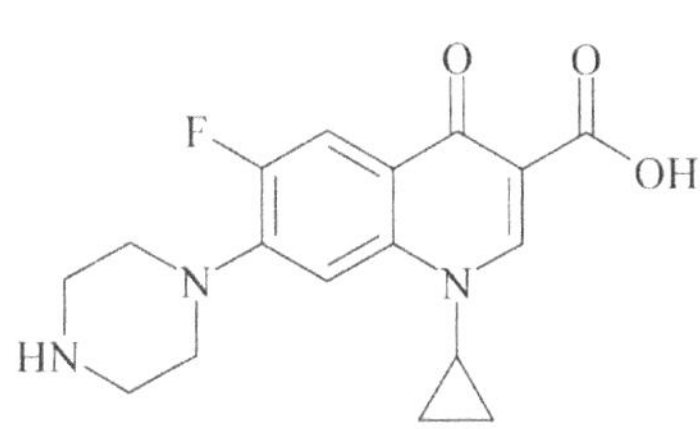

Ciprofloxacin

Figure 16.3 Chemical structures of atorvastatin, fluoxetine and ciprofloxacin.

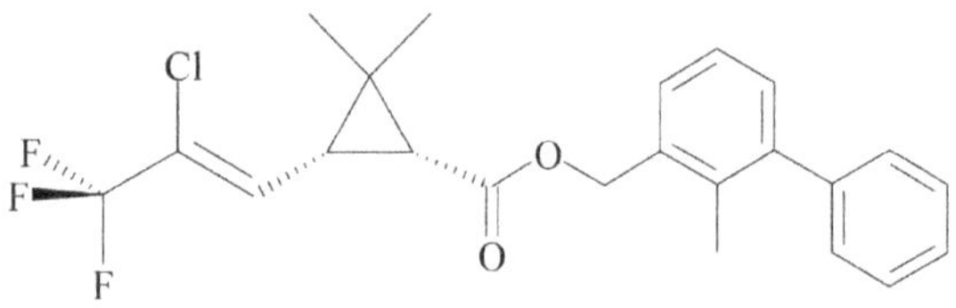

Figure 16.4 Chemical structure of bifenthrin.

(Azarpazhooh and Main, 2008). Moreover, the use of fluorinated water and mouth rinses are the most important and effective strategies against dental caries in the third world.

In the agrochemical industry, fluorine compounds are used for pesticide production. They have used for preparation of wood preservatives and insecticides such as Bifenthrin (Figure 16.4) (Bouaziz *et al.*, 2006). Fluorinated compounds can reach the soil and rocks through different anthropogenic activities, either directly *via* phosphate fertilizers and/or indirectly *via* industrial pollution, fossil fuels burning and environmental organic chemicals (Bouaziz *et al.*, 2006).

16.3 Fluoride Accumulation

Fluorine is the most electronegative element among the Periodic Table of the Elements (Roesky, 2010). It occurs naturally as a negatively charged anion. In acidic conditions fluoride ion changes to HF form (Whitford *et al.*, 1994). Membrane permeability of HF is similar to that of water and about seven times higher than that of fluoride (Gutknecht and Walter, 1981). Recent studies also demonstrate that about 40% and 45% of ingested fluoride is absorbed through stomach and intestine, respectively (Whitford and Pashley, 1984). Intestinal absorption of HF is insensitive to pH and happens *via* facilitated diffusion (He *et al.*, 1998). Another study, which was performed with G292 osteoblastic cells model, suggests that fluoride ions stimulates $K^{(+)}$-selective ion channels activity (Gofa and Davidson, 1996). Fluoride-induced $K^{(+)}$ and $Ca^{(2+)}$ channels activation may influence a sum of early events in affected cells (Gofa and Davidson, 1996). Besides HF diffusion mechanism; there are numerous carrier mechanisms for fluoride transport, *i.e.* fluoride can go through the cellular membrane by a F^-–H^+ cotransporter or F^-–OH^- exchangers; however, additional studies are needed to get a better understanding of these mechanisms (Barbier *et al.*, 2010). There are several cations such as calcium, magnesium, aluminum, *etc.* whose complexes with fluoride are insoluble and depend on the amount of fluoride, that may lead to hypocalcemia, trace-element deficiency, inhibition of Mn-dependent enzymes, *etc.* as well as decrease of fluoride absorption (ŞİRELİ and Bülbül, 2009).

Absorbed fluoride quickly spreads *via* bloodstream across the body and part of it is stored in bone and teeth (calcium rich area of body) (Fagin, 2008).

In adults the level of absorbed fluoride retention in the calcium-rich area is about 60%, however, this level in children is between 80 and 90%. Urine is the main way of fluoride excretion (Buzalaf and Whitford, 2011).

Fluoride–inorganic complexes can interact with structural proteins, for example actin and troponin C (Barbier *et al.*, 2010).

16.4 Fluoride Toxicity

The accidental intoxication with fluorinated compounds (such as sodium fluoride, sodium fluorosilicate, fluosilicic acid, or hydrogen fluoride) may injury bones, cardiac muscle, liver, skin, kidney, brain, and erythrocytes and may cause thyroid dysfunction (Figure 16.5) (Nabavi *et al.*, 2012).

Chronic long-term exposure to high levels of fluoride mainly through drinking water leads to fluorosis, a serious health problem (Nabavi *et al.*, 2012). Although excessive consumption of fluoride can alter the normal function of different organs, its main effects are generally manifested through two disorders: skeletal and dental fluorosis. Skeletal fluorosis occurs following long-term excessive fluoride exposures and the characteristic symptoms are axial skeletal pain and occasionally spinal rigidity with kyphosis (Whyte *et al.*, 2005). Radiographic analyses have demonstrated axial osteosclerosis, exostoses, periostalgia, and pelvis ligament calcification in skeletal fluorosis. A histopathological study also reports osteomalacia in patients with skeletal fluorosis (Whyte *et al.*, 2005). Up to now, there is not a clear treatment protocol for skeletal fluorosis (Whyte *et al.* 2005). Dental fluorosis is a fluoride-induced disorder that disturbs enamel development through hypomineralization. In dental fluorosis, fluoride changes Ca^{2+} activities and, therefore, induces a change in Ca-dependent proteases activity in the fluid surrounding protein-crystal conglomerates, resulting in an inhibition of the protein degradation and elimination from enamel (Aoba and Fejerskov, 2002). consequently; it can influence the enamel

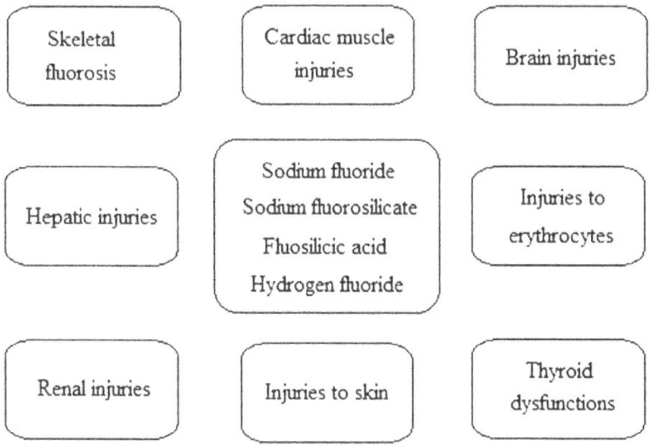

Figure 16.5 Pathological effects of fluoride toxicity.

formation by increasing the porosity and the intercrystalline spaces width. The most crucial period for dental fluorosis is comprised from birth to 8 years of age (Mascarenhas 2000).

Fluorosis is a widespread disease in different areas of the world mainly affecting developing countries such as India and Pakistan. In India, there are more than 60 million people in different areas with dental and skeletal fluorosis (Nabavi *et al.*, 2012).

Numerous scientific reports have evidenced a fluoride-induced toxicity to living systems, because of the progressive increase in the amount of fluoride pollutants released into water resources (Podder *et al.*, 2011). Fluorides accumulated in the body lead to numerous metabolic disorders (Nabavi *et al.*, 2012). Exposure to fluorides affects several enzymatic reactions and may increase reactive oxygen species (ROS) production that can result in oxidative stress and cytotoxicity (Nabavi *et al.*, 2013). Fluoride intoxication leads to organ dysfunction through a disturbance in the oxidative balance, increases the generation of nitric oxide, lipid peroxides, resulting in cellular injury that affects cellular structures and their function (Nabavi *et al.*, 2012).

16.4.1 Molecular Mechanism of Fluoride-Induced Toxicity

A plethora of information shows that fluoride intoxication affects normal organelle functions of the cells, especially mitochondria and endoplasmic reticulum, through different molecular pathways such as p53, mitogen-activated protein kinase, nuclear factor kappa beta, activator protein-1, *etc.* (Figure 16.6) (Wang *et al.*, 2005; Strunecka *et al.*, 2007; Zhang *et al.*, 2008; Barbier *et al.*, 2010).

Due to the crucial role of mitochondria-rich tissues such as those of kidney and liver in body homeostasis, fluoride intoxication causes more damage to these tissues (Karaoz *et al.*, 2004). Fluoride can inhibit the enzymes involved in the Krebs cycle (such as Na^+/K^+-ATPases) (Iukhnovets and Bachinskiĭ, 1981) and phosphatases, (Wergedal *et al.*, 1988) and increases cytochrome C release (Lee *et al.*, 2008), inducing the caspase cascade that leads to toxic energy depletion and damage to vita cellular process (Anuradha *et al.*, 2001). Na^+/K^+-ATPases inhibition also stimulates the synthesis of nitric oxide that leads to suppression of mitochondrial respiration (Liu *et al.*, 2003). It is believed that cytochrome C release is the main mechanism of fluoride-induced opening of the permeability transition pore (Anuradha *et al.*, 2001). Previous studies demonstrate a close correlation between loss of mitochondrial membrane potential and increasing in superoxide generation (García-Montalvo *et al.*, 2009).

He *et al.*, (2012) reported that chronic fluoride exposure causes alteration in the expression level of functional and structural genes and expressed sequence tags in rat liver mitochondria, including ATP synthase H^+ transporting mitochondrial F_0 complex subunit C, mitochondrial succinyl-CoA synthetase alpha subunit, mitochondrial IF1 protein, selenoprotein S, microsomal glutathione S-transferase.

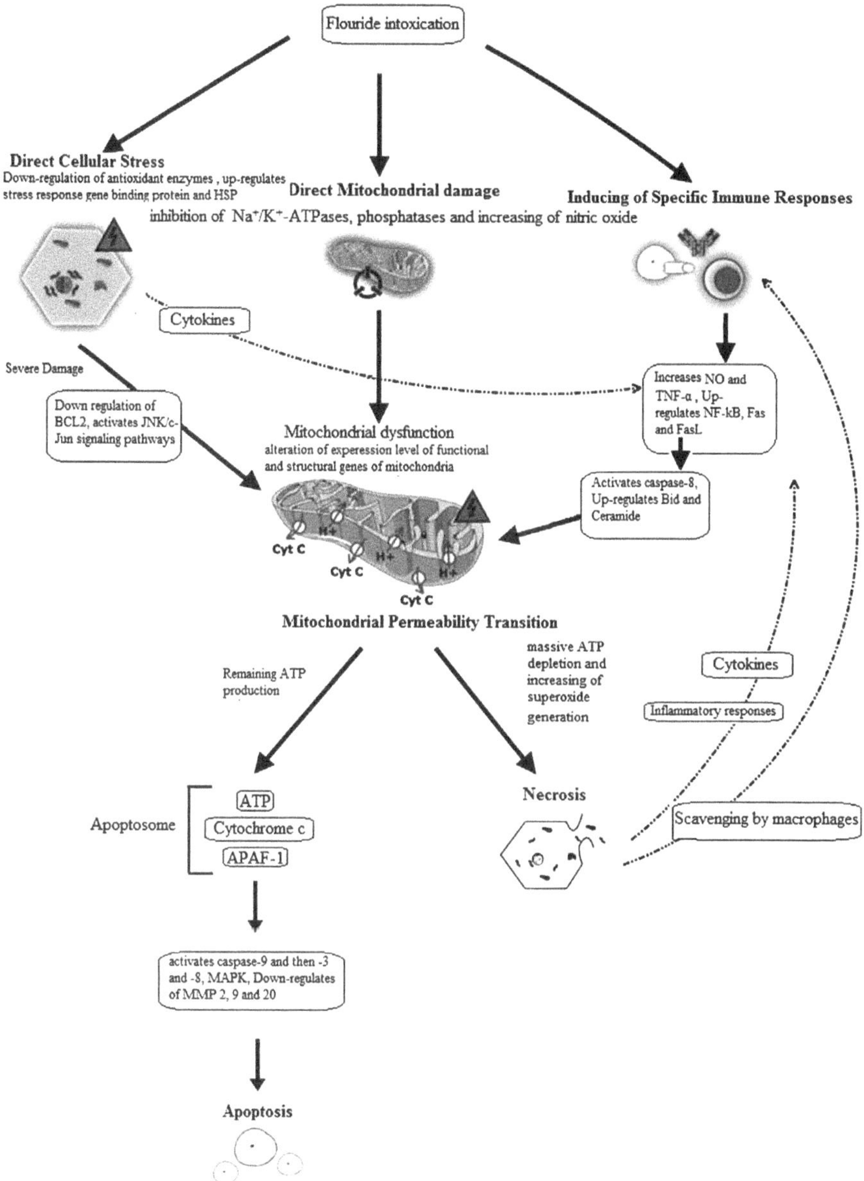

Figure 16.6 Molecular mechanism of fluoride toxicity. FasL: Fas ligand; NF-kB: Nuclear factor kappa beta; Bid: BH3 interacting-domain death agonist; No: nitric oxide; JNK: c-Jun N-terminal kinase; BCl2: B-cell lymphoma 2; MMP: Matrix metalloproteinase; MAPK: Mitogen-activated protein kinase; APAF-1: Apoptotic protease activating factor 1; ATP: Adenosine triphosphate.

Available evidence supports the hypothesis that fluoride-induced oxidative stress has detrimental effects on different soft tissues such as brain, lung, kidney, liver, testes and thyroid in animal and in humans who suffer from fluoride intoxication (Nabavi *et al.*, 2013). Fluoride accumulation in tissue also elevates the formation of ROS and reactive nitrogen species (RNS) including superoxide onion, hydrogen peroxide, nitric oxide, *etc.* Hydrogen peroxide accumulation in tissues generates hydroxyl radical through Fenton reaction (Nabavi *et al.*, 2012). Furthermore, fluoride-induced nitric oxide can react with superoxide and both thiols and metal-center proteins in order to form peroxynitrite and nitrosyl adducts, respectively (Barbier *et al.*, 2010). Fluoride exposure also may cause protein misfolding by interfering with disulfide-bond formation, which leads to endoplasmic reticulum stress (Barbier *et al.*, 2010).

It is well known that fluoride-induced oxidative stress suppresses the activity of the antioxidant enzymes such as catalase, superoxide dismutase, glutathione peroxidase, glutathione reductase, glucose 6-phosphate dehydrogenase, glutathione transferase, *etc.* (Flora *et al.*, 2012, Nabavi *et al.*, 2012). Moreover, it affects the level of intracellular nonenzymatic antioxidant such as reduced glutathione thus increasing protein carbonyl content, conjugated dienes and DNA damages (Thangapandiyan and Miltonprabu, 2013). Membrane phospholipids are other targets of fluoride-induced oxidative stress leading to lipid peroxidation and mitochondrial membrane depolarization (Nabavi *et al.*, 2012). Previous reports show that fluoride induces apoptosis through activation of mitogen-activated protein kinases (Lu *et al.*, 2010) and downregulation of matrix metalloproteinase 2, 9 and 20 *via* mediation of JNK/c-Jun signaling pathways (Zhang *et al.*, 2007). Ras/MEKK/MEK-mediated pathways play important role in activation of stress-activated protein kinases such as plevin (Paul *et al.*, 1997). Moreover, protein kinase C dependent pathways and change in tyrosine phosphorylation are involved in fluoride-induced mitogen-activated protein kinase activation. Fluoride exposure can also change gene expression of target cells (Refsnes *et al.*, 2003). According to previous findings, fluoride exposure upregulates the expression of genes encoding a signal transduction component, stress-induced factors and proapoptotic proteins (Barbier *et al.*, 2010).

It has been reported that fluoride exposure reduces hepatic zinc/copper superoxide dismutase mRNA (Zhan *et al.*, 2006a), upregulates stress response genes encoding binding protein and heat shock protein (HSP)-70 (Chen *et al.*, 2009). The binding protein acts as a molecular chaperone and suppressor of protein aggregation in the endoplasmic reticulum (Xu *et al.*, 2007). HSP70, as part of chaperone machinery, is activated as a cellular response to fluoride (Chen *et al.*, 2009). The results obtained from investigations on Humans who live in fluoride-contaminated areas and on fluoride exposed HepG2 cells demonstrated that HSP70 level increases in response to fluoride, as a mediator tolerance (Barbier *et al.*, 2010).

The sperm cation channels (CatSper) is another oxidative stress-related gene target of fluoride exposure. It has been reported that fluoride exposure downregulation of CatSper and antioxidant treatments lead to upregulation of CatSper (Mohammadi *et al.*, 2009). This family is essential for sperm mortality and male fertility.

It has been reported that during fluoride-induced apoptosis, intracellular calcium content plays an important role either as a direct target of fluoride toxicity or as a result of other cellular processes (Barbier *et al.*, 2010). Fluoride exposure can also alter expression of proapoptotic and antiapoptotic genes. Fluoride exposure at 20 Mm to human gingival fibroblasts (Lee *et al.*, 2008) and 2–4 Mm to human neuroblast cells induces (Xu *et al.*, 2011) apoptosis through down-regulation of Bcl-2, activates caspase -3, -8 and -9, stimulates the release of cytochrome C and the cleavage of poly (ADP-ribose) polymerase (PARP). Xu *et al.*, (2011) also found that fluoride exposure causes the upregulation of Fas and Fas ligand and NF-Kb expression. Salgado-Bustamante *et al.*, (2010) report that chronic fluoride exposure at 2.16–7.30 mg of F g^{-1} creatinine leads to significant downregulation of caspase-6, CD40L, HVEM, LT-b, TRAF-2, TRAF-5 genes in peripheral blood mononuclear cells in comparison with healthy group. They also found that fluoride exposure downregulate TNF/TNFR superfamily and CIDE family genes as well as several pro- and anti-inflammatory genes. In contrast, Ridley and Matsuoka, (2009) reported that fluoride exposure to A549 human pulmonary epithelial cells up-regulate cyclooxygenase-2 expression through p38 and ERK pathways.

Hirano and Ando, (1997) showed that fluoride induced cytotoxicity depending on pH. They found that acidic conditions increase cytotoxicity while alkaline ones decrease fluoride cytotoxicity.

16.4.2 Fluoride-Induced Hepatotoxicity

Fluoride hepatotoxicity is one of the most important complications of fluoride intoxication that alters the oxidant–antioxidant balance systems in hepatic tissue thus inducing liver oxidative injuries (Dubey *et al.*, 2013). Fluoride exposure causes several abnormalities concerning hepatic antioxidant factors (superoxide dismutase, catalase, malondialdehyde, reduced glutathione), DNA, RNA and protein contents, mitochondrial dysfunction, caspase-9 and caspase-3, hydroxyproline and collagen as well as changes of serum markers such as alkaline phosphatase, aspartate aminotransferase, alanine transaminase, triacylglycerol, high-density lipoprotein, low-density lipoprotein, total bilirubin, direct bilirubin, total protein and albumin (Zhan *et al.*, 2006b; Inkielewicz-Stêpniak and Knap, 2012; Jha *et al.*, 2012; Siddiqi, 2012). We also demonstrated that in experimental animals one week pretreatment with quercetin and gallic acid reduces sodium fluoride-induced oxidative stress and liver dysfunction (Nabavi *et al.*, 2013). We showed that intraperitoneally administration of quercetin and gallic acid modified oxidative stress (superoxide dismutase, catalase, malondialdehyde, reduced glutathione) and serum levels of alkaline phosphatase, aspartate aminotransferase, alanine transaminase, triacylglycerol, high-density lipoprotein, low-density lipoprotein, total bilirubin, direct bilirubin, total protein and albumin (Nabavi *et al.*, 2012, 2013). We concluded that the beneficial role of quercetin and gallic acid in fluoride-induced hepatoxicity is the result of their antioxidant effects (Nabavi *et al.*, 2012, 2013).

It is well known that ROS play crucial roles in development of hepatic dysfunction and liver injuries. Chinoy *et al.*, (2004) showed that one-month administration of fluoride at 10 mg kg^{-1} significantly increased the levels of lipid peroxide and dehydroascorbic acid. They also reported that fluoride intoxication affects the level of glutathione and the activity of superoxide dismutase, catalase, as well as glutathione peroxidase. Chinoy *et al.*, (2005) reported that a protein-rich diet normalizes fluoride-induced hepatotoxicity and oxidative stress *via* restoration of antioxidant enzymes activities and amount of nonenzymatic antioxidants. Wang *et al.*, (2004) reported that 12 h incubation of normal human primary hepatocytes with sodium fluoride at 80 μg ml^{-1} causes hepatocyte damage increasing the amount of apoptotic cells, DNA damage, and lipid peroxide as well as the number of cells in S phase. They also reported that fluoride toxicity is associated with an increased activity of aspartate transaminase and lactate dehydrogenase, and a decreased level of glutathione in cultured human hepatocytes. Thangapandiyan and Miltonprabu, (2013) examined the hepatoprotective role of epigallocatechingallate (EGCG) on DNA damage in sodium fluoride-induced oxidative stress in rats. They showed that oral administration of sodium fluoride at 25 mg kg^{-1} body weight for 4 weeks increases the serum levels of hepatic markers, total bilirubin, and also the levels of lipid peroxidation products, *i.e.* thiobarbituric acid reactive substances and lipid hydroperoxides. Also, they showed that sodium fluoride intoxication increases the levels of protein carbonyl content and conjugated dienes. According to the study of Thangapandiyan and Miltonprabu, (2013), sodium fluoride causes aberration of the activity of antioxidant enzymes and of nonenzymatic antioxidants level in liver and also increased DNA damage and fragmentation in hepatocytes of experimental animals. Moreover, they demonstrated that the potent antioxidant and antigenotoxic effect of EGCG significantly mitigates sodium fluoride-induced hepatotoxicity and oxidative stress in rats. Panneerselvam *et al.*, (2013) investigated the beneficial role of ferulic acid against fluoride-induced hepatotoxicity and oxidative stress in rats. They indicated that a 12 week oral administration of fluoride at 25 mg l^{-1} alters different serum marker enzymes such as aspartate transaminase, alanine transaminase, alkaline phosphatase, acid phosphatase, gamma glutamyltransferase, lactate dehydrogenase, bilirubin, lipid hydroperoxides, nitric oxide, antioxidant enzymes, total protein content levels. They also reported that fluoride induces histopathological changes in liver tissues. Panneerselvam *et al.*, (2013) observed that ferulic acid significantly mitigated sodium fluoride-induced hepatotoxicity *via* restoration of oxidative stress and hepatic serum markers as well as reducing the extent of histopathological abnormalities.

Manna *et al.*, (2007) reported that treatment with 43Kd protein (isolated from the herb *Cajanus indicus* L.) normalized the fluoride-induced oxidative stress in mice hepatic and renal tissues. They showed that 43Kd protein reduces fluoride-induced alterations of the activity of antioxidant enzymes (superoxide dismutase, catalase, glutathione-S-transferase, glutathione reductase) and of nonenzymatic antioxidants (reduced glutathione, oxidized glutathione, total thiols, lipid peroxidation end products) in liver and kidney (Manna *et al.*, 2007).

Histopathological examination of fluoride-induced hepatotoxicity showed that fluoride intoxication is associated with the increase of hepatocellular and centrilobular necrosis, degenerative changes, extensive vacuolization in rabbit hepatocytes, caspase-dependent apoptosis, fatty acid deposition, and hyperplasia in hepatic tissues (Shashi and Thapar, 2001). Nair *et al.* (2004) also reported that treatment with vitamin C, or calcium phosphate and vitamin E reduced fluoride-induced abnormalities of serum transaminases and glycogen and phosphorylase activity as well as histopathological changes. Stawiarska-Pięta *et al.* (2012) demonstrated that 5 weeks treatment with vitamin E and methionine partially modified fluoride-induced pathomorphological changes and also restored the abnormalities in other biochemical markers fructose-1, 6-biphosphate aldolase, malate dehydrogenase, lactate dehydrogenase, sorbitol dehydrogenase, glutathione peroxidase activity, glutathione transferase and glutathione reductase. Chawla *et al.* (2008) reported that daily *intraperitoneal* injection of melatonin (10 mg kg^{-1} body weight) for one month counters the fluoride-induced decrease of body and liver weight in female mice. The treatment with melatonin also normalizes the activities of succinate dehydrogenase, acid phosphatase, alkaline phosphatase, serum glutamate pyruvate transaminase, serum glutamate oxaloacetate transaminase, and the amount of total protein (Chawla *et al.* 2008). Another report by Ekambaram *et al.* (2010) showed that *Tamarindus indica* mitigated fluoride-induced hepatotoxicity modifying the activity of antioxidant enzymes, serum aspartate transaminase and serum alanine transaminase, and also restoring the levels of hepatic malondialdehyde and ascorbic acid as well as serum levels of calcium and fluoride in female rats (Ekambaram *et al.*, 2010). Pawłowska-Góral *et al.* (2013) showed that N-acetylcysteine decreased fluoride-induced oxidative stress *via* normalization of the activity of antioxidant enzymes, *i.e.* lactate dehydrogenase, superoxide dismutase, glutathione peroxidase, glutathione reductase and of the level of reduced glutathione in primary rat hepatocytes.

Hassan and Yousef, (2009) also showed that sodium fluoride induces hepatotoxicity and oxidative stress through alteration of the levels of thiobarbituric acid reactive substances and glutathione and nitric oxide, superoxide dismutase, catalase, lactate dehydrogenase, creatine kinase activities, total antioxidant capacity, alanine transaminase and aspartate aminotransferase. They also reported that fluoride affects the level of total lipids, cholesterol, triglycerides, low density lipoprotein–cholesterol (LDL–c) as well as high-density lipoprotein–cholesterol (HDL–c). Hassan and Yousef, (2009) found that black berry juice treatment induces a decrease of hepatotoxicity and oxidative stress related to sodium fluoride (NaF) intoxication.

According to the study of Xiong *et al.*, (2007), there is a correlation between fluoride consumption and liver and kidney dysfunction in children. Molecular investigation showed that fluoride intoxication increases the content of both mRNA and protein expression of p53, GRP78, Fas and FasL, while decreasing the expression of apolipoprotein E gene and therefore fluoride induces apoptosis in hepatocytes (Wang *et al.*, 2005; Da Silva Pereira *et al.*, 2013; Miao *et al.*, 2013). Miao *et al.* (2013) showed that selenium reduced hepatocyte apoptosis *via* inhibition of the expression of Fas and FasL. Lu *et al.* (2010) performed proteomics analysis on hepatic tissues of *Takifugu rubripes* intoxicated with fluoride. They identified 8 proteins that are affected by fluoride intoxication, namely disulfide isomerase ER-60, 4SNc-Tudor domain, SMC3, Cyclin D1, mitogen-activated protein kinase 10, and also three unknown proteins.

Overall, the above-mentioned results demonstrate that fluoride induces oxidative stress and apoptosis in liver, through stimulation or suppression of gene expression and consequently the content of cognate proteins, giving out different signaling messages through several mechanisms.

Summary Points

- This chapter focuses on hepatotoxicity induced by fluoride.
- Humans accumulate fluorine, mainly in bones and teeth but also in soft tissues through both oral and inhalation exposure.
- Fluorinated compounds can improve the binding efficacy and selectivity of drugs and prevent dental caries.
- Chronic exposure to high levels of fluoride leads to fluorosis.
- Fluoride intoxication affects the normal function of cells and organelles.
- Fluoride alters the oxidant–antioxidant balance systems of hepatic tissue resulting in liver oxidative injuries.
- Treatment with antioxidants reduce fluoride-induced hepatotoxicity.

Key Facts

Key Features of Fluoride Toxicity

1. Fluoride is an essential mineral necessary for the maintenance of healthy bone and teeth.
2. Excessive consumption of fluoride can alter the normal function of different organs and tissues.
3. Excess of fluoride alters accretion and resorption of bone tissue and also affects the homeostasis of bone and dental mineral metabolism resulting in fluorosis.
4. Skeletal and dental fluorosis is a widespread disease mainly affecting developing countries with more than 60 million of affected people.
5. In nonskeletal structures fluoride affects the normal function of the cells *via* reactive oxygen and nitrogen species overproduction.

Key Features of Fluoride-Induced Hepatotoxicity

1. Liver plays a major role in metabolism and has a number of functions in the body, including glycogen storage, decomposition of red blood cells, plasma protein synthesis, hormone production and detoxification.
2. Because of its strategic functions, the liver is vital for survival and it also prone to many diseases.
3. It is well documented that the excess of reactive oxygen species are related with hepatic dysfunction and liver injuries.
4. Fluoride accumulation can alter the oxidant–antioxidant balance of the inducing oxidative damage and affecting its functionality.
5. Antioxidants such as quercetin, gallic acid or vitamin C mitigate the fluoride-induced hepatotoxicity reducing the oxidative stress.

Definitions of Words and Terms

Antioxidant system: The antioxidant system comprises molecules produced by the body or consumed by the diet that can safely interact with reactive species avoiding oxidative damage to cells and tissues.
Apoptosis: A natural process of self-destruction in certain cells that are genetically programmed and induced by a stimulus, such as irradiation, toxic drugs or an excess of cellular damage.
Fluorinated compounds: Compounds with fluorine in their chemical structure.
Fluorosis: Disease caused by exposure to excessive amounts of fluorine or its compounds affecting mainly the skeletal system.
Hepatotoxicity: Liver damage induced by chemical compounds.
Homeostasis: Capability to maintain internal stability in an organism to compensate for environmental changes.
Molecular pathway: A series of step-by-step actions among molecules in a cell that leads to a certain end point or cell function.
Oxidative stress: Disturbances in the normal redox state of cells caused by an imbalance between the reactive oxygen species production and the systems to detoxify these reactive intermediates.
Reactive species: Chemically reactive molecules that are able to react with cellular components inducing changes in their function and cellular damage.
Xenobiotic: Chemical compound foreign to a living organism that is not normally naturally produced by or expected to be present within that organism.

Abbreviations

Epigallocatechingallate	(EGCG)
Heat shock protein	(HSP)
High-density lipoprotein-csholesterol	(HDL-c)

Low-density lipoprotein-cholesterol	(LDL-c)
Poly (ADP-ribose) polymerase	(PARP)
Reactive oxygen species	(ROS)
Sperm cation channels	(CatSper)
Superoxide dismutase	(SOD)

References

Allibone, R., Cronin, S. J., Charley, D. T., Neall, V. E., Stewart, R. B. and Oppenheimer, C., 2012. Dental fluorosis linked to degassing of Ambrym volcano, Vanuatu: a novel exposure pathway. *Environmental Geochemistry and Health.* **34**: 155–170.

Anuradha, C. D., Kanno, S. and Hirano, S., 2001. Oxidative damage to mitochondria is a preliminary step to caspase-3 activation in fluoride-induced apoptosis in HL-60 cells. *Free Radical Biology and Medicine.* **31**: 367–373.

Aoba, T. and Fejerskov, O., 2002. Dental fluorosis: chemistry and biology. *Critical Reviews in Oral Biology and Medicine.* **13**: 155–170.

Azarpazhooh, A. and Main, P. A., 2008. Pit and fissure sealants in the prevention of dental caries in children and adolescents: a systematic review. *Journal of the Canadian Dental Association.* **74**: 171–177.

Barbier, O., Arreola-Mendoza, L. and Del Razo, L. M., 2010. Molecular mechanisms of fluoride toxicity. *Chemico-Biological Interactions.* **188**: 319–333.

Bégué, J. P. and Bonnet-Delpon, D., 2006. Recent advances (1995–2005) in fluorinated pharmaceuticals based on natural products. *Journal of Fluorine Chemistry.* **127**: 992–1012.

Belitz, H. D., Grosch W. and Schieberle, P., 2009. Food Chemistry. Springer Berlin Heidelberg 4th edn.

Bouaziz, H., Ketata, S., Jammoussi, K., Boudawara, T., Ayedi, F., Ellouze, F. and Zeghal, N., 2006. Effects of sodium fluoride on hepatic toxicity in adult mice and their suckling pups. *Pesticide Biochemistry and Physiology.* **86**: 124–130.

Buzalaf, M. A. R. and Whitford, G. M., 2011. Fluoride and the Oral Environment. *Monographs in Oral Science.* Basel, Karger. vol. **22**, 20–36 pp.

Chawla, S. L., Yadav, R., Shah, D. and Rao, M. V., 2008. Protective action of melatonin against fluoride-induced hepatotoxicity in adult female mice. *Fluoride.* **41**: 44–51.

Chen, Q., Wang, Z., Xiong, Y., Xue, W., Kao, X., Gao, Y., Muhammad, N. and Song, D., 2009. Selenium increases expression of HSP70 and antioxidant enzymes to lessen oxidative damage in Fincoal-type fluorosis. *Journal of Toxicological Sciences.* **34**: 399–405.

Chinoy, N. J., Mehta, D. and Jhala, D. D., 2005. Beneficial effects of a protein rich diet on fluoride induced free radical toxicity in the liver of male mice. *Fluoride.* **38**: 276–283.

Chinoy, N. J., Sharma, A. K., Patel, T. N., Memon, R. and Jhala, D. D., 2004. Recovery from fluoride and aluminium induced free radical liver toxicity in mice. *Age.* **12**: 14–16.

Dasarathy, S., Das, T. K., Gupta, I. P., Susheela, A. K. and Tandon, R. K., 1996. Gastroduodenal manifestations in patients with skeletal fluorosis. *Journal of Gastroenterology.* **31**: 333–337.

Da Silva Pereira, H. A. B., De Lima Leite, A., Charone, S., Lobo, J. G. V. M., Cestari, T. M., Peres-Buzalaf, C. and Buzalaf, M. A. R., 2013. Proteomic Analysis of Liver in Rats Chronically Exposed to Fluoride. *PloS One.* **8**: e75343.

Dubey, N., Khan, A. M. and Raina, R., 2013. Sub-acute deltamethrin and fluoride toxicity induced hepatic oxidative stress and biochemical alterations in rats. *Bulletin of Environmental Contamination and Toxicology.* **91**: 334–338.

Ekambaram, P., Namitha, T., Bhuvaneswari, S., Aruljothi, S., Vasanth, D. and Saravanakumar, M., 2010. Therapeutic efficacy of Tamarindusindica (L) to protect against fluoride-induced oxidative stress in the liver of female rats. *Fluoride.* **43**: 134–140.

Fagin, D., 2008. Second thoughts about fluoride. *Scientific American.* **298**: 74–81.

Flora, S. J., Mittal, M., Pachauri, V. and Dwivedi, N., 2012. A possible mechanism for combined arsenic and fluoride induced cellular and DNA damage in mice. *Metallomics.* **4**: 78–90.

García-Montalvo, E. A., Reyes-Pérez, H. and Del Razo, L. M., 2009. Fluoride exposure impairs glucose tolerance *via.* decreased insulin expression and oxidative stress. *Toxicology.* **263**: 75–83.

Gao, H., Zhang, Z. and Wan, X., 2012. Influences of charcoal and bamboo charcoal amendment on soil-fluoride fractions and bioaccumulation of fluoride in tea plants. *Environmental Geochemistry and Health.* **34**(5): 551–562.

Gofa, A. and Davidson, R. M., 1996. NaF potentiates a K+-selective ion channel in G292 osteoblastic cells. *The Journal of Membrane Biology.* **149**: 211–219.

Gutknecht, J. and Walter, A., 1981. Hydrofluoric and nitric acid transport through lipid bilayer membranes. *Biochimica et Biophysica Acta (BBA)-Biomembranes.* **644**: 153–156.

Hassan, H. A. and Yousef, M. I., 2009. Mitigating effects of antioxidant properties of black berry juice on sodium fluoride induced hepatotoxicity and oxidative stress in rats. *Food and Chemical Toxicology.* **47**: 2332–2337.

He, H., Ganapathy, V., Isales, C. M. and Whitford, G. M., 1998. pH-dependent fluoride transport in intestinal brush border membrane vesicles. *Biochimica et Biophysica Acta (BBA)-Biomembranes.* **1372**: 244–254.

He Y. Q., Pan Y., Ying L. J., Zhao R., 2012. Differential expression ESTs associated with fluorosis in rats liver. *International Journal of Genomics.* 208–390.

Hirano, S. and Ando, M., 1997. Fluoride mediates apoptosis in osteosarcoma UMR 106 and its cytotoxicity depends on the pH. *Archives of Toxicology.* **72**: 52–58.

Inkielewicz-Stêpniak, I. and Knap, N., 2012. Effect of exposure to fluoride and acetaminophen on oxidative/nitrosative status of liver and kidney in male and female rats. *Pharmacological Reports.* **64**: 902–911.

Iukhnovets, R. A. and Bachinskiĭ, P. P., 1981. Effect of fluoride and insulin on cation-dependent ATPase activity of the enterocytes during threonine absorption. *Voprosymeditsinskoikhimii.* **28**: 46–50.

Jha, S. K., Nayak, A. K. and Sharma, Y. K., 2009. Fluoride toxicity effects in onion (Alliumcepa L.) grown in contaminated soils. *Chemosphere.* **76**: 353–356.

Jha, A., Shah, K. and Verma, R. J., 2012. Effects of sodium fluoride on DNA, RNA and protein contents in liver of mice and its amelioration by Camellia sinensis. *Acta Poloniae Pharmaceutica.* **69**: 551–555.

Karaoz, E., Oncu, M., Gulle, K., Kanter, M., Gultekin, F., Karaoz, S. and Mumcu, E., 2004. Effect of chronic fluorosis on lipid peroxidation and histology of kidney tissues in first-and second-generation rats. *Biological Trace Element Research.* **102**: 199–208.

Krishnamachari, K. A. V. R., 1987. *Fluorine in Trace Elements in Human and Animal Nutrition*. Mertz. 5th edn. vol. 1, pp. 356–415.

Lee, J. H., Jung, J. Y., Jeong, Y. J., Park, J. H., Yang, K. H., Choi, N. K., Kim, S. H. and Kim, W. J., 2008. Involvement of both mitochondrial-and death receptor-dependent apoptotic pathways regulated by Bcl-2 family in sodium fluoride-induced apoptosis of the human gingival fibroblasts. *Toxicology.* **243**: 340–347.

Liu, G., Chai, C. and Cui, L., 2003. Fluoride causing abnormally elevated serum nitric oxide levels in chicks. *Environmental Toxicology and Pharmacology.* **13**: 199–204.

Lu, J., Zheng, J., Liu, H., Li, J., Xu, Q. and Chen, K., 2010. Proteomics analysis of liver samples from puffer fish Takifugurubripes exposed to excessive fluoride: an insight into molecular response to fluorosis. *Journal of Biochemical and Molecular toxicology.* **24**: 21–28.

Manna, P., Sinha, M. and Sil, P. C., 2007. A 43 kD protein isolated from the herb Cajanusindicus L attenuates sodium fluoride-induced hepatic and renal disorders in vivo. *Journal of Biochemistry and Molecular Biology.* **40**: 382–395.

Mascarenhas, A. K., 2000. Risk factors for dental fluorosis: a review of the recent literature. *Pediatric Dentistry.* **22**: 269–277.

Miao, K., Zhang, L., Yang, S., Qian, W. and Zhang, Z., 2013. Intervention of selenium on apoptosis and Fas/FasL expressions in the liver of fluoride-exposed rats. *Environmental Toxicology and Pharmacology.* **36**: 913–920.

Mohammadi, S., Movahedin, M. and Mowla, S. J., 2009. Up-regulation of CatSper genes family by selenium. *Reproductive Biology and Endocrinolology.* 7: 126.

Müller, K., Faeh, C. and Diederich, F., 2007. Fluorine in pharmaceuticals: looking beyond intuition. *Science.* **317**: 1881–1886.

Nabavi, S. M., Nabavi, S. F., Eslami, S. and Moghaddam, A. H., 2012. In vivo protective effects of quercetin against sodium fluoride-induced oxidative stress in the hepatic tissue. *Food Chemistry.* **132**: 931–935.

Nabavi, S. F., Nabavi, S. M., Habtemariam, S., Moghaddam, A. H., Sureda, A., Jafari, M. and Latifi, A. M., 2013. Hepatoprotective effect of gallic acid isolated from Peltiphyllumpeltatum against sodium fluoride-induced oxidative stress. *Industrial Crops and Products.* **44**: 50–55.

Nair, S. B., Jhala, D. D. and Chinoy, N. J., 2004. Beneficial effects of certain antidotes in mitigating fluoride and/or arsenic induced hepatotoxicity in mice. *Fluoride.* **37**: 60–70.

Panneerselvam, L., Subbiah, K., Arumugam, A. and Senapathy, J. G., 2013. Ferulic acid modulates fluoride-induced oxidative hepatotoxicity in male Wistar rats. *Biological Trace Element Research.* **151**: 85–91.

Paul, A., Wilson, S., Belham, C. M., Robinson, C. J., Scott, P. H., Gould, G. W. and Plevin, R., 1997. Stress-activated protein kinases: activation, regulation and function. *Cellular Signaling.* **9**: 403–410.

Pawłowska-Góral, K., Kurzeja, E. and Stec, M., 2013. N-acetylcysteine protects against fluoride-induced oxidative damage in primary rat hepatocytes. *Toxicology in Vitro.* **27**: 2279–2282.

Podder, S., Chattopadhyay, A. and Bhattacharya, S., 2011. Reduction in fluoride-induced genotoxicity in mouse bone marrow cells after substituting high fluoride-containing water with safe drinking water. *Journal of Applied Toxicology.* **31**: 703–705.

Qin, X., Wang, S., Yu, M., Zhang, L., Li, X., Zuo, Z., Zhang, X. and Wang, L., 2009. Child skeletal fluorosis from indoor burning of coal in Southwestern China. *Journal of Environmental and Public Health.* 1–7.

Refsnes, M., Schwarze, P. E., Holme, J. A. and Laêg, M., 2003. Fluoride-induced apoptosis in human epithelial lung cells (A549 cells): role of different G protein-linked signal systems. *Human and Experimental Toxicology.* **22**: 111–123.

Ridley, W. and Matsuoka, M., 2009. Fluoride-induced cyclooxygenase-2 expression and prostaglandin E2 production in A549 human pulmonary epithelial cells. *Toxicology Letters.* **188**: 180–185.

Roesky, H. W., 2010. A flourish of fluorine. *Nature Chemistry.* **2**: 240.

Roos, J., Dumolard, A., Bourget, S., Grange, L., Rousseau, A., Gaudin, P., Calop, J. and Juvin, R., 2005. Ostéosefluorée due à uneconsommation excessive de dentifrice fluoré. *La Presse Médicale.* **34**: 1518–1520.

Ruan, J., Ma, L., Shi, Y. and Han, W., 2004. The impact of pH and calcium on the uptake of fluoride by tea plants (Camellia sinensis L.). *Annals of Botany.* **93**: 97–105.

Salgado-Bustamante, M., Ortiz-Pérez, M. D., Calderón-Aranda, E., Estrada-Capetillo, L., Niño-Moreno, P., González-Amaro, R. and Portales-Pérez, D., 2010. Pattern of expression of apoptosis and inflammatory genes in humans exposed to arsenic and/or fluoride. *Science of the Total Environment.* **408**: 760–767.

Scientific Committee on Health and Environmental Risks., 2011. Critical Review of Any New Evidence on the Hazard Profile, Health Effects, and Human Exposure to Fluoride and the Fluoridating Agents of Drinking Water.

Shashi, A. and Thapar, S. P., 2001. Histopathology of fluoride-induced hepatotoxicity in rabbits. *Fluoride.* **34**: 34–42.

Siddiqi, N. J., 2012. Effect of sodium fluoride and magnesium chloride on different hydroxyproline fractions in rat liver. *Indian Journal of Biochemistry and Biophysics.* **49**: 130–133.

ŞİRELİ, M. and Bülbül, A., 2009. The effect of acute fluoride poisoning on nitric oxide and methemoglobin formation in the guinea pig. *Turkish Journal of Veterinary and Animal Sciences.* **28**: 591–595.

Shivarajashankara, Y. M. and Shivashankara, A. R., 2012. Neurotoxic effects of fluoride in endemic skeletal fluorosis and in experimental chronic fluoride toxicity. *Journal of Clinical and Diagnostic Research.* 740–744.

Soriano, M. and Manchon, F., 1966. Radiological aspects of a new type of bone fluorosis, periostitisdeformans. *Radiology.* **87**: 1089–1094.

Stawiarska-Pięta, B., Bielec, B., Birkner, K. and Birkner, E., 2012. The influence of vitamin E and methionine on the activity of enzymes and the morphological picture of liver of rats intoxicated with sodium fluoride. *Food and Chemical Toxicology.* **50**: 972–978.

Strunecka, A., Patocka, J., Blaylock, R. L. and Chinoy, N. J., 2007. Fluoride interactions: from molecules to disease. *Current Signal Transduction Therapy.* **2**: 190–213.

Thangapandiyan, S. and Miltonprabu, S., 2013. Epigallocatechingallate effectively ameliorates fluoride-induced oxidative stress and DNA damage in the liver of rats. *Canadian Journal of Physiology and Pharmacology.* **91**: 528–537.

Wang, A. G., Chu, Q. L., He, W. H., Xia, T., Liu, J. L., Zhang, M., Nussler, A. K., Chen, X. M. and Yang, K. D., 2005. Effects on protein and mRNA expression levels of p53 induced by fluoride in human embryonic hepatocytes. *Toxicology Letters.* **158**: 158–163.

Wang, A. G., Xia, T., Ru, R., Yuan, J., Chen, X., Yang, K. and Yang, K. D., 2004. Antagonistic effect of selenium on oxidative stress, DNA damage, and apoptosis induced by fluoride in human hepatocytes. *Fluoride.* **37**: 107–116.

Wergedal, J. E., Lau, K. H. W. and Baylink, D. J., 1988. Fluoride and bovine bone extract influence cell proliferation and phosphatase activities in human bone cell cultures. *Clinical Orthopaedics and Related Research.* **233**: 274–282.

Whitford, G. M., 1992. Acute and chronic fluoride toxicity. *Journal of Dental Research.* **71**: 1249–1254.

Whitford, G. M. and Pashley, D. H., 1984. Fluoride absorption: the influence of gastric acidity. *Calcified Tissue International.* **36**: 302–307.

Whitford, G. M., Bawden, J. W., Bowen, W. H., Brown, L. J., Ciardi, J. E., Clarkson, T. W., Imrey, P. B., Kleerekoper, M., Marthaler, T. M., McGuire, S., Ophaug, R. H., Robinson, C., Schultz, J. S., Stookey, G. K., Tochman, M. S., Venkateswarlu, P. and Zero, D. T., 1994. Report for Working Group I: Strategies for improving the assessment of fluoride accumulation in body fluids and tissues. *Advance Dental Research.* **8**: 113–115.

Whyte, M. P., Essmyer, K., Gannon, F. H. and Reinus, W. R., 2005. Skeletal fluorosis and instant tea. *The American Journal of Medicine.* **118**(1): 78–82.

Xiong, X., Liu, J., He, W., Xia, T., He, P., Chen, X., Yang, K. and Wang, A., 2007. Dose-effect relationship between drinking water fluoride levels and damage to liver and kidney functions in children. *Environmental Research.* **103**: 112–116.

Xu, B., Xu, Z., Xia, T., He, P., Gao, P., He, W., Zhang, M., Guo, L., Niu, Q. and Wang, A., 2011. Effects of the Fas/Fas-L pathway on fluoride-induced apoptosis in SH-SY5Y cells. *Environmental Toxicology.* **26**: 86–92.

Xu, H., Zhou, Y. L., Zhang, J. M., Liu, H., Jing, L. and Li, G. S., 2007. Effects of fluoride on the intracellular free Ca2+ and Ca2+-ATPase of kidney. *Biological Trace Element Research.* **116**(3): 279–287.

Zhan, X. A., Wang, M., Xu, Z. R., Li, W. F. and Li, J. X., 2006a. Effects of fluoride on hepatic antioxidant system and transcription of Cu/Zn SOD gene in young pigs. *Journal of Trace Elements in Medicine and Biology.* **20**(2): 83–87.

Zhan, X. A., Wang, M., Xu, Z. R., Li, W. F. and Li, J. X., 2006b. Evaluation of caspase-dependent apoptosis during fluoride-induced liver lesion in pigs. *Archives of Toxicology.* **80**: 74–80.

Zhang, Y., Li, W., Chi, H. S., Chen, J. and DenBesten, P. K., 2007. JNK/c-Jun signaling pathway mediates the fluoride-induced down-regulation of MMP-20 in vitro. *Matrix Biology.* **26**: 633–641.

Zhang, M., Wang, A., Xia, T. and He, P., 2008. Effects of fluoride on DNA damage, S-phase cell-cycle arrest and the expression of NF-κB in primary cultured rat hippocampal neurons. *Toxicology Letters.* **179**: 1–5.

Zheng, B. S. and Huang, R. G., 1989. Human fluorosis and environmental geochemistry in Southwest China. Developments in Geoscience: Contribution to 28th International Geological Congress. Washington, D.C., USA, Science Press, Beijing, China: 171.

CHAPTER 17

Effect of Fluoride on the Sensitivity and Secretion of Insulin

MERCEDES LOMBARTE[a], ALFREDO RIGALLI*[a], FERNANDO YAMAMOTO CHIBA[b], AND DORIS HISSAKO SUMIDA[c]

[a]Bone Biology Laboratory, School of Medicine, Rosario National University, Santa Fe 3100, 2000. Rosario, Argentina; [b]Department of Child and Social Dentistry, Dental School of Araçatuba, UNESP - Univ Estadual Paulista, José Bonifácio street, 1193, CEP 16015-050, Araçatuba, São Paulo, Brazil; [c]Department of Basic Sciences, Dental School of Araçatuba, UNESP - Univ Estadual Paulista, José Bonifácio street, 1193, CEP 16015-050, Araçatuba, São Paulo, Brazil

*E-mail: arigalli@conicet.gov.ar

Fluoride (F^-) has many toxic effects on the endocrine system, both in animals and human beings. Table 17.1 displays the most important effect on this system.

Although the effect of fluoride on glucose metabolism has been known for decades (Wang and Himoe, 1974), the effects of fluoride on insulin secretion and insulin action on target tissues were described twenty years ago (Rigalli *et al.*, 1990). Since then, the knowledge about the effect of fluoride on insulin and glucose metabolism has increased constantly.

As the effect of fluoride on insulin resistance and secretion resembles type 1 and type 2 diabetes, a short revision of them is done below in this chapter. Fluoride makes the effects after reaching the blood stream, and the main

Food and Nutritional Components in Focus No. 6
Fluorine: Chemistry, Analysis, Function and Effects
Edited by Victor R Preedy

Published by the Royal Society of Chemistry, www.rsc.org

Table 17.1 Effect of fluoride on the endocrine system.

System	Effects	References
Reproductive system	Decrease in estrogen and progesterone	Zhoum *et al.* (2013)
Parathyroid glands	Fluoride induces changes in ultra-structure of the glands.	Faccini and Care (1965)
Thyroid function	Fluoride has stimulatory effect on thyroid gland in iodine deficiency	Zhao *et al.* (1998)
Adrenal glands	Decrease in RNA, DNA and proteins in the gland.	Shashi (2003)
Insulin secretion	Decrease in plasma insulin levels	Rigalli *et al.* (1990)

way to enter the body is through the gastrointestinal tract. A brief revision of fluoride intake and distribution in the body is done below in this chapter.

17.1 Diabetes Mellitus

Diabetes mellitus is a chronic disease with a high and growing prevalence, costly in social-economic terms, and considered a severe public health problem in several countries of the world (Georg *et al.*, 2005). Progression of the disease leads to the development of systemic complications such as alteration in the carbohydrates, lipids and proteins metabolism, microangiopathy, neuropathy, nephropathy and macrovascular diseases.

Diabetes mellitus can be divided into two main types: type 1 diabetes and type 2 diabetes. In type 1 diabetes, there is a deficiency in insulin production, and the amount of the hormone produced is insufficient to the maintenance of survival. In type 2 diabetes it is observed that insulin is incapable of carrying out its functions properly.

The increase of extracellular glucose is the main determinant of tissue damage caused by diabetes.

Insulin resistance is a genetic or acquired status in which physiological plasma insulin levels have a subnormal biological response, such as a decrease in glucose uptake by muscle and adipose tissues. As a consequence, there is an increase in insulin production, without normal effect on tissues or resistance to the hormone (Cesaretti and Kohlmann Junior, 2006).

Studies have shown that among the chemical elements, fluoride can promote alteration in the secretion of insulin and in consequence in the metabolism of lipids and carbohydrates (Handler *et al.*, 1946; Dost *et al.*, 1977; Shashi *et al.*, 1988; Rigalli *et al.*, 1990; Rigalli *et al.*, 1995; Grucka-Mamczar *et al.*, 2004).

17.2 Fluoride Metabolism

Fluoride enters the body mainly by the gastrointestinal tract. The absorption of fluoride occurs in the stomach and small intestine, and can be modified by other elements, such as calcium, aluminum, *etc.*, which reduce fluoride absorption (da Silva Cardoso and Buzalaf, 2008; Ekstrand and Ehrnebo, 1979; Whitford, 1999). In the gastrointestinal tract, fluoride is absorbed by passive diffusion as hydrogen fluoride in the stomach and as fluoride in the intestine (da Silva Cardoso and Buzalaf, 2008). There is also evidence that an active component is present in the gastrointestinal absorption (Rigalli *et al.*, 2001). Once absorbed, fluoride is distributed in plasma and soft tissues, retained in calcified tissues and excreted mainly by the kidney (da Silva Cardoso and Buzalaf, 2008).

Over the last few decades, industrialized and developing countries have shown a decline of dental caries and an increase of dental fluorosis. This fact is due mainly to increased exposure to fluoridated products, among which the use of fluoridated toothpaste stands out (Bratthall *et al.*, 1996). Dental fluorosis is a change of tooth enamel that is related to prolonged and excessive fluoride intake during tooth formation (Ishii and Suckling, 1991, Buzalaf *et al.*, 2001). The limit dose of fluoride intake to prevent dental caries and to avoid undesirable degrees of dental fluorosis is estimated at approximately 0.05–0.07 mg F^-/kg bw/day (Burt, 1992). As most of the fluoride intake comes from dentifrice ingestion, some researchers have investigated formulas to reduce the fluoride concentration in toothpaste without interfering with their anticariogenic property (Vilhena *et al.*, 2010).

Fluoride is also present in significant amounts in chocolate bars, biscuits (da Silva Cardoso and Buzalaf, 2008), breakfast cereals and snacks (Cardoso *et al.*, 2003), black tea and tea-based drinks (Hayacibara *et al.*, 2004). Another important aspect is the fact that the benefits of fluoridation of the public water supply have a diffuse action, so that products and beverages processed in fluoridated areas are consumed for residents in nonfluoridated areas (Ramires and Buzalaf, 2007).

17.3 Fluoride and Glucose Homeostasis

Chronic or acute treatment with sodium fluoride (NaF) generate hyperglycemia. This condition is similar to diabetes (Rigalli *et al.*, 1992; Sakurai *et al.*, 1993; Appleton, 1995; Rigalli *et al.*, 1995; Grucka-Mamczar *et al.*, 2004). In some cases changes resemble type 1 and in other type 2 diabetes mellitus.

The mechanism by which fluoride induces hyperglycemia involves several factors, and they are summarized in Table 17.2.

17.3.1 Effect of Fluoride on Glycogen and Glucose Metabolism

Fluoride increases the conversion of adenosine-triphosphate (ATP) to cyclic adenosine-monophosphate (cAMP) by stimulation of adenylyl cyclase. The cAMP activates cAMP-dependent protein kinase (PKA), which activates glycogen phosphorylase–the key enzyme of glycogenolysis – by activation of a cascade of phosphorylations of proteins. This activated enzyme promotes

Table 17.2 Effect of fluoride on metabolic pathways or hormones related to glucose metabolism.

Metabolic pathway or hormone	Effect of fluoride	References
Glycolysis	Inhibition of enzymes	Wang and Himoe (1974)
Glycogenolysis	Increase in cAMP	Allmann and Kleiner (1980)
Epinephrine	Increase in release	McGown and Suttie (1977)
Insulin	Inhibition of secretion	Rigalli *et al.* (1990)
Insulin effect	Secretion of polypeptides without hypoglycemic effect on target cells	Menoyo *et al.* (2008)
Insulin effect	Changes in phosphorylation of proteins downstream insulin receptor on target cells	Chiba *et al.* (2010)

glycogen breakdown, which leads to the hyperglycemia due to release of glucose in the bloodstream. It should be noted that fluoride can also decrease cAMP, and the different effects could be the consequence of a difference in fluoride concentrations. Fluoride inhibits enolase, an enzyme of the glycolytic pathway, as a consequence a decrease in glucose metabolism is observed (Wang and Himoe, 1974). However, the effect of fluoride on the glycolytic enzymes is not observed at fluoride concentrations that usually are found in people living in areas of endemic fluorosis. Figure 17.1 displays the effect of fluoride on the metabolism of glucose and glycogen.

17.3.2 Inhibition of the Secretion of Insulin

A simultaneous increase in plasma glucose levels and a decrease in plasma insulin levels has been observed after an oral dose of fluoride both in rats and human beings. Experiments carried out with isolated Langerhans islets stimulated with glucose showed that fluoride at 5–20 μmol L^{-1} inhibited insulin secretion (Rigalli *et al.*, 1990). These authors suggest that fluoride produces changes in insulin secretion by modification of some stage in the cascade of intracellular events that involves calmodulin, protein kinase C and cAMP (Menoyo *et al.*, 2005). Fluoride inhibits insulin secretion when it is stimulated with glucose, forskolin (an activator of adenylyl cyclase), phorbol ester (activator of C-kinase) and calcium ionophore A23187 (Figure 17.2).

In further studies, the authors also demonstrated that fluoride causes a decrease in insulin secretion, while the fluoride concentration is above 10 μmol L^{-1} and that the insulin secretion is restored when plasma fluoride concentration return to basal levels (Menoyo *et al.*, 2008). However, the insulin effect is not normal because insulin is secreted as molecules with a lower hypoglycemic effect, probably as a modification of the intracellular processing of insulin precursors. The effect of fluoride on insulin secretion is at least in part a direct effect on beta cells, as the effect was observed in isolated Langerhans' islets and with perfused rat pancreas (Menoyo *et al.*, 2005).

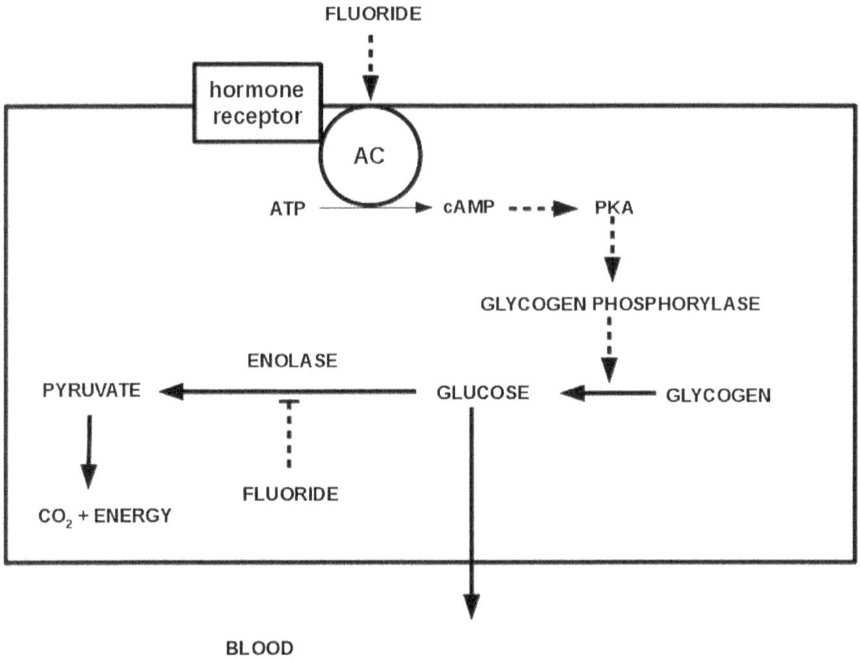

Figure 17.1 Effect of fluoride on glucose metabolism. Effect of fluoride on glucose and glycogen metabolism. Arrows indicate biochemical processes, dashed arrows indicate stimulation and dashed lines with a hyphen indicate inhibition. AC: adenylyl cyclase, PKA: protein kinase A.

Data obtained from glucose tolerance tests performed in patients living in areas of endemic fluorosis in Argentina showed an inverse relationship between plasma fluoride levels and the area under the curve of insulin concentration (de la Sota *et al.*, 1997). Rigalli *et al.* in 1990 found in people with high fluoride intake an increase in plasma glucose levels and a delay in the peak of plasma insulin concentration after a glucose-tolerance test (Rigalli *et al.*, 1990).

Studies performed in rats demonstrated that there is an alteration on the glucose homeostasis only when plasma fluoride level is above 5 μmol L^{-1} (Rigalli *et al.*, 1995). These researchers also reported a decrease of insulin secretion both in patients and in rats after ingesting 0.4 and 7.6 mg F^-/kg bw as NaF, respectively (Rigalli *et al.*, 1990). This difference in the dose to reach similar peak plasma fluoride levels in both models is probably due to the large difference in the metabolism and rate of bone remodeling.

Hyperglycemia was observed in rats after one dose of fluoride (7.6 mg F^-/kg bw) as NaF or disodium monofluorophosphate (MFP), without differences between treatments (Rigalli *et al.*, 1995). However, when rats drank water for 90 days with 100 ppm NaF or MFP, while NaF-treated rats showed abnormal glucose tolerance test, MFP-treated rats showed normal response to glucose load. The difference was attributed to low plasma fluoride levels in

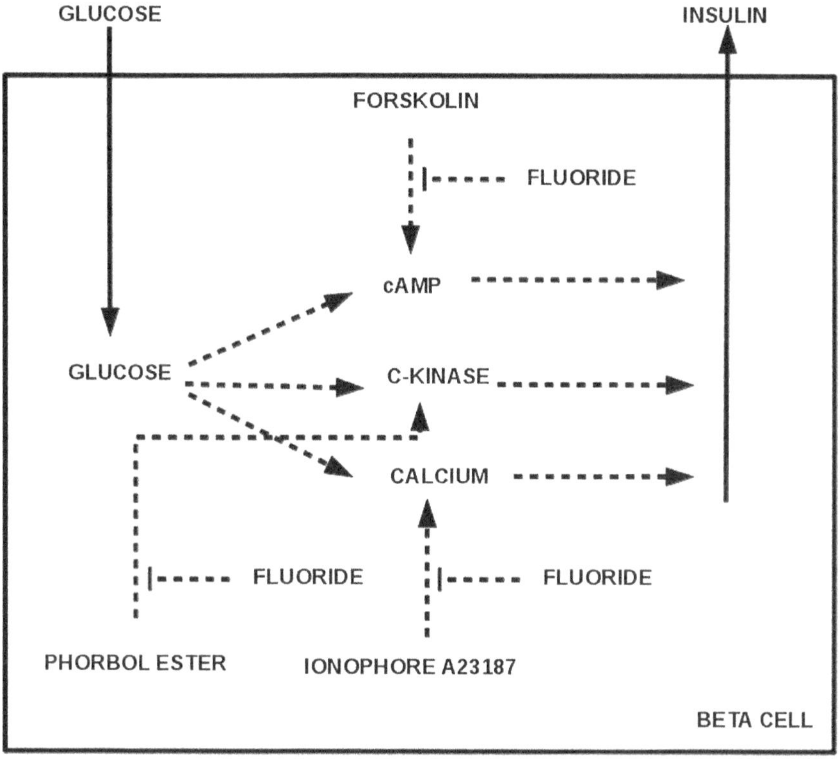

Figure 17.2 Effect of fluoride on insulin secretion. Glucose-stimulated insulin secretion and fluoride effect. Arrows indicate biochemical processes, dashed arrows indicate stimulation and dashed lines with a hyphen indicate inhibition.

MFP-treated rats compared to NaF-treated rats. In rats treated with MFP fluorine is bound to proteins, with low ionic fluoride levels. Interestingly, NaF-treated rats that increased bone mass as a consequence of fluoride stimulus had normal glucose tolerance test. The increase in bone mass was concomitant with a return to normal plasma fluoride levels as a consequence of the increase in fluoride clearance by bone tissue (Rigalli *et al.*, 1992).

17.3.3 Fluoride-Induced Insulin Resistance

Studies carried out in rats showed that the treatment with fluoride 100 ppm in the drinking water for one year promoted an increase of plasma insulin levels (Hu *et al.*, 2012). Further studies confirmed that the chronic ingestion of 4 mg F^{-}/kg bw/day causes insulin resistance (Chiba *et al.*, 2010; Chiba *et al.*, 2012). This was demonstrated by the HOMA-IR index (homeostatic model assessment for insulin resistance), and by a short intravenous insulin tolerance test.

The difference in HOMA-IR index is a consequence of a high fasting plasma insulin level while the fasting plasma glucose level is unchanged. Table 17.3

Table 17.3 Glycemia, insulinemia and HOMA-IR index in rats treated with water with 15 ppm fluoride (NaF treatment) or water without fluoride (controls). Mean ± Standard error, $n = 8$ per group.

Treatment	**Glycemia ($mmol\ l^{-1}$)**	**Insulinemia ($\mu UI\ ml^{-1}$)**	**HOMA-IR index**
Controls	5.5 ± 0.2	19.1 ± 6.3	4.4 ± 1.3
NaF treatment	5.4 ± 0.4	83.4 ± 18.0[a]	20.2 ± 4.4[a]

[a]indicates significant differences compared to controls, Student's *t* test.

shows data of fasting glycemia and insulinemia, and HOMA-IR index in healthy rats (controls) and rats that drank water with NaF 15 ppm for 30 days (NaF treatment).

However, Chehoud *et al.* did not observe any changes in insulin sensitivity or pp185 tyrosine phosphorylation status in muscle or white adipose tissues in rats treated with a single dose of 1.0 mg F^-/kg bw as NaF by gavage (Chehoud *et al.*, 2008). The difference can be explained by the lower dose of fluoride and the treatment as a single dose. Moreover, Trivedi *et al.* observed that 40% of patients with endemic fluorosis had impaired glucose tolerance, and verified that this alteration was reversed with the removal of excessive amount of fluoride in the drinking water (Trivedi *et al.*, 1993).

Lupo *et al.* evaluated the effect of the treatment with drinking water containing fluoride 0, 1, 5, and 15 ppm as NaF for 60 days on the glucose metabolism in rats with normal and chronic renal failure. They found that plasma glucose levels were not different among rats with normal and deficient renal function, and among rats with different intake of fluoride. However, plasma insulin levels were higher in animals that drank fluoridated water, showing insulin resistance both in normal and in rats with chronic renal disease (Lupo *et al.*, 2011). Lombarte *et al.* demonstrated that daily running on a treadmill for 60 min can reverse the insulin resistance in rats treated with drinking water with fluoride 15 ppm for 30 days. The rats subjected to physical exercise also showed a decrease in plasma fluoride levels and an increase in the amount of fluoride in bone tissue. The authors concluded that the reduction of insulin resistance could be the consequence of the decrease in plasma fluoride levels due to bone fluoride uptake and the increase of glucose transporters GLUT4 at the level of insulin-sensitive tissues (Lombarte *et al.*, 2013).

The binding of insulin to the receptor activates a tyrosine-specific protein kinase activity. This activation results in autophosphorylation followed by tyrosine phosphorylation of several intracellular substrates, such as IRS-1 (Kasuga *et al.*, 1982). These processes start several intracellular events that mediate the effects of insulin. In this manner, the effect of insulin at the level of insulin-dependent tissues depends on the phosphorylation of proteins downstream insulin receptor. Viñals *et al.* observed a reduction on insulin-induced tyrosine phosphorylation of insulin receptors purified from rat skeletal muscle and human placenta, when exposed to fluoride 190 ppm (Viñals *et al.*, 1997). Chronic fluoride intake (4.0 mg F^-/kg bw/day) in castrated rats promotes a decrease in insulin receptor substrate (pp185 - IRS-1/2) tyrosine phosphorylation status in muscle and white adipose tissue (Chiba *et al.*, 2010) and an increase in the IRS-1 serine phosphorylation status in white adipose tissue (Chiba *et al.*, 2012). This serine phosphorylation promotes a decrease in insulin signal transduction by impairment in the capacity of the insulin receptor to phosphorylate tyrosine residues after insulin stimulation (Figure 17.3). This inhibitory phosphorylation causes negative feedback in the insulin pathway and can lead to insulin resistance. The authors suggest that the decrease in insulin signal observed in muscle and white adipose tissue can be related to increased plasma concentrations of tumor necrosis factor alpha (TNF-α) as it was observed in these animals chronically treated with fluoride (Chiba *et al.*, 2012).

Lang *et al.* (1992) demonstrated that chronic administration of TNF-α in animals was able to induce a state of insulin resistance. The relationship between TNF-α and insulin resistance has been demonstrated by experiments where the treatment with TNF-α causes a decrease in tyrosine phosphorylation of insulin receptor and IRS-1 and glucose uptake, *in vivo* and in adipocytes culture. This reduction in tyrosine phosphorylation of IRS-1 may be due to a higher rate of serine phosphorylation of this insulin substrate (Figure 17.3) (Kanety *et al.*, 1995). On the other hand, neutralization of TNF-α by specific antibodies results in the improvement in tyrosine

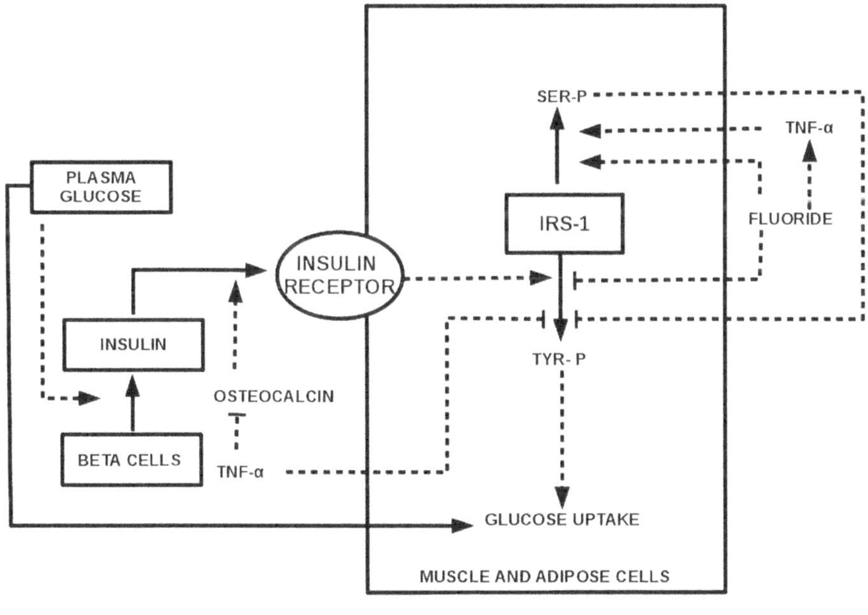

Figure 17.3 Effect of fluoride on target cells of insulin. Response of target cells to insulin and relationship with TNF-α, fluoride and osteocalcin. Arrows indicate biochemical processes, dashed arrows indicate stimulation and dashed lines with a hyphen indicate inhibition. SER-P and TYR-P indicate serine and tyrosine phosphorylation of IRS-1.

phosphorylation of insulin receptor and IRS-1 in muscle and adipose tissue from rats and causes reduction of plasma insulin concentration (Hotamisligil *et al.*, 1994).

Proinflammatory cytokines, such as TNF-α and interleukin-1 beta (IL-1β) are able to activate the pathways of c-Jun N-terminal kinase (JNK) and inhibitor of kappa β kinase/nuclear factor kappa β (IKKβ/NF-kβ) through interaction with specific receptors found in plasma membrane and thereby promote a cascade of intracellular signaling that culminates in insulin resistance (Shoelson *et al.*, 2006).

Studies have been developed to investigate the role of skeleton on endocrine regulation of energetic metabolism and thereby to contribute to the understanding of this process and metabolic disorders. Osteocalcin is a bone-specific protein expressed by osteoblasts, which undergoes post-translational modification by vitamin-K-dependent gamma-carboxylation. This protein strongly binds to hydroxyapatite, is stored in bone matrix, and can be used as marker of osteoblast differentiation and bone formation (Hauschka *et al.*, 1989). Studies performed in osteocalcin knockout mice showed that these animals had higher blood glucose level and lower plasma insulin levels than control mice (Lee *et al.*, 2007). It was also verified that insulin secretion and sensitivity as well as glucose tolerance (analyzed by glucose stimulated insulin secretion tests, glucose tolerance tests, insulin tolerance tests, and hyperinsulinemic euglycemic clamps) were decreased in osteocalcin knockout mice. Moreover, islets size and number, beta-cell mass, pancreas insulin content and insulin immunoreactivity were significantly decreased in osteocalcin knockout mice. These animals also showed decreased beta-cell proliferation, an increase in fat mass, and the number of adipocytes in comparison to control mice.

Fluoride decreases the osteocalcin levels as demonstrated by Hu *et al.* (2012) who observed that treatment with high doses of fluoride causes a decrease in osteocalcin expression in cells derived from newborn mouse calvaria. The effect of fluoride on osteocalcin depends on the dose of fluoride. In rats that drank water with fluoride 400 ppm a decrease in osteocalcin was observed (Song *et al.*, 2011). However, *in vitro* studies with fluoride 4 ppm on Saos-2 cells demonstrated an increase in osteocalcin (Huo *et al.*, 2013). In the same paper a reduction in cell growth was observed with fluoride concentration higher than 0.2 ppm. Probably a reduction in osteocalcin was present, although it was not measured. The normal metabolism of glucose observed in rats with 100 ppm fluoride in the drinking water after 100 days and the increase in bone mass (Rigalli *et al.*, 1992) could be the consequence of the increase in osteoblast function and osteocalcin production.

Based on these data, one can conclude that osteocalcin exhibits an important correlation with insulin sensitivity, as observed in clinical studies in which the decrease in osteocalcin was inversely related to body mass index, fat mass, plasma glucose levels, fasting plasma insulin levels, HOMA-IR, and high-sensitivity C-reactive protein (Kindblom *et al.*, 2009). Furthermore, the decrease in osteocalcin levels (Song *et al.*, 2011) and the increase in TNF-α levels (Lee *et al.*, 2007) promoted by treatment with fluoride may also have been two of the reasons for the decrease in insulin sensitivity.

The increase in TNF-α levels is related to diseases involving loss of bone mass and bone microarchitecture changes, such as osteoporosis and osteopenia, and can be related to the increase of osteoclasts and thus to increased bone resorption. Serum osteocalcin levels are significantly lower in patients with osteoporosis (Pietschmann *et al.*, 1991), and this change could be related to an increase in TNF-α levels. Gilbert *et al.*, (2000) demonstrated that TNF-α treatment inhibits cellular differentiation of osteoblasts, reduces mineralizing nodules and decreases the secretion of osteocalcin.

Summary Points

- After an oral dose of fluoride there is a decrease in plasma insulin levels as a consequence of an inhibitory effect on insulin secretion.
- Chronic treatment with fluoride produces insulin resistance as a consequence in the state of phosphorylation of insulin receptor substrate-1.
- Drinking water with 15 ppm induced insulin resistance in normal rats and in rats with chronic renal failure.
- Human beings living in areas of endemic fluorosis have insulin resistance.
- Tumor necrosis alpha levels are positively related with insulin resistance. The chronic treatment with fluoride produces increase in tumor necrosis alpha levels.
- The increase in bone fluoride uptake reduce the effect of fluoride on glucose homeostasis.
- Ostecalcin levels are inversely related to insulin resistance. The chronic treatment with fluoride produces a decrease in osteocalcin levels.
- Physical exercise ameliorates fluoride-induced insulin resistance.

Key Facts of Insulin Resistance

- Drinking fluoridated water may produce insulin resistance.
- Type 2 diabetes mellitus is characterized by insulin resistance.
- Physical exercise ameliorates insulin resistance induced by chronic fluoride intake.
- Osteocalcin levels are negatively related to insulin resistance and fluoride decreases osteocalcin levels.
- Tumor necrosis factor alpha is positively related to insulin resistance and fluoride increases tumor necrosis alpha levels.
- Insulin resistance depends on tyrosine and serine phosphorylation of insulin receptor substrate-1. Fluoride modifies the phosphorylation of tyrosine and serine in insulin receptor substrate-1.

Definitions of Words and Terms

Cyclic adenosine-monophosphate (cAMP): Cyclic nucleotide involved in the response of target cells to some hormones.
Euglycemia: A condition in which plasma glucose level is in the normal range (70–110 mg dl^{-1} for human beings).
Fluorosis: Chronic endemic illness produced by the ingestion of high amounts of fluoride. Fluorosis can affect teeth but also it can affect the function of different organs and systems.
Glucose homeostasis: The processes whereby the plasma glucose levels of an organism tends to remain balanced and stable.
Glucose metabolism: Reactions that involve glucose and other carbohydrates, such as glycogen.
Gluconeogenesis: Synthesis of glucose from other substances, such as pyruvate, aminoacids and glycerol.
Glucose tolerance test: Clinical test in which an amount of glucose is given orally to test glucose metabolism.
GLUT4: A glucose transport protein found in muscle and adipose tissue that promotes transport of glucose from the blood into target tissues. The inactive form is localized in the cytoplasm of the target cells and in response to insulin, it is translocated to the plasma membrane.
Glycogen: Carbohydrate that stores glucose in liver and muscle.
Glycogenolysis: Metabolic pathway of liver and muscle that transforms glycogen into glucose-6-phosphate.
Glycolysis: Metabolic pathway that transforms glucose into pyruvate.
Glycolytic enzymes: Enzymes of the metabolic pathway (glycolysis) that transform glucose into pyruvate.
HOMA-IR index: Abbreviation of homeostatic model assessment for insulin resistance index. HOMA-IR index is calculated with fasting plasma glucose and insulin. HOMA-IR values positively correlates with insulin resistance.
Hyperglycemia: A condition in which the plasma glucose level is higher than the normal value.
Hyperinsulinemic euglycemic clamp: Test in which plasma insulin levels are maintained over the normal value and glucose is infused to hold this constant at the normal value. In this condition, the rate of glucose infusion is a measure of glucose uptake by insulin-dependent tissues.
Hypoglycemia: A condition in which the plasma glucose level is lower than the normal value.
Insulin: Hormone produced by the pancreas that regulates the metabolism of carbohydrates and lipids.
Insulin tolerance test: Test in which insulin is intravenously injected to assess the insulin effect.
Osteocalcin: Protein produced by osteoblast that is a marker of bone formation.

Osteoporosis: An illness associated with a decrease in bone mineral content and density, changes in bone microarchitecture and an increase in the risk of bone fracture.

List of Abbreviations

AC	Adenylyl cyclase
ATP	Adenosine-triphosphate
bw	Body weight
cAMP	Cyclic adenosine-monophosphate
F^-	Fluoride
GLUT4	glucose transporter type 4
HOMA-IR	Homeostatic model assessment for insulin resistance
IKKβ/NF-kβ	Inhibitor of kappa β kinase/nuclear factor kappa β
IL-1β	Interleukin-1 beta
IRS-1	Insulin receptor substrate 1
JNK	c-Jun N-terminal kinase
kg	kilogram
l	liter
MFP	disodium monofluorophosphate
NaF	Sodium fluoride
PKA	cAMP-dependent protein kinase
ppm	parts per million
pp185	alternative name of IRS-1
SER-P	phosphorylated serine
TYR-P	phosphorylated tyrosine
TNF-α	Tumor necrosis factor alpha

References

Allmann, D. W. and Kleiner, H. S., 1980. Effect of NaF on rat tissue cAMP levels in vivo. *Pharmacology and Therapeutics in Dentistry.* **5**: 73–78.

Appleton, J., 1995. Changes in the plasma electrolytes and metabolites of the rat following acute exposure to sodium fluoride and strontium chloride. *Archives of Oral Biology.* **40**: 265–268.

Bratthall, D., Hänsel-Petersson, G. and Sundberg, H., 1996. Reasons for the caries decline: what do the experts believe? *European Journal of Oral Sciences.* **104**: 416–422.

Burt, B. A., 1992. The changing patterns of systemic fluoride intake. *Journal of Dental Research.* **71**: 1228–1237.

Buzalaf, M. A., Granjeiro, J. M., Damante, C. A. and de Ornelas, F., 2001. Fluoride content of infant formulas prepared with deionized, bottled mineral and fluoridated drinking water. *ASDC Journal of Dentistry for Children.* **68**: 37–41.

Chehoud, K. A., Chiba, F. Y., Sassaki, K. T., Garbin, C. A. S. and Sumida, D. H., 2008. Effects of fluoride intake on insulin sensitivity and insulin. signal transduction. *Fluoride.* **41**: 270–275.

Cardoso, V. E., Olympio, K. P., Granjeiro, J. M. and Buzalaf, M. A., 2003. Fluoride content of several breakfast cereals and snacks found in Brazil. *Journal of Applied Oral Science.* **11**: 306–310.

Cesaretti, M. L. R. and Kohlmann Junior, O., 2006. Experimental models of insulin resistance and obesity: lessons learned. *Arquivos brasileiros de Endocrinologia e Metabologia.* **50**: 190–197.

Chiba, F. Y., Colombo, N. H., Shirakashi, D. J., de Souza Gomes, W. D., Saliba Moimaz, S. A., Saliba Garbin, C. A., Silva, C. A. and Sumida, D. H., 2010. Insulin signal decrease in muscle but not in the liver of castrated male rats from chronic exposure to fluoride. *Fluoride.* **43**: 25–30.

Chiba, F. Y., Colombo, N. H., Shirakashi, D. J., Silva, V. C., Moimaz, S. A. S., Garbin, C. A. S., Antoniali, C. and Sumida, D. H., 2012. NaF treatment increases TNF-α and resistin concentrations and reduces insulin signal in rats. *Journal of Fluorine Chemistry.* **136**: 3–7.

Dost, F. N., Knaus, R. M., Johnson, D. E. and Wang, C. H., 1977. Fluoride impairment of glucose utilization: nature of effect in rats during and after continuous NaF infusion. *Toxicology and Applied Pharmacology.* **41**: 451–458.

Ekstrand, J. and Ehrnebo, M., 1979. Influence of milk products on fluoride bioavailability in man. *European Journal of Clinical Pharmacology.* **16**: 211–215.

Faccini, J. M. and Care, A. D., 1965. Effect of sodium fluoride on the ultrastructure of the parathyroid glands of the sheep. *Nature.* **207**: 1399–1401.

Georg, A. E., Duncan, B. B., Toscano, C. M., Schmidt, M. I., Mengue, S., Duarte, C. and Polanczyk, C. A., 2005. Economic analysis of a screening program for diabetes mellitus in Brazil. *Revista de Saúde Pública.* **39**: 452–460.

Gilbert, L., He, X., Farmer, P., Boden, S., Kozlowski, M., Rubin, J. and Nanes, M. S., 2000. Inhibition of osteoblast differentiation by tumor necrosis factor-alpha. *Endocrinology.* **141**: 3956–3964.

Grucka-Mamczar, E., Birkner, E., Kasperczyk, S., Kasperczyk, A., Chlubek, D., Samujło, D. and Cegłowska, A., 2004. Lipid balance in rats with fluoride-induced hyperglycemia. *Fluoride.* **37**: 195–200.

Handler, P., Herring, H. E. Jr. and Hebb, J. H., 1946. The effects of insulin in fluoride-poisoned rats. *The Journal of Biological Chemistry.* **164**: 679–683.

Hauschka, P. V., Lian, J. B., Cole, D. E. and Gundberg, C. M., 1989. Osteocalcin and matrix Gla protein: vitamin K-dependent proteins in bone. *Physiological Reviews.* **69**: 990–1047.

Hayacibara, M. F., Queiroz, C. S., Tabchoury, C. P. and Cury, J. A., 2004. Fluoride and aluminum in teas and tea-based beverages. *Revista de saúde pública.* **38**: 100–105.

Hotamisligil, G. S., Budavari, A., Murray, D. and Spiegelman, B. M., 1994. Reduced tyrosine kinase activity of the insulin receptor in obesity-diabetes. Central role of tumor necrosis factor-alpha. *The Journal of Clinical Investigation.* **94**: 1543–1549.

Hu, C. Y., Ren, L. Q., Li, X. N., Wu, N., Li, G. S., Liu, Q. Y. and Xu, H., 2012. Effect of fluoride on insulin level of rats and insulin receptor expression in the MC3T3-E1 cells. *Biological Trace Element Research.* **150**: 297–305.

Huo, L., Liu, K., Pei, J., Yang, Y., Ye, Y., Liu, Y., Sun, J., Han, H., Xu, W. and Gao, Y., 2013. Flouride promotes viability and differentiation of osteoblast-like Saos-2 cells *via.* BMP/Smads signaling pathway. *Biological Trace Element Research.* **155**: 142–149.

Ishii, T. and Suckling, G., 1991. The severity of dental fluorosis in children exposed to water with a high fluoride content for various periods of time. *Journal of Dental Research.* **70**: 952–956.

Kanety, H., Feinstein, R., Papa, M. Z., Hemi, R. and Karasik, A., 1995. Tumor necrosis factor alpha-induced phosphorylation of insulin receptor substrate-1 (IRS-1). Possible mechanism for suppression of insulin-stimulated tyrosine phosphorylation of IRS-1. *The Journal of Biological Chemistry.* **270**: 23780–23784.

Kasuga, M., Karlsson, F. A. and Kahn, C. R., 1982. Insulin stimulates the phosphorylation of the 95,000-dalton subunit of its own receptor. *Science.* **215**: 185–187.

Kindblom, J. M., Ohlsson, C., Ljunggren, O., Karlsson, M. K., Tivesten, A., Smith, U. and Mellström, D., 2009. Plasma osteocalcin is inversely related to fat mass and plasma glucose in elderly Swedish men. *Journal of Bone and Mineral research.* **24**: 785–791.

Lang, C. H., Dobrescu, C. and Bagby, G. J., 1992. Tumor necrosis factor impairs insulin action on peripheral glucose disposal and hepatic glucose output. *Endocrinology.* **130**: 43–52.

Lee, N. K., Sowa, H., Hinoi, E., Ferron, M., Ahn, J. D., Confavreux, C., Dacquin, R., Mee, P. J., McKee, M. D., Jung, D. Y., Zhang, Z., Kim, J. K., Mauvais-Jarvis, F., Ducy, P. and Karsenty, G., 2007. Endocrine regulation of energy metabolism by the skeleton. *Cell.* **130**: 456–469.

Lombarte, M., Fina, B. L., Lupo, M., Buzalaf, M. A. and Rigalli, A., 2013. Physical exercise ameliorates the toxic effect of fluoride on the insulin-glucose system. *The Journal of Endocrinology.* **218**: 99–103.

Lupo, M., Buzalaf, M. A. and Rigalli, A., 2011. Effect of fluoridated water on plasma insulin levels and glucose homeostasis in rats with renal deficiency. *Biological Trace Element Research.* **140**: 198–207.

Menoyo, I., Rigalli, A. and Puche, R. C., 2005. Effect of fluoride on the secretion of insulin in the rat. *Arzneimittel-Forschung.* **55**: 455–460.

Menoyo, I., Puche, R. C. and Rigalli, A., 2008. Fluoride-induced resistance to insulin in the rat. *Fluoride.* **41**: 260–269.

McGown, E. L. and Suttie, J. W., 1977. Mechanism of fluoride-induced hyperglycemia in the rat. *Toxicology and Applied Pharmacology.* **40**: 83–90.

Pietschmann, P., Resch, H., Krexner, E., Woloszczuk, W. and Willvonseder, R., 1991. Decreased serum osteocalcin levels in patients with postmenopausal osteoporosis. *Acta Medica Austriaca.* **18**: 114–116.

Ramires, I. and Buzalaf, M. A., 2007. Fifty years of fluoridation of public water supplies in Brazil: benefits for the control of dental caries. *Ciência & Saúde Coletiva.* **12**: 1057–1065.

Rigalli, A., Ballina, J. C., Roveri, E. and Puche, R. C., 1990. Inhibitory effect of fluoride on the secretion of insulin. *Calcified Tissue International.* **46**: 333–338.

Rigalli, A., Ballina, J. C. and Puche, R. C., 1992. Bone mass increase and glucose tolerance in rats chronically treated with sodium fluoride. *Bone and Mineral.* **16**: 101–108.

Rigalli, A., Alloatti, R., Menoyo, I. and Puche, R. C., 1995. Comparative study of the effect of sodium fluoride and sodium monofluorophosphate on glucose homeostasis in the rat. *Arzneimittel-Forschung.* **45**: 289–292.

Rigalli, A., Beinlich, A. and Puche, R. C., 2001. Intestinal absorption of fluoride at high luminal concentration of fluoride. *Arzneimittel-Forschung.* **51**: 151–155.

Sakurai, T., Suzuki, K., Taki, T. and Suketa, Y., 1993. The mechanism of changes in metabolism and transport of glucose caused by fluoride administration to rats. *Fluoride.* **26**: 210.

Shashi, A., Singh, J. P. and Thapar, S. P., 1988. Changes in glycogen content in some tissues during fluorosis—An experimental study on rabbits. *Fluoride.* **21**: 82–86.

Shashi, A., 2003. Fluoride and adrenal gland function in rabbits. *Fluoride.* **36**: 241–251.

Shoelson, S. E., Lee, J. and Goldfine, A. B., 2006. Inflammation and insulin resistance. *The Journal of Clinical Investigation.* **116**: 1793–1801.

da Silva Cardoso, V. E. and Buzalaf, M. A., 2008. Destinos do fluoreto no organismo. In: Buzalaf, M. A. (ed.) *Fluoretos e Saúde bucal. Livraria Santos Editora*, Santos, Brazil, pp. 45–67.

Song, Y. E., Tan, H., Liu, K. J., Zhang, Y. Z., Liu, Y., Lu, C. R., Yu, D. L., Tu, J. and Cui, C. Y., 2011. Effect of fluoride exposure on bone metabolism indicators ALP, BALP, and BGP. *Environmental Health and Preventive Medicine.* **16**: 158–163.

de la Sota, M., Puche, R., Rigalli, A., Fernández, L. M., Benassati, S. and Boland, R., 1997. Changes in bone mass and in glucose homeostasis in subjects with high spontaneous fluoride intake. *Medicina.* **57**: 417–420.

Trivedi, N., Mithal, A., Gupta, S. K. and Godbole, M. M., 1993. Reversible impairment of glucose tolerance in patients with endemic fluorosis. Fluoride Collaborative Study Group. *Diabetologia.* **36**: 826–828.

Vilhena, F. V., Olympio, K. P., Lauris, J. R., Delbem, A. C. and Buzalaf, M. A., 2010. Low-fluoride acidic dentifrice: a randomized clinical trial in a fluoridated area. *Caries Research.* **44**: 478–484.

Viñals, F., Camps, M., Testar, X., Palacín, M. and Zorzano, A., 1997. Effect of cations on the tyrosine kinase activity of the insulin receptor: inhibition by fluoride is magnesium dependent. *Molecular and Cellular Biochemistry.* **171**: 69–73.

Wang, T. and Himoe, A., 1974. Kinetics of the rabbit muscle enolase-catalyzed dehydration of 2-phosphoglycerate. Fluoride and phosphate inhibition. *The Journal of Biological Chemistry.* **249**: 3895–3902.

Whitford, G. M., 1999. Fluoride metabolism and excretion in children. *Journal of Public Health Dentistry.* **59**: 224–228.

Zhao, W., Zhu, H., Yu, Z., Aoki, K., Misumi, J. and Zhang, X., 1998. Long-term effects of various iodine and fluorine doses on the thyroid and fluorosis in mice. *Endocrine Regulations.* **32**: 63–70.

Zhoum, Y., Qiu, Y., He, J., Chen, X., Ding, Y., Wang, Y. and Liu, X., 2013. The toxicity mechanism of sodium fluoride on fertility in female rats. *Food and Chemical Toxicology.* **62**: 566–572.

CHAPTER 18

Preventing Fluoride Toxicity with Selenium

SWARAN J. S. FLORA*[a] AND MEGHA MITTAL[b]

[a]Defence Research and Development Establishment, Jhansi Road, Gwalior - 474 002, India; [b]Manav Rachna College of Engineering, Faridabad, Haryana, 121001, India
*E-mail: sjsflora@drde.drdo.in

18.1 Introduction

The mechanism of fluoride toxicity is mediated mainly through oxidative stress, apoptosis, and increased intracellular calcium level. The hypothesis suggesting oxidative stress as the key mechanism in fluoride-induced toxic effects has been confirmed in the patients of fluorosis, animals models of chronic fluorosis or cultured cells treated with varying concentration of fluoride (Figure 18.1). Oxidative stress is a condition generated due to the imbalance between the normal pro-oxidant and antioxidant status of the cell and results in the excessive production of free radicals. There have been copious reports suggesting downregulation of antioxidant capacity of cellular defense mechanism due to fluoride exposure (Mittal and Flora, 2006; Chouhan *et al.*, 2010; Reddy *et al.*, 2014). These findings have shown reduced antioxidant status, apoptosis and altered neurological and cardiorespiratory variables (Flora *et al.*, 2009, 2012).

Trace elements or ultratrace elements like selenium are required in the body in minute quantities (mg/day or ng/day) for maintaining normal redox cellular status. Owing to its antioxidant potential it might exhibit antagonistic effects on fluoride-induced altered activities of antioxidant enzymes,

Food and Nutritional Components in Focus No. 6
Fluorine: Chemistry, Analysis, Function and Effects
Edited by Victor R Preedy

Published by the Royal Society of Chemistry, www.rsc.org

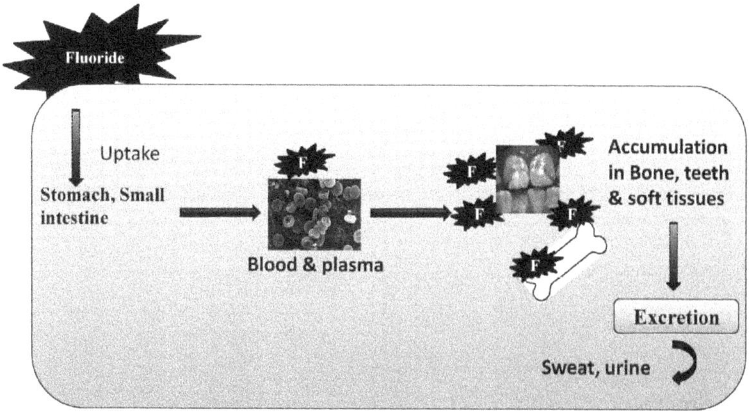

Figure 18.1 Toxicokinetic of fluoride.

metabolic pathways, increased concentration of reactive oxidant species and cellular damage (Figure 18.2). Selenium (Se) in a small quantity might inhibit or delay oxidation of a substrate. The most important source of selenium intake is through nutritional diet. It is reported to play a critical role in counteracting free-radical-induced damage to macromolecules and also cellular injury. It may act through different mechanisms and in different cellular compartments but its role as a free-radical scavenger is well defined. It may act through three possible mechanism, (i) by directly neutralizing free radicals, (ii) by reducing peroxide concentrations and repairing oxidized membranes, and (iii) by quenching iron to decrease ROS production (*via* lipid metabolism, short-chain free fatty acids and cholesterol esters neutralize ROS) (Flora *et al.*, 2009). Thus, supplementation of selenium is useful in the recovery of fluoride-induced altered biochemical variables suggestive of oxidative stress and organ damage.

18.2 Selenium: An Essential Component of an Antioxidant System

Selenium is an important component of the antioxidant defense system and generally exists in the environment in both inorganic and organic forms (Figure 18.3).

During the process of converting inorganic form to organic form of selenium or *vice versa*, various selenium metabolites are formed. Methylation is the major pathway for selenium metabolism in plants, humans and microbes, while demethylation occurs only in animals. Inorganic forms of Se, selenite (Se^{+4}) or selenate (Se^{+6}), are reduced by glutathione (GSH) to yield selenodiglutathione (GS–Se–SG), which is converted to hydrogen selenide (H_2Se) in a reaction catalyzed by glutathione reductase (GR). Hydrogen selenide is the putative intermediary metabolite that either serves as a precursor for the synthesis of

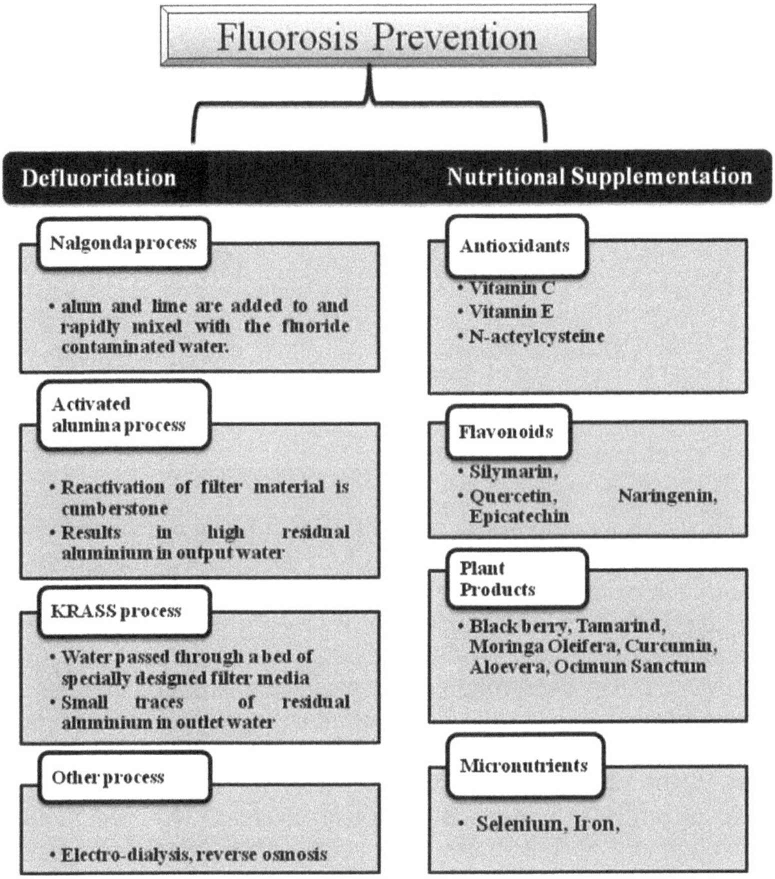

Figure 18.2 Various strategies to prevent fluorosis.

selenocysteine or is methylated to methylselenol (CH_3SeH), dimethylselenide, $(CH_3)_2Se$, and trimethylselenonium cation, $(CH_3)_3Se^+$ (Figure 18.4).

Selenium is a cofactor for various selenoenzymes involved in the maintenance of redox homeostasis and stress response. It is an essential dietary trace element and a potent antioxidant that prevents damage to cell organelle, oxidation of protein, lipid and DNA by the attack of free radicals or hydroperoxides. It is active in a variety of selenoproteins that support the immune system, prevent specific cancers, arthritis, atherosclerosis, maintain normal functioning of cellular and subcellular organelles, and alters immunological functions. Selenium is an integral part of specific proteins called selenoproteins in which it is incorporated as selenocysteine or selenomethionine. In all selenoproteins, selenium is co-translationally incorporated into polypeptides in the form of selenocysteine, the 21st amino acid. Selenium is also incorporated in a nonspecific manner in protein *via* substitution of selenomethionine for methionine. In selenium-tolerant plants, selenocysteine is converted to

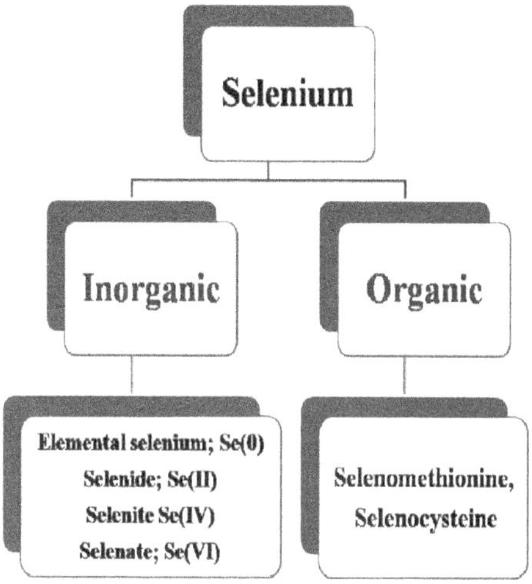

Figure 18.3 Environmental occurrence of selenium in both organic and inorganic form.

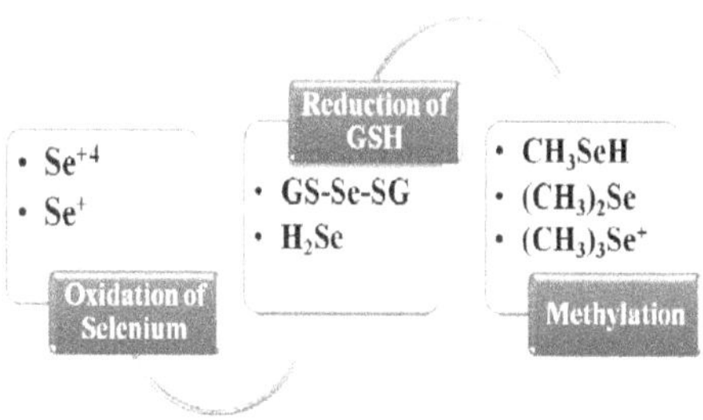

Figure 18.4 Various metabolic fates of selenium inside the cell in order to neutralize free radicals and warfare fluoride toxicity.

Se-methylselenocysteine in the presence of selenocysteine methyltransferase. It thus prevents the competition of selenocysteine with cysteine to be incorporated with protein. These methylated metabolites of selenium have been shown as potent anticarcinogenic agents (Ip *et al.*, 1991). There are also few other selenium binding proteins that tightly bind selenium for their activity.

The beneficial effects of selenium supplementation on human health have been studied extensively. Selenium acts as an antioxidant along with anti-inflammatory and antiviral activities. Researchers further evaluated the potential of selenium as an antiangiogenesis or antimetastasis agent. Beneficial effects of selenium are implicated in the protection of cellular structure against oxidative stress, maintenance of the antioxidative defense system, modulation of growth and development and anticarcinogenic properties (Zeng *et al.*, 2005).

Selenium exhibits antagonistic effects against high fluoride concentrations by depleting the fluoride burden from blood, soft and hard tissues, restoring antioxidant enzyme activities, repairing lipid metabolism, inhibiting lipid per oxidation, and promoting antioxidation reactions in organisms (Feng *et al.*, 2011). Selenium as a cofactor is required for the activity of numerous selenoenzymes (peroxidase, Selenoprotein P, thioredoxin reductase and iodothyronine deiodinase) involved in the maintenance of tissue antioxidant levels and stress.

The glutathione pathway is considered as a major defense mechanism against cellular oxidative stress condition. Glutathione peroxidase (GPx) is an important enzyme in this pathway containing four Se cofactors that catalyze the breakdown of hydrogen peroxide and organic hydroperoxides and converts reduced glutathione (GSH) to oxidized glutathione (GSSG). It also reduces phospholipid hydroperoxides in cell membrane and prevents cellular apoptosis.

$$2GSH + H_2O_2 \xrightarrow{GPx(Se)} GSSG + 2H_2O$$

Chemical Reaction 1.1: Oxidation of GSH in the presence of GPx.

Selenium alters the activity of GPx in cytosol and mitochondria facilitating fluoride excretion. Decreased GPx activity reflects the higher concentration of malonaldehyde (MDA) and selenium may antagonize lipid peroxidation induced by high fluoride concentration. Selenium forms an active center of GPx and protects the cells against deleterious effects of fluoride by bolstering its activity.

Selenoprotein P is a plasma selenoprotein, rich in selenocysteine and acts as a selenium transporter due to its plasma location and predominant origin in the liver. It is associated with endothelial cells, through its heparin-binding properties and contains more than one selenium atom per polypeptide chain. There are few controversies about the functions of selenoprotein P. If selenium is covalently bound to protein then it might not release selenium inside the cells. Burk *et al.*, (1995) reported that selenoprotein P transports selenium to the brain *via* labeling selenoprotein P with selenium and injecting into rats. Selenoprotein P has been postulated to protect against oxidative injury and transport selenium from the liver to peripheral tissues. Selenoproteins are responsible for most of the beneficial effects of selenium including preventing cancer and heart diseases, delaying aging process, and its role in the developmental and immunological function.

The thioredoxin/thioredoxin reductase (TRx/TRx-R) system is also known to provide a key role in reducing oxidative stress induced by fluoride. It is mainly located and expressed in endothelial and medial smooth muscle cells and targeted by free radicals with insinuation in cellular pathways. The characteristic feature of this antioxidant is the presence of a two cysteine residue at the catalytic site Trp–Cys–Gly–Pro–Cys–Lys along with two or three more cysteine residues. The disulfide group in thioredoxin is reduced to dithiol in the presence of the selenium-dependent enzyme TRx-R with NAPDH as an electron donor.

$$\text{Thioredoxin}-\overset{S}{\underset{S}{|}} + NADPH + H^+ \underset{}{\overset{TRx\text{-}R}{\rightleftarrows}} \text{Thioredoxin}-\overset{SH}{\underset{SH}{+}} NADP^+$$

Chemical Reaction 1.2: Reduction of disulfide form to dithiol form with NADPH as an electron source.

Thioredoxin reductase is known to recycle lipoic acid and vitamin C, regulate the metabolism of vitamin K_3, and acts to regulate cell growth and the activity of tumor-suppressing protein p53. The level of TRx is shown to be overexpressed in tumors because it is secreted by tumor cells and enhances the growth of cancerous tissues. Selenium being an essential component of TRx-R is also responsible for the reduction of TRx. Under the condition of oxidative stress, TRx is translocated from cytoplasm to the nucleus, suggesting a possible role in sensing and transducing oxidative stress signals. Another enzyme thioredoxin peroxidase (TRxP) leads to degradation of peroxides and hydroperoxides, responsible for DNA damage, tissue deformities and cell death. Thiol moieties when they react with free radicals form thiyl species, which are highly reactive and capable of triggering oxidative damage to the biological molecules like lipids, proteins and DNA. Thioredoxin peroxidase is involved in the detoxification of these thiyl radicals, thus explicitly a protective action. Excessive exposure to fluoride leads to the generation of reactive nitrogen species (RNS) such as nitric oxide (NO). A study on activated human neutrophils suggests that NO attacks thiols and leads to the formation of nitrosothiols. Thioredoxin reductase cleaves the molecule of nitrosothiols to NO and thiols and thereby maintains the normal antioxidant status of the body. It is evident that mammalian Trx-R possesses the ability to reduce several oxidants directly like, quinines and hydrogen peroxides due to the presence of a highly reactive selenocysteine moiety. They are also known to regulate the number of transcription factors affecting cell proliferation and death, through redox regulation.

In addition to providing indirect beneficial effects against fluoride toxicity *via* maintaining normal pro-oxidant/antioxidant status, selenium can directly bind to fluoride. The interaction of selenium with fluoride is highly dependent on pH. Adsorption of fluoride primarily takes place in acidic medium, however, the presence of selenium under a specific concentration can reduce the cellular fluoride uptake (Jinadasa and Dissanayake, 1992).

18.3 Effect of Selenium on Apoptosis and Cell Death

Excessive generation of free radicals may also lead to apoptosis. Accumulation of fluoride in cells attacks oxygen and influences oxygen metabolism leading to the massive generation of reactive oxygen species (ROS) that mediates apoptosis *via* various pathways (Yang *et al.*, 2011). Apoptosis is programmed cell death controlled by various receptors and mediated by three major pathways: (a) death receptor pathway (b) mitochondrial pathway and (c) endoplasmic reticulum pathways. The mitochondrial pathway is triggered by the release of cytochrome c from mitochondria into the cytosol, which activate caspase-9 followed by activation of caspase-3. However, the death receptor pathway is initiated by receptors located at the cellular surface. Tumor necrosis factor (TNF), tumor necrosis factor receptor (TNFR) and Fas/FasL are the important membrane proteins that mediate apoptosis. These receptors can transduce signals during apoptosis, activate the caspase cascade and induce apoptosis.

Numerous reports suggest upregulation of Fas/FasL pathways during high fluoride intake (Lee *et al.*, 2008). Oxidative stress or genotoxicity may be the crucial factors underlying the mechanism in fluoride-induced toxic manifestations. Fluoride-induced apoptosis by oxidative stress-mediated lipid peroxidation may lead to the mitochondrial dysfunction and the activation of downstream pathways. Fluoride may also cause injury to the cell membrane and affect the activity of membrane-bound enzymes like ATPase by disturbing membrane fluidity and membrane integrity and altering its permeability. The increased activity of ATPase on fluoride exposure could impart excessive breakdown of ATP in order to maintain the energy requirement of cells.

Selenium, being a cofactor for various selenoenzymes is involved in stress response, the antiapoptotic mechanism and maintenance of high tissue antioxidant levels. On the other hand, excessive accumulation of selenium may lead to selenosis and show clinical symptoms of "Keshan and Kashin Beck diseases" (Whanger *et al.*, 1996). Thus, an appropriate dose of selenium can reduce the expression of the death receptor and prevent fluoride-induced apoptosis.

Apoptosis is mediated by certain apoptotic molecules namely Bax and Bcl-2. Bcl-2 is responsible for the inhibition of apoptosis while Bax forms heterologous dimeric complexes with Bcl-2 and promotes cellular apoptosis. Cells easily undergo apoptosis on fluoride exposure that is indicated by an imbalance between the expression of Bax to Bcl-2 ratio. Selenium is reported to inhibit the expression of Bax to Bcl-2 ratio *via* (a) enhancing antioxidant enzymes activities (b) reducing expression of death receptors (c) removing fluoride from the body and (d) recovering normal cellular functions.

18.4 Selenium as an Anticarcinogenic Agent

Fluoride is an equivocally recognized carcinogen by the National Cancer Institute Toxicological Program and number of mutagenic studies carried out using sodium fluoride (Slamenova *et al.*, 1996). Using mammalian cells *in vitro*, fluoride caused genetic damage (including chromosomal aberrations) at cytotoxic concentrations (10 mg liter^{-1}), while various incidences of tumor growth were reported in workers exposed to fluoride.

Selenium is a potent antitumorigenic agent against various types of cancers such as lung, bladder, colorectal, esophageal, gastric cardiac, prostate and thyroid cancers (Chen *et al.*, 2013). Multiple studies have demonstrated the antiangiogenic effects of selenium, a potential mechanism of cancer prevention. Selenium supplementation with methylselenocysteine inhibits microvessel formation and tumor growth. Selenium not only suppresses tumor initiation and primary tumor growth but also shows promising results in the metastasis condition. Anticarcinogenic effects of Se are illustrated in Figure 18.5.

Monomethylated Se compounds at very low concentration are effective in providing chemopreventive effects. Selenium compounds inhibit cell growth

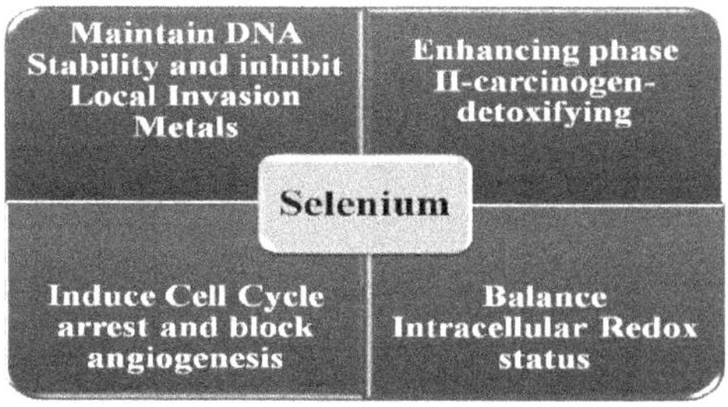

Figure 18.5 Anticarcinogenic effects of selenium mediated *via* various pathways suggesting multidirectional effects to reduce occurrence of fluorosis.

at micromolar levels by inactivating transcription factors, NF-κB and AP-1 that are known to be modulated by a redox mechanism. It has been reported that the effects of selenium are more pronounced in the early stages of carcinogenesis due to normalization of regulatory pathways. Both pro- and antiapoptotic effects are associated with selenium supplementation. Methylselenic acid has been shown to promote apoptosis in vascular endothelial cells, suggesting antiangiogenic effects, while selenite is reported to attenuate oxysterol-induced apoptosis in vascular smooth muscle cells (Tang *et al.*, 2005).

Selenium inhibits tumor growth both in the initial and later stages of carcinogenesis. Selenium supplementation enhances the activity of GPx, which upregulates the expression of interleukin-2 receptors and acts as immunomodulator. It also inhibits tumor cell proliferation and suppresses tumor angiogenesis. The anticarcinogenic activity of selenium does not depend upon the cellular level of selenium, rather on specific metabolic form of selenium. It has been suggested that the inorganic form of Se (selenite) is more effective compared to its organic form (selenomethionine) in preventing tumor growth (Ip and Hayes 1989). Selenium exerts an inhibitory effect upon fluoride-induced carcinogenesis *via* modulating the rate of cell division, stimulating DNA repair and protecting chromosomal damage.

Selenium affects DNA methylation and gene transcription both *in vivo* and *in vitro*. It causes global DNA hypomethylation associated with hypermethylation of the promoter of the tumor suppressor gene p53. It increases the availability of *S*-adenosyl methionine, the methyl donor for all DNA methylation reactions, and affects the concentration of *S*-adenosyl homocysteine, a potent inhibitor of DNA methyltransferases (Zeng *et al.*, 2011).

It has also been reported that selenium administration increases the expression of HSP70 in fluoride exposed conditions. HSPs are a family of stress-induced proteins responsible for preserving cell viability and offer cytoprotection against oxidative stress. HSP 70 is the most widely used family of all HSPs and are used as a biomarker due to increased expression

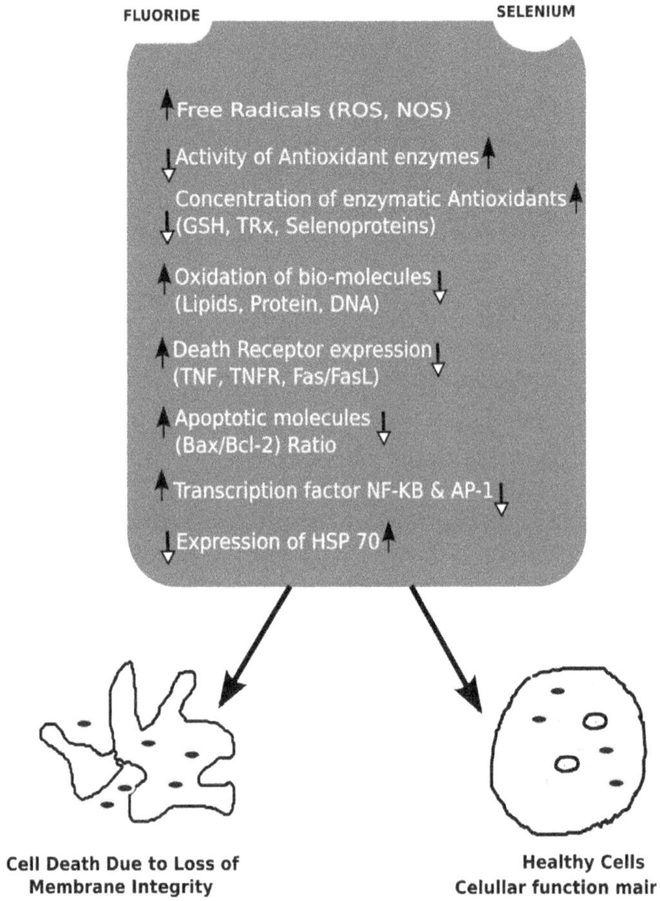

Figure 18.6 Major functions of selenium against fluoride-induced cellular death. The figure depicts possible site of action where selenium antagonizes adverse effects of fluoride. Abbreviations: ROS – Reactive oxygen species; RNS – Reactive nitrogen species; GSH – glutathione; TRx – Thioredoxin; DNA – Deoxyribonucleic acid; TNF – Tumor necrosis factor; TNFR – tumor necrosis factor receptor; Fas/FasL – Fas ligand, a type-II transmembrane protein that belongs to the tumor necrosis factor family; Bax/Bcl-2 – ratio of proapoptotic (BAX) to antiapoptotic (BCL-2) members; NF-κB – nuclear factor kappa-light-chain-enhancer of activated B cells is a protein complex that controls transcription of DNA; AP-1 – activator protein 1; Hsp70 – 70 kilodalton heat shock proteins (Hsp70 s) are a family of conserved ubiquitously expressed heat shock proteins.

during stress conditions. HSP70 is primarily known to repair protein damage and plays a key role in cytoprotection. Chen *et al.*, (2009) suggested selenium supplementaion during chronic fluorosis led to an increased expression of HSP70 and offered resistance against chronic fluorosis. Possible sites of action of selenium are depicted in Figure 18.6.

18.5 Protective Effects of Selenium on Organ Damage

Specific concentration of selenium has the potential to ameliorate fluoride concentration from soft tissues, maintain normal pro-oxidant–antioxidant status and promote better clinical recovery. Moreover, it may also restrict or decrease entry of fluoride in the hard tissues, thereby leading to decreased occurrence of skeletal fluorosis. Selenium ameliorates high concentrations of fluoride, promotes urinary excretion of fluoride and provides clinical recovery. The effects of selenium on enzymatic or nonenzymatic defense mechanisms may be related to sensitivity, dose, duration, temperature and many other factors. Free radicals generated due to the exposure of fluoride cause lipid peroxidation and changes membrane fluidity thus damaging cellular structure and their function (Sun *et al.*, 2011). Excessive generation of free radicals leads to the oxidation of biomacromolecules such as, lipid, protein and DNA, subsequently leading to cell death. Various mechanisms for DNA damage induced by fluoride have been suggested and they include, (i) fluoride being a strong nucleophile have a direct effect on DNA as it shows strong affinity for uracil and amide bonds by –NH-----F– interactions that can induce the rupture of hydrogen bonds in the base pairing of adenine and thymine. This results in the disturbance of DNA synthesis and increases the error frequency of linkage between basic groups in the process of DNA replication (Li *et al.*, 1987), (ii) fluoride may bind stably with DNA by covalent bonding and thus affecting the normal structure of DNA, (iii) fluoride induces the production of free radicals, which may damage DNA strands directly or by lipid peroxidation initiated by free radicals and, (iv) fluoride may decrease enzyme activity, such as with DNA polymerase, which might further affect the process of DNA replication or repair and thus leading to damaged DNA.

Supplementation of selenium in fluoride-exposed animals may also improve the activity of GPx and enhance cellular antioxidant status. Evidently, a preventive and therapeutic efficacy of selenium has been reported against acute and chronic fluoride toxicity and the antioxidant mechanism for the same can be illustrated *via* various site of action. The protective efficacy of selenium on fluoride-induced alterations mainly relies on its efficacy to enhance activity of antioxidant enzymes, free-radical scavenging potential and depletion of body fluoride burden. The alteration in the enzyme vitality depends upon their consumption and induction during the neutralization of free radicals or conversion into less harmful radicals.

18.6 Neurological Manifestations

The central nervous system (CNS) is highly vulnerable to free radical induced damage owing to the utilizaton of large amounts of oxygen, the presence of postmitotic cells, easily oxidizable fatty acid and low antioxidant defense system. Children are more susceptible to the attack of free radicals due to the heterogenity of the developing nervous system with different cell types and function. Thus, the supplementation of antioxidants is a better choice against fluoride intoxication. Fluoride is known to produce neuronal dysfunction and synaptic injury and reduce learning-memory ability in rats (Basha *et al.*, 2011). Exposure to high concentrations of fluoride in drinking water is found to be associated with lower intelligence in children (Xiang *et al.*, 2003). A positive correlation has been reported between neurological impairment and oxidative stress and other brain biochemical variables. Antioxidant suplementation can restore perturbation in redox status and maintain antioxidant homeostasis of the cell, thereby supressing fluoride toxicity.

Selenium supplementation could supress the formation of lipid peroxides (LPO), thus delaying the inception of LPO intrusion with neutrophic regulation. Selenium supplementation can also decrease protein oxidation in CNS by reducing the fragmentation and aggregation of amino acids. The oxidative effects of fluoride are different in various brain regions due to the difference in cellular formation and maturation and thus the antioxidant effects also vary for various brain regions. Fluoride exposure is also known to cause severe degenerative and necrotic changes followed by decreased neuronal density. Fluoride has been found to alter the levels of dopamine, serotonin, 5-hydroxyindoleacetic acid, homovanillic acid, nor epinephrine, and epinephrine in the hippocampus and neocortex regions of the rat brain (Reddy *et al.*, 2014). The decreased levels of neurotransmitters may either be due to the decreased activity of enzymes involved in their synthesis like tyrosine hydroxylase, DOPA decarboxylase and dopamine β-hydroxylase or to the enhanced release of catechol-O-methyl transferase caused by increased neuronal activity. Fluoride also exhibits developmental toxicity as it has the ability to cross the placental barrier, thus leading to fetal toxicity and congenital birth defects.

Fluoride-induced DNA damage in the brain follows two basic mechanisms, (i) an indirect mechanism, where exposure to fluoride results in the generation of free radicals, which in turn attack the hydrogen bonds of the DNA molecule to give DNA adducts and, (ii) the direct mechanism, where fluoride targets free –NH group. Fluoride accumulation in the brain is known to alter levels of neurotransmitters thus causing neurobehavioral impairments. However, these altered levels can significantly be restored by selenium supplementation. Protection against neuronal damage by selenium is attributed to the enhanced antioxidant capacity of cells by increasing GSH content and restoring activity of superoxide dismutase (SOD) and GPx. Selenium supplementation along with fluoride was also found to reverse pro- and antioxidant systems towards normal. Selenium facilitates excretion of fluoride from soft and hard tissues due to the altered activity of GPx and GST in cytosol and mitochondrial matrix.

Qian *et al.* (2013) studied the effect of selenium on fluoride-induced changes in the synaptic plasticity in rat hippocampus and suggested the possible role of selenium in combating fluoride induced neurotoxicity. They suggested that fluoride-induced neurotoxicity is mediated *via* an increase in postsynaptic density-93 (PSD-93) expression and selenium supplementation could also revert this expression back to normal. Thus, PSD-93 expression might be the key factor to fluoride-induced central nervous toxicity and the possible site of selenium intervention. The optimal dose of selenium pertaining to its beneficial effects was reported to be 0.75–1.5 mg l^{-1}. Administration of sodium selenite at a dose of 5 mg kg^{-1} body weight has been shown to have promising clinical benefits while, 10 mg kg^{-1} body weight of selenite administration may significantly reduce fluoride concentration from soft tissues.

Sialic acids (SA) are substituted neuraminic acid derivatives and important components of membrane glycoprotein and glycolipid. Massive generation of free radicals influences sialic acid decomposition from membranes, thereby altering the membrane configuration and affecting many enzymatic and transporter activities. Degradation of cellular membrane results in the decreased fluidity and increased permeability and eventually collapse of the membrane structure (Zhang *et al.*, 2007). Selenium supplementation to fluoride-exposed animals promotes recoveries of damaged membrane and the degeneration of cytomembranes and descending fluidity.

18.7 Renal System

Kidney is an important target organ for fluoride accumulation, drainage, and toxic effect. The influence of selenium on cellular apoptosis against fluorosis has been explored along with expression of Bcl-2 and Bax proteins and genes. There are numerous studies suggesting alteration in kidney tissue morphology and ultrastructural changes in animals and patients with chronic fluorosis. Studies have found that the chronic fluorosis may lead to kidney dysfunction in human patients as well as animal models. Increased expression of Bax and Bax/Bcl-2 indicated apoptosis to be one of the key mechanisms in the occurrence of fluorosis in renal tissue. The molecular mechanism behind beneficial role of selenium against fluorosis to inhibit apoptosis may be due to the upregulation of Bcl-2 expression and downregulation of Bax expression. However, due to a narrow range between nutritional and toxic dose of selenium, the antagonistic effects are primarily dependent upon the dose of selenium. The dose of 1.5 mg L^{-1} of selenium against chronic fluorosis has been reported to be the most promising at which it can intervene in fluoride-induced cellular apoptosis.

18.8 Cardiovascular System

Cardiac effects following acute and chronic fluorosis are well reported (Shashi and Thapar, 2001). Animals when exposed to high doses of fluoride shows cardiac dysfunction and low blood pressure while in humans changes in the electrocardiogram and heart enlargement has also been found to be positively correlated to the presence of fluoride in the drinking water. Flora *et al.*, (2012) reported a positive relationship between oxidative stress and physiological variables indicative of cardiorespiratory dysfunction. They also suggested that cardiorespiratory impairment was mediated *via* generation of ROS and depletion of the antioxidant defense system. Oxidative stress has been found to be the key mechanism in atherogenesis, myocardial infarction, and heart failure. Free-radical-mediated damage induces oxidative changes in low-density cholesterol particles that initiate and promote atherosclerotic changes. Thus, antioxidant therapy is the first line of treatment for the prevention and management of atherosclerotic cardiovascular disease. Multiple preclinical studies have demonstrated that during myocardial injury produced by ischemia and the combination of ischemia and reperfusion, selenoproteins, particularly GPx, reduces oxidative stress. Selenium deficiency worsens, and selenium supplementation abrogates, myocardial injury (Rakotovao *et al.*, 2005).

Selenium metabolism is intricately linked to cardiovascular function, and deficiency of selenium is associated with cardiac pathology. There are no reports that suggest that selenium supplementation protects against fluoride-induced cardiomyopathy however, on the basis of the antioxidant action of selenium and its proven efficacy to decrease the risk of impaired cardiac function, it can also be recommended for fluoride toxicity. Selenium supplementation has been found to improve coronary flow reserve, endothelial dysfunction and atherosclerosis. Long-term supplementation of selenium reduces the occurrence of impaired cardiac function due to excessive generation of reactive species. Increase in selenium supplementation may be associated with decreased risk of coronary artery disease. A small pilot study conducted in the UK demonstrated that selenium supplementation had modest benefits on reducing total and non-HDL cholesterol (Rayman *et al.*, 2011). In cases of cardiomyopathy, the activity of GPx and zinc gets reduced significantly, selenium supplementation on the other hand was beneficial to restore the activity of GPx thereby improving antioxidant defense system. Subjects were randomized to receive additional selenium and zinc supplements or to continue on routine heart-failure therapy.

Selenium supplementation has been shown to effect functions of immune system *via* redox balance and calcium flux, which may be imperative for cardiovascular biology. Matrix composition of the vascular wall and myocardium is vital to cardiovascular function and selenium may modulate cardiovascular health *via* effects on matrix metabolism.

18.9 Interaction with Vitamin E and Essential Elements

Supplementation of selenium in combination with vitamin E have shown better clinical recovery against fluoride-induced oxidative stress. They are known to prevent muscular dystrophy and restore normal functioning of cellular pathways. Both vitamin E and Se has specific functions in preventing tissue oxidative stress. Vitamin E decreases formation of hydroperoxide and selenium being an essential component of GPx catalyzes the breakdown of peroxides and hydroperoxides. Selenium in combination with vitamin E also enhances antibody formation. Cobalt, copper and vitamin B_6 also act synergistically with selenium and facilitate conversion of selenomethionine to GPx thereby bolstering the antioxidant defense system.

18.10 Efficacy of Nanosized Selenium

In recent years with the emergence of nanochemistry, nanosized elemental selenium has been synthesized that offers higher efficacy as compared to selenomethionine. It has shown better efficacy in enhancing antioxidant potential by upregulating GPx expression (Peng *et al.*, 2007). Nanoparticles of selenium interact with $-NH_2$, $-COO$, $-CO$ and $-CN$ group of proteins and shows higher biological activity. The size and shape of nanosized selenium plays a critical role in safeguarding oxidative stress. It has been suggested that hollow spherical nanoparticles of selenium exert better antioxidant potential. Moreover, 5–200 nm sized particles of selenium can directly scavenge reactive oxygen species in a size-dependent manner (Wang *et al.*, 2007). Selenium nanoparticles enhances the activities of several antioxidant enzymes like thioredoxin reductase, glutathione peroxidase and glutathione reductase by acting at their active sites (Zhang *et al.*, 2001). Nanosized selenium demonstrates comparable efficacy to selenite in upregulation of selenoenzyme and was also found to be less toxic.

18.11 Conclusion

Administration of antioxidants during fluoride exposure can significantly overcome fluoride toxicity and therefore may be a promising therapeutic strategy for cases of chronic fluorosis. Selenium plays a critical role in the maintenance and proper functioning of cellular defense mechanism and acts as a potent protective agent *via* expression of selenoproteins. An appropriate amount of selenium can improve the body's immunity, thereby reducing cellular damage. Interference of selenium in balancing cellular redox status and restoring the level of antioxidant enzymes to a normal value, is the key factor in ameliorating fluoride toxicity. Thus, the conclusion can be drawn that selenium is significantly important in reducing chronic fluorosis and detoxification of tissues against toxic effects of fluoride by preventing fluoride induced oxidative stress.

Summary Points

- This chapter focuses on the effects of selenium on fluoride-induced toxic effects including oxidative stress, organ toxicity, exercise ventilation and pulmonary function.
- Selenium is an essential dietary trace element and a potent antioxidant that prevents damage of cell organelle, oxidation of protein, lipid and DNA by the attack of free radicals or hydroperoxides.

- Selenium acts as an antagonist against fluoride by promoting removal of fluoride from blood, and tissues, improves antioxidant enzymes system, repairs lipid metabolism and inhibits lipid peroxidation.
- Selenium also suppresses tumor initiation and primary tumor growth and also shows promise in metastasis conditions.
- The positive effect of selenium on fluoride-induced changes in the synaptic plasticity in rat hippocampus suggest a possible role of selenium in combating fluoride-induced neurotoxicity.
- Selenium is significantly important in reducing chronic fluorosis and detoxification of tissues against the toxic effects of fluoride by preventing oxidative stress.

Key Facts

- **Fluorosis**: Excessive intake of fluorine compounds leads to mottling of the teeth and, if severe, calcification of the ligaments.
- **Keshan disease**: The disease known to be linked to selenium deficiency, is a cardiomyopathy found primarily in children.
- **Beneficial effects of selenium**: Selenium individually and in combination with other nutrients such as vitamin E, vitamin C, and glutathione prevent oxygen molecules from becoming too reactive, thereby prevent cellular damage.
- **Apoptosis, or programmed cell death**: This is a naturally occurring process in the body and involves a controlled sequence of steps in which cells signal self-termination. Apoptosis works to keep the body's natural process of cell division or mitosis in check.
- **Oxidative stress**: This is a condition generated when there is disturbance in the balance between the generation of reactive oxygen species (free radicals) and antioxidant defenses. Important free radicals are described and biological sources of origin discussed, together with the major antioxidant defense mechanisms.
- **Body selenium requirement**: Upper limit for selenium of 400 μg per day for men and women 19 years and above. One of the body's most powerful antioxidant enzymes is glutathione peroxidase and Se is required for its functioning.

Definition and Explanation of Key Terms

Reactive oxygen species: Molecules generated due to incomplete one-electron reduction of oxygen and reactive oxygen intermediates include singlet oxygen; superoxides; peroxides; hydroxyl radical; and hypochlorous acid.
Trace elements: Those chemicals that are generally essential for body and are required in minute quantity.
Apoptosis: Apoptosis, or programmed cell death, is a naturally occurring process in the body and involves a controlled sequence of steps in which

cells signal self termination. Apoptosis works to keep the body's natural process of cell division or mitosis in check.

Antioxidants: Antioxidants are man-made or natural substances that may prevent or delay some types of cell damage. Antioxidants are found in many foods, including fruits and vegetables. They are also available as dietary supplements.

Lipid peroxidation: It is a process where free radicals take electrons from the lipids in our cell membranes, resulting in the cell damage and increased production of free radicals.

DNA methylation: It is a biological process in which an organic functional group like methyl group (CH_3), is added to DNA nucleotide.

Antioxidant enzyme system: There are several enzyme systems that catalyze reactions to neutralize free radicals and reactive oxygen species. These form the body's endogenous defence mechanisms to help protect against free-radical-induced cell damage.

Carcinogenesis: This is also known as tumorigenesis, and is the molecular process by which cancer develops.

Cardiomyopathy: Chronic disease of the myocardium, the heart muscle. The muscle gets abnormally enlarged, thickened, and/or stiffened and loses the ability to pump blood effectively leading to disturbed heart beat or even heart failure.

Neurotransmitter: The brain chemicals that communicate information throughout our brain and body. They relay signals between nerve cells, called "neurons." They affect mood, sleep, concentration, weight, and can cause adverse symptoms when they are out of balance.

Nanoparticle: A nanoparticle is a microscopic particle with at least one dimension less than 100 nm.

List of Abbreviations

GSH	glutathione (GSH)
TRx/TRx-R	Thioredoxin/thioredoxin reductase
NO	Nitric oxide
ROS	Reactive oxygen species
RNS	Reactive nitrogen species
TNF	Tumor necrosis factor
TNFR	Tumor necrosis factor receptor
CNS	Central nervous system
LPO	Lipid peroxides
GPx	Glutathione peroxidase
GST	Glutathione S-transferase
PSD-93	Postsynaptic density-93
SA	Sialic acids
b.w.	Body weight
Ca	calcium

References

Basha, P. M., Rai, P. and Begum, S., 2011. Fluoride toxicity and status of serum thyroid hormones, brain histopathology, and learning memory in rats: a multigenerational assessment. *Biological Trace Element Research.* **144**: 1083–1094.

Burk, R. F., Hill, K. E., Awad, J. A., Morrow, J. D., Kato, T., Cockell, K. A. and Lyons, P. R., 1995. Pathogenesis of diquat-induced liver necrosis in selenium-deficient rats: assessment of the roles of lipid peroxidation and selenoprotein P. *Hepatology.* **21**: 561–569.

Chen, Q., Wang, Z., Xiong, Y., Xue, W., Kao, X., Gao, Y., Muhammad, N. and Song, D., 2009. Selenium increases expression of HSP70 and antioxidant enzymes to lessen oxidative damage in Fincoal-type fluorosis. *Journal of Toxicological Sciences.* **34**: 399–405.

Chen, Y. C., Prabhu, K. S. and Mastro, A. M., 2013. Is selenium a potential treatment for cancer metastasis? *Nutrients.* **5**: 1149–1168.

Chouhan, S., Lomash, V., Flora, S. J. S., 2010. Fluoride-induced changes in haem biosynthesis pathway, neurological variables and tissue histopathology of rats. *Journal of Applied Toxicology.* **30**: 63–73.

Feng, P., Wei, J. and Zhang, Z., 2011. Intervention of selenium on chronic fluorosis-induced injury of blood antioxidant capacity in rats. *Biological Trace Element Research.* **144**: 1024–1031.

Flora, S. J. S., Mittal, M. and Mishra, D., 2009. Co-exposure to arsenic and fluoride on oxidative stress, glutathione linked enzymes, biogenic amines and DNA damage in mouse brain. *Journal of Neurological Sciences.* **285**: 198–205.

Flora, S. J. S., Mittal, M., Pachauri, V. and Dwivedi, N., 2012. A possible mechanism for combined arsenic and fluoride induced cellular and DNA damage in mice. *Metallomics.* **4**: 78–90.

Fora, S. J. S., 2009. Structural, chemical and biological aspects of antioxidants for strategies against metal and metalloid exposure. *Oxidative Medicine and Cellular Longevity.* **2**: 191–206.

Ip, C. and Hayes, C., 1989. Tissue selenium levels in selenium-supplemented rats and their relevance in mammary cancer protection. *Carcinogenesis.* **10**: 921–925.

Ip, C., Hayes, C., Budnick, R. M. and Ganther, H. E., 1991. Chemical form of selenium, critical metabolites, and cancer prevention. *Cancer Research.* **51**: 595–600.

Jinadasa, K. B. and Dissanayake, C. B., 1992. The effect of selenium on fluoride - clay interactions: Possible environmental health implications. *Environmental Geochemistry & Health.* **14**: 3–7.

Lee, J. H., Jung, J. Y., Jeong, Y. J., Park, J. H., Yang, K. H., Choi, N. K., Kim, S. H. and Kim, W. J., 2008. Involvement of both mitochondrial- and death receptor-dependent apoptotic pathways regulated by Bcl-2 family in sodium fluoride-induced apoptosis of the human gingival fibroblasts. *Toxicology.* **243**: 340–347.

Li, Y. M., Heerema, N. A., Dunipace, A. J. and Stookey, G. K., 1987. Genotoxic effects of fluoride evaluated by sister-chromatid exchange. *Mutation Research.* **192**: 191–201.

Mittal, M., Flora, S. J. S., 2006. Effects of individual and combined exposure to sodium arsenite and sodium fluoride on tissue oxidative stress, arsenic and fluoride levels in male mice. *Chemico- Biological Interaction.* **162**: 128–139.

Peng, D., Zhang, J., Liu, Q. and Taylor, E. W. Size effect of elemental selenium nanoparticles (Nano-Se) at supranutritional levels on selenium accumulation and glutathione S-transferase activity. *Journal of Inorganic Biochemistry.* **101**: 1457–1463.

Qian, W., Miao, K., Li, T. and Zhang, Z., 2013. Effect of selenium on fluoride-induced changes in synaptic plasticity in rat hippocampus. *Biological Trace Element Research.* **155**: 253–260.

Rakotovao, A., Tanguy, S., Toufektsian, M. C., Berthonneche, C., Ducros, V., Tosaki, A., de Leiris, J. and Boucher, F., 2005. Selenium status as determinant of connexin-43 ephosphorylation in ex vivo ischemic/reperfused rat myocardium. *Journal of Trace Element in Medicine & Biology.* **19**: 43–47.

Rayman, M. P., Stranges, S., Griffin, B. A., Pastor-Barriuso, R. and Guallar, E., 2011. Effect of supplementation with high-selenium yeast on plasma lipids: a randomized trial. *Annal of Internal Medicine.* **154**: 656–665.

Reddy, Y. P., Tiwari, S. K., Shaik, A. P., Alsaeed, A., Sultana, A. and Reddy, P. K., 2014. Effect of sodium fluoride on neuroimmunological parameters, oxidative stress and antioxidative defenses. *Toxicology Mechanism & Methods.* **24**: 31–36.

Shashi, A. and Thapar, S. P., 2001. Histopathology of myocardial damage in experimental fluorosis in rabbits. *Fluoride.* **34**: 43–50.

Slamenová, D., Ruppová, K., Gábelová, A. and Wsólová, L., 1996. Evaluation of mutagenic and cytotoxic effects of sodium fluoride on mammalian cells influenced by an acid environment. *Cell Biology & Toxicology.* **12**: 11–17.

Sun, Z., Niu, R., Wang, B., Jiao, Z., Wang, J., Zhang, J., Wang, S. and Wang, J., 2011. Fluoride-induced apoptosis and gene expression profiling in mice sperm in vivo. *Archives of Toxicology.* **85**: 1441–1452.

Tang, R., Liu, H., Wang, T. and Huang, K., 2005. Mechanisms of selenium inhibition of cell apoptosis induced by oxysterols in rat vascular smooth muscle cells. *Archives of Biochemistry & Biophysics.* **441**: 16–24.

Wang, H., Zhang, J. and Yu, H., 2007. Elemental selenium at nano size possesses lower toxicity without compromising the fundamental effect on selenoenzymes: comparison with selenomethionine in mice. *Free Radical Biology & Medicine.* **42**: 1524–1533.

Whanger, P., Vendeland, S., Park, Y. C. and Xia, Y., 1996. Metabolism of subtoxic levels of selenium in animals and humans. *Annal of Clinical & Laboratory Sciences.* **26**: 99–113.

Xiang, Q., Liang, Y., Chen, L., Wang, C., Chen, B., Chen, X. and Zhou, M., 2003. Effect of fluoride in drinking water on children's intelligence. *Fluoride.* **36**: 84–94.

Yang, S., Wang, Z., Farquharson, C., Alkasir, R., Zahra, M., Ren, G. and Han, B., 2011. Sodium fluoride induces apoptosis and alters bcl-2 family protein expression in MC3T3-E1 osteoblastic cells. *Biochemistry Biophysics Research Communication.* **410**: 910–915.

Zeng, H., Uthus, E. O. and Combs Jr, G. F., 2005. Mechanistic aspects of the interaction between selenium and arsenic. *Journal of Inorganic Biochemistry.* **99**: 1269–1274.

Zeng, H., Yan, L., Cheng, W. H. and Uthus, E. O., 2011. Dietary selenomethionine increases exon-specific DNA methylation of the p53 gene in rat liver and colon mucosa. *Journal of Nutrition.* **141**: 1464–1468.

Zhang, J. S., Gao, X. Y., Zhang, L. D. and Bao, Y. P., 2001. Biological effects of a nano red elemental selenium. *Biofactors.* **15**: 27–38.

Zhang, M., Wang, A., He, W., He, P., Xu, B., Xia, T., Chen, X. and Yang, K., 2007. Effects of fluoride on the expression of NCAM, oxidative stress, and apoptosis in primary cultured hippocampal neurons. *Toxicology.* **236**: 208–216.

CHAPTER 19

Fluoride and Effects on Caspases

JUNDONG WANG*[a] AND RUIYAN NIU[a]

[a]Shanxi Key Lab of Ecological Animal Science and Environmental Veterinary Medicine, Shanxi Agricultural University, Taigu, Shanxi 030801, PR China
*E-mail: wangjd@sxau.edu.cn

19.1 Introduction

Exposure to elevated fluoride has been found to induce apoptosis *in vitro* in different cultured cell types, as well as *in vivo* in various tissues from animals and human beings. Apoptosis was first proposed to be involved in the toxic mechanism of fluoride by Hirano and Ando who observed the apoptotic cell death in rat alveolar macrophages incubated with sodium fluoride (Hirano and Ando, 1996). To get further evidence for fluoride-induced apoptosis, this research group later applied the osteosarcoma cell line (UMR 106) and found that after 5 mM fluoride exposure for 8 h caused significant UMR 106 apoptosis (Hirano and Ando, 1997). In human osteogenic sarcoma cells, 48 h sodium fluoride exposure (10, 50 or 250 ppm) induced the cell apoptosis that was detected by flow cytometric analysis, accompanied by the appearance of nucleosomal DNA fragmentation and apoptotic bodies, and the increase in cytochrome c release and Caspase-3 activity (Song *et al.*, 2005). Through establishing animal models, our group also found the apoptosis in mouse sperm (Sun *et al.*, 2011), rat thymus (Wang *et al.*, 2009) and kidney (Chen *et al.*, 2009).

Apoptosis is classified as the programmed cell death, which is a complex and highly regulated phenomenon normally functioning in unnecessary cell elimination, proliferation, differentiation, tissue homeostasis, and aging

Food and Nutritional Components in Focus No. 6
Fluorine: Chemistry, Analysis, Function and Effects
Edited by Victor R Preedy

Published by the Royal Society of Chemistry, www.rsc.org

(Lauber *et al.*, 2004). During the process of apoptosis, typically the cell undergoes shrinkage, the chromatin is condensed, and DNA is fragmented further forming apoptotic bodies. These above changes together with mitochondrial depolarization and membrane blebbing form the morphological features for apoptosis, which are always observed in cells from different organs and tissues exposed to fluoride. Our earlier investigation showed that 150 mg F ion/L in drinking water for 60 days resulted in condensed and lessened chromatins, nuclear envelope, swelling mitochondria and significant DNA damage, as well as the increased amounts of apoptotic cells in rat thymus (Wang *et al.*, 2009). The consistent apoptotic characteristics were also depicted in other animal experiments. 50 mg L^{-1} sodium fluoride in drinking water for 6 months promoted the renal apoptosis and decreased the DNA relative content in rats (Yu *et al.*, 2006). In pig liver, both apoptotic hepatocytes and the obvious DNA ladder were observed after 400 mg kg^{-1} fluoride treatment for 50 days (Zhan *et al.*, 2006). It is worth noting that a large number of investigations, which focused on the fluoride-induced apoptosis, also provided the evidence for oxidative stress including overproduced reactive oxygen species (ROS) and weakened status of intracellular antioxidant systems. Oxidative stress is considered to be an accepted mode of fluoride action, and plays a triggered role in apoptosis induction through the intrinsic pathway in a concentration-dependent manner (Barbier *et al.*, 2010). Many works concluded that oxidative stress involved in the mechanism of fluoride-induced apoptosis.

The initiation and process of apoptosis at least in mammal cells require delicate cooperation of signaling proteins *via* three main pathways. The first one is the extrinsic pathway, in which apoptosis is trigged by ligation of death receptors at the plasma membrane, followed by the activation of caspase-8 (Boatright and Salvesen, 2003). The second is intrinsic pathway, also called mitochondria-initiated death pathway (Denault and Salvesen, 2008). When mitochondria is exposed to intracellular stress such as ROS, DNA damage, ultraviolet (UV) radiation, γ-irradiation, heat, viral virulence factors and the actions of tumor suppressor genes, the permeability of mitochondria membrane increases, resulting in the release of proapoptotic proteins like cytochrome c and AIF that activates the caspase-9. The third one is related to the cytotoxic cell protease granzyme B (Nakagawa *et al.*, 2000). In this pathway, granzyme B is transported to target cells further executing apoptosis. During the execution of these pathways mentioned above, activation of caspases-3 and/or -7 are determined to be the sharing phase of apoptosis (Salvesen and Dixit, 1997; Thornberry and Lazebnik, 1998). The requirement of Caspases in apoptosis was first implicated by the findings that *Caenorhabditis elegans death* gene 3 acts as cysteine proteases in the initiation of programmed cell death (Yuan *et al.*, 1993). Defects in apoptosis induced by the caspase inhibitors further certified the involvement of caspases in apoptosis (Chang and Yang, 2000). Many *in vitro* and *in vivo* studies showed that the activities of caspases-3, -9, and -8 were enhanced by fluoride exposure, which suggests an ability of fluoride to induce caspase-dependent apoptosis. However, besides apoptosis, caspases participate in various biological actions. Here, we focus on the effects of fluoride on caspases proteolytic system through summarizing the recent related literature data, and update our understanding of the mechanism underling alterations in caspases-dependent functions induced by fluoride.

19.2 Caspase Family

After the first official nomenclature of caspase-1 that was originally discovered to be a homolog of CED-3 in Caenorhabditis elegans (C elegans) cells and function in the activation of interleukin-1β-converting enzyme, there are 14 members of cysteine-dependent aspartate-specific proteases (caspases) family have been identified in different cell types and tissues from different species (Nakagawa and Yuan, 2000). Among the already discovered caspases, 12 are determined to express in human beings (Fuentes-Prior and Salvesen, 2004), whilst caspase-13 is proved to be expressed only in bovine that is a homolog of human caspase-4 (Koenig *et al.*, 2001), and caspase-11 (also a homolog of human caspase-4) found only in murine, indicating species specificity (Degterev *et al.*, 2003).

Caspase members share some common structural features including an NH_2-terminal prodomain of variable length, a catalytic domain composed by a large subunit (17–20 kDa) and a small subunit (10–12 kDa), and a linker region connecting these subunits (Fuentes-Prior and Salvesen, 2004). The length and the amino acid sequence in NH_2-terminal prodomain vary in different caspases. For instance, caspase-8 and -10 possess long prodomains containing a special motif such as DED that is essential for the interaction with adaptor proteins and linkage with signaling pathways. Others like caspase-1, -2, -4, -5, -9, -11 and -12 have caspase recruitment domains (CARD) (Kang *et al.*, 2004). The structural differences highly influence caspase functions. In other words, apoptosis is not the only biological functions in which caspases participate. Several investigations reported that caspases also take part in cell survival, proliferation, differentiation or inflammation (Los *et al.*, 2001). According to the phylogenetic analysis, caspases are subdivided into two groups: caspases-1, -4, -5, -11 and -13 mainly responsible for the inflammatory response system (Martinon and Tschopp, 2004), and caspases-2, -3, -6, -7, -8, -9, -10, and -12 involving in apoptosis (Degterev *et al.*, 2003). Based on their placement within the cascade of apoptosis pathway, the later apoptotic caspases can are further divided into initiator caspases including caspases-2, -8, -9, -10, and -12, and effector caspases such as caspases-3, -6 and -7 (Earnshaw *et al.*, 1999). Although previous studies had reported that elevated fluoride exposure contributed to the inflammation, few mentioned the involvement of inflammatory caspases (caspases-1, -4, -5, -11 and -13). By contrast, the numbers of reports on fluoride affecting apoptotic caspases increased greatly. Therefore, the following mainly introduce the caspase activation in apoptotic pathways.

19.3 Apoptotic Caspases Activation

In the process of caspase-dependent apoptosis, the extrinsic pathway relying on the death receptor ligation at cell membrane and intrinsic pathway at the mitochondria are the most extensively researched signaling pathways that are briefly described as follows (see Figure 19.1).

In extrinsic pathway, extracellular ligands stimulate the death receptors located in the cell membrane. Then, the receptors begin aggregation and recruit the adaptor molecule Fas-associated protein with death domain (FADD). When the death domain (DD) of FADD binds to the DD of death receptor (Ashkenazi and Dixit, 1998), the death-inducing signaling complex (DISC) is formed (Boldin *et al.*, 1995). Subsequently procaspase-8, one of the initiator caspases, is recruited to the DISC and cleaved into activated caspase-8 (Bodmer *et al.*, 2000). The activated caspase-8 then initiates apoptosis by directly cleaving procaspase-3 into activated caspase-3 (Denault and Salvesen, 2008), an effector caspase in the downstream of apoptosis cascade.

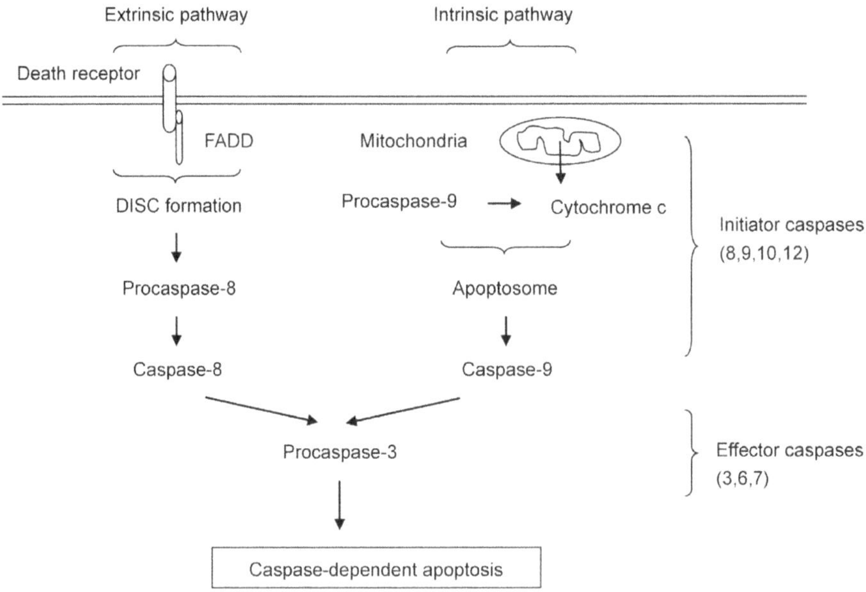

Figure 19.1 Simplified process of caspase-dependent apoptosis, including the intrinsic and extrinsic pathways. In an extrinsic pathway, the death receptor binds to Fas-associated protein with death domain (FADD) to form the death-inducing signaling complex (DISC), which cleaves procaspase-8 into activated caspase-8. In the intrinsic pathway, the extra- and intracellular stresses increase mitochondrial membrane permeabilization, leading to the release of proapoptotic proteins such as cytochrome c. The cytochrome c binds to the cytoplasm Apaf1 and recruits procaspase-9, finally forming apoptosome. Then, procaspase-9 is cleaved into caspase-9. Either caspase-8 or caspase-9 then activates caspase-3 to evoke the apoptosis.

In intrinsic pathways, various extra- and intracellular stresses result in mitochondrial membrane permeabilization, leading to the release of proapoptotic proteins such as cytochrome c, AIF, and noncaspase-dependent (Endo G). The cytochrome c binds to the cytoplasm Apaf1 and recruits procaspase-9 (an initiator caspase), finally forming the multiprotein complex (cytochrome c- Apaf1-procaspase 9) called apoptosome (Ly *et al.*, 2003). After activation, caspase-9 then activates caspase-3 to evoke the apoptosis (Li *et al.*, 1997).

It is easy to find that caspase-3 activation is an overlap phase of both extrinsic and intrinsic pathways. Besides, another acknowledged link of the two pathways is that the proapoptotic proteins in mitochondria can also be released by the truncated Bid (tBid) that is cleaved by caspase-8 (Ghavami *et al.*, 2009).

Reviewing studies on fluoride-induced apoptosis, it is obvious that caspase-3, -9 and -8 got more attention, and thus here we mainly introduce the functions of these three caspases in apoptosis.

19.4 Effects on Caspases

Caspase-3 is studied extensively (Ghavami *et al.*, 2009) and has been considered to be an important effector caspase in executing apoptosis due to at least two reasons: first, its activation involves in both extrinsic and intrinsic pathways; secondly, the inhibition of most downstream affairs in apoptosis occurs when caspase-3, rather than caspase-6 and/or -7 (other effector caspases), is decreased (Slee *et al.*, 2001). Therefore, caspases-3 is usually selected as the target molecular in fluoride-induced apoptosis.

Exposure of human leukemia (HL-60) cells to 100–250 mg L^{-1} sodium fluoride (NaF) (Song *et al.*, 2002) or 2–5 mM NaF (Anuradha *et al.*, 2000) resulted in increased activation of caspase-3. In human osteogenic sarcoma cells, caspase-3 activity was significantly enhanced by sodium fluorosilicate (Song *et al.*, 2005). 50 mg L^{-1} NaF also increased the protein expression of caspase-3 in rat leukocytes (Gutierrez-Salinas *et al.*, 2010). Besides the change in caspase-3, the significant apoptosis was also observed in all the above mentioned studies. However, it is still difficult to conclude that Caspase-3 is a necessary effector involved in fluoride-induced apoptosis. Because 10^{-2} mol L^{-1} NaF exposure for 24 h significantly increased caspase-3 activity in murine osteoblast-like MC3T3-E1 cells (Ren *et al.*, 2011), whilst in the same cell 4 mM NaF treatment for 4 h failed to induce caspase-3 activation (Chien *et al.*, 2009).

Actually, caspase-3, -8, and -9 were reported to be involved in cytotoxicity of fluoride in diverse cell types and organs. Dental and skeletal fluorosis were the two classical symptom of fluoride poisoning. The incubation of 20 mM NaF for 6 h increased the cytochrome c release in human gingival fibroblasts, and upregulated the caspase-3, -8, and -9 activities, indicating the mitochondria-mediated pathways regulated by a death receptor-mediated pathway (Lee *et al.*, 2008). The exposure of 5 mM NaF for 24 h enhanced the expression of caspase-3, and -8 in first molar germ (Jacinto-Aleman *et al.*). 5 mM NaF also activated the caspase-3 in mouse odontoblast-lineage cells (Karube *et al.*, 2009). In human ameloblast-lineage cells incubated with NaF, the caspase-3-mediated DNA fragmentation was found (Kubota *et al.*, 2005). All these evidence suggested that caspase-3 might participate in apoptosis in dental fluorosis. In our previous study, the increased mRNA levels of caspase-3 and -9, and osteoblast apoptosis induced by 72 h NaF (5 mg L^{-1}) treatment provided evidence for capases-mediated apoptosis participating in skeletal fluorosis (Yan *et al.*, 2009).

The results of *in vivo* studies are consistent with those of *in vitro* ones. After 400 mg L^{-1} NaF treatment for 50 days, significant increases in apoptotic hepatocytes plus activated caspase-3 and -9 in pig liver indicated fluoride induced caspase-dependent aopotosis (Zhan *et al.*, 2006). In thymus, 150 mg L^{-1} F^{-} (338 mg L^{-1} NaF) in drinking water leads to the high mRNA expression of caspase-3 and -9 and thymus apoptosis in rats (Wang *et al.*, 2009). The same dose and ingestion way of NaF also resulted in renal apoptosis and increased caspase-3 expression in rats (Chen *et al.*, 2009). In addition, a significant increase in protein expressions of both cytochrome c and caspase-3 were observed in sperm of mice exposed to 150 mg L^{-1} NaF for 8 weeks (Sun *et al.*, 2011).

19.5 Conclusions

Activated caspases, included caspase-3, -9, and -8, have been reported to be involved in fluoride-induced apoptosis in diverse tissues and organs such as bone, tooth, kidney, thyroid, liver, and sperm. However, the cascade events in apoptotic caspases in fluorosis models have not been identified exactly. There still exists incomplete and controversial data about molecular mechanisms of the fluoride-induced apoptosis. Therefore, future investigations are required to clarify the intracellular pathways in caspase-dependent apoptosis caused by fluoride exposure.

Summary Points

- The chapter is an overview of the effects of fluoride on caspases in human, animals, and cell lines.
- High-fluoride exposure has been found to induce apoptosis *in vitro* in different cultured cell types, as well as *in vivo* in various tissues from animals and human beings.
- Caspase members, caspases-2, -3, -6, -7, -8, -9, -10, and -12 participating in apoptosis, which can induce the caspase-dependent apoptosis.
- In the process of caspase-dependent apoptosis, the extrinsic pathway relying on the death receptor ligation at cell membrane and intrinsic pathway at the mitochondria are the most extensively researched signaling pathways, involving the activation of caspase-3, -9, -6, -7and -8.
- The mechanism by which fluoride induces cell apoptosis involves the activated caspases, included caspase-3, -9, and -8, which been reported in diverse tissues and organs such as bone, tooth, kidney, thyroid, liver, and sperm.

Definition of Words and Terms

Apoptosis: A complex and highly regulated and programmed cell death, during which typically the cell becomes shrinkage, the chromatin is condensed, and DNA is fragmented further forming apoptotic bodies.

Caspases: Cysteine-dependent aspartate-specific proteases family. At present, there are 14 members that share some common structural features including an NH_2-terminal prodomain of variable length, a catalytic domain composed by a large subunit (17–20 kDa) and a small subunit (10–12 kDa), and a linker region connecting these subunits.

Caspase 3: A member of the cysteine-aspartic acid protease (caspase) family is formed from a 32 kDa zymogen that is cleaved into 17 kDa and 12 kDa subunits. It plays a central role in the execution-phase of cell apoptosis that cleaves and activates caspases 6 and 7 and activated by caspases 8, 9, and 10.

Caspase recruitment domains (CARD): Interaction motifs found in a wide array of proteins including helicases, kinases, mitochondrial proteins, caspases, and other cytoplasmic factors, which involve in processes relating to inflammation and apoptosis.

Death domain (DD): A protein interaction module composed of a bundle of six alpha-helices. The C terminal death domain is a part of FADD.

Death-inducing signaling complex (DISC): A multiprotein complex formed by members of the death receptor family of apoptosis-inducing cellular receptors.

Fas-associated protein with death domain (FADD): An adaptor protein that bridges tumor necrosis factor receptor to procaspases, forming the death-inducing signaling complex during apoptosis.

Mitochondria-initiated death pathway: The essential role of mitochondria in signaling apoptosis. In this pathway, proapoptotic proteins target mitochondria and cause mitochondrial swelling, and increase the permeability of the mitochondrial membrane, resulting in apoptotic effectors to leak out.

Oxidative stress: A disturbance in the balance between the production of reactive oxygen species (free radicals) and antioxidant defenses.

The death-effector domain (DED): A protein interaction domain found in inactive procaspases (cysteine proteases) and proteins that regulate caspase activation in the apoptosis cascade such as FADD.

List of Abbreviations

Caspases	Cysteine-dependent aspartate-specific proteases
ROS	Reactive oxygen species
CARD	Caspase recruitment domains
FADD	Fas-associated protein with death domain
DD	Death domain
DISC	Death-inducing signaling complex
HL	Human leukemia

References

Anuradha, C. D., Kanno, S. and Hirano, S., 2000. Fluoride induces apoptosis by caspase-3 activation in human leukemia HL-60 cells. *Archives of Toxicology.* **74**: 226–230.

Ashkenazi, A. and Dixit, V. M., 1998. Death receptors: signaling and modulation. *Science.* **281**: 1305–1308.

Barbier, O., Arreola-Mendoza, L. and Del Razo, L. M., 2010. Molecular mechanisms of fluoride toxicity. *Chemico-biological Interactions.* **188**: 319–333.

Boatright, K. M. and Salvesen, G. S., 2003. Mechanisms of caspase activation. *Current Opinion in Cell Biology.* **15**: 725–731.

Bodmer, J. L., Holler, N., Reynard, S., Vinciguerra, P., Schneider, P., Juo, P., Blenis, J. and Tschopp, J., 2000. TRAIL receptor-2 signals apoptosis through FADD and caspase-8. *Nature Cell Biology.* **2**: 241–243.

Boldin, M. P., Varfolomeev, E. E., Pancer, Z., Mett, I. L., Camonis, J. H. and Wallach, D., 1995. A novel protein that interacts with the death domain of Fas/APO1 contains a sequence motif related to the death domain. *Journal of Biological Chemistry.* **270**: 7795–7798.

Chang, H. Y. and Yang, X., 2000. Proteases for cell suicide: functions and regulation of caspases. *Microbiology and Molecular Biology Reviews.* **64**: 821–846.

Chen, H. C., Hong, J. H., Niu, R. Y., Sun, Z. L. and Wang, J. D., 2009. Influence of fluoride on caspase-3 expression in kidney of rats fed a normal or a low calcium and protein diet and its reversal by protein and calcium supplementation. *Fluoride.* **42**: 121–126.

Chien, C. H., Sakagami, H., Kouhara, M., Sasaki, A., Matsumoto, K. and Kanegae, H., 2009. Effect of simulated orthodontic forces on fluoride-induced cytotoxicity in MC3T3-E1 osteoblast-like cells. *In Vivo.* **23**: 259–265.

Degterev, A., Boyce, M. and Yuan, J., 2003. A decade of caspases. *Oncogene.* **22**: 8543–8567.

Denault, J. B. and Salvesen, G. S., 2008. Apoptotic caspase activation and activity. *Methods in Molecular Biology.* **414**: 191–220.

Earnshaw, W. C., Martins, L. M. and Kaufmann, S. H., 1999. Mammalian caspases: structure, activation, substrates, and functions during apoptosis. *Annual Review of Biochemistry.* **68**: 383–424.

Fuentes-Prior, P. and Salvesen, G. S., 2004. The protein structures that shape caspase activity, specificity, activation and inhibition. *Biochemical Journal.* **384**: 201–232.

Ghavami, S., Hashemi, M., Ande, S. R., Yeganeh, B., Xiao, W., Eshraghi, M., Bus, C. J., Kadkhoda, K., Wiechec, E., Halayko, A. J. and Los, M., 2009. Apoptosis and cancer: mutations within caspase genes. *Journal of Medical Genetics.* **46**: 497–510.

Gutierrez-Salinas, J., Morales-Gonzalez, J. A., Madrigal-Santillan, E., Esquivel-Soto, J., Esquivel-Chirino, C., Gonzalez-Rubio, M. G., Suastegui-Dominguez, S. and Valadez-Vega, C., 2010. Exposure to sodium fluoride produces signs of apoptosis in rat leukocytes. *International Journal of Molecular Sciences.* **11**: 3610–3622.

Hirano, S. and Ando, M., 1996. Apoptotic cell death following exposure to fluoride in rat alveolar macrophages. *Archives of Toxicology.* **70**: 249–251.

Hirano, S. and Ando, M., 1997. Fluoride mediates apoptosis in osteosarcoma UMR 106 and its cytotoxicity depends on the pH. *Archives of Toxicology.* **72**: 52–58.

Jacinto-Aleman, L. F., Hernandez-Guerrero, J. C., Trejo-Solis, C., Jimenez-Farfan, M. D. and Fernandez-Presas, A. M., In vitro effect of sodium fluoride on antioxidative enzymes and apoptosis during murine odontogenesis. *Journal of Oral Pathology and Medicine.* **39**: 709–714.

Kang, T. B., Ben-Moshe, T., Varfolomeev, E. E., Pewzner-Jung, Y., Yogev, N., Jurewicz, A., Waisman, A., Brenner, O., Haffner, R., Gustafsson, E., Ramakrishnan, P., Lapidot, T. and Wallach, D., 2004. Caspase-8 serves both apoptotic and nonapoptotic roles. *The Journal of Immunology.* **173**: 2976–2984.

Karube, H., Nishitai, G., Inageda, K., Kurosu, H. and Matsuoka, M., 2009. NaF activates MAPKs and induces apoptosis in odontoblast-like cells. *Journal of Dental Research.* **88**: 461–465.

Koenig, U., Eckhart, L. and Tschachler, E., 2001. Evidence that caspase-13 is not a human but a bovine gene. *Biochemical and Biophysical Research Communications.* **285**: 1150–1154.

Kubota, K., Lee, D. H., Tsuchiya, M., Young, C. S., Everett, E. T., Martinez-Mier, E. A., Snead, M. L., Nguyen, L., Urano, F. and Bartlett, J. D., 2005. Fluoride induces endoplasmic reticulum stress in ameloblasts responsible for dental enamel formation. *The Journal of Biological Chemistry.* **280**: 23194–23202.

Lauber, K., Blumenthal, S. G., Waibel, M. and Wesselborg, S., 2004. Clearance of apoptotic cells: getting rid of the corpses. *Molecular Cell.* **14**: 277–287.

Lee, J. H., Jung, J. Y., Jeong, Y. J., Park, J. H., Yang, K. H., Choi, N. K., Kim, S. H. and Kim, W. J., 2008. Involvement of both mitochondrial- and death receptor-dependent apoptotic pathways regulated by Bcl-2 family in sodium fluoride-induced apoptosis of the human gingival fibroblasts. *Toxicology.* **243**: 340–347.

Li, P., Nijhawan, D., Budihardjo, I., Srinivasula, S. M., Ahmad, M., Alnemri, E. S. and Wang, X., 1997. Cytochrome c and dATP-dependent formation of Apaf-1/caspase-9 complex initiates an apoptotic protease cascade. *Cell.* **91**: 479–489.

Los, M., Stroh, C., Janicke, R. U., Engels, I. H. and Schulze-Osthoff, K., 2001. Caspases: more than just killers? *Trends in Immunology.* **22**: 31–34.

Ly, J. D., Grubb, D. R. and Lawen, A., 2003. The mitochondrial membrane potential (deltapsi(m)) in apoptosis; an update. *Apoptosis.* **8**: 115–128.

Martinon, F. and Tschopp, J., 2004. Inflammatory caspases: linking an intracellular innate immune system to autoinflammatory diseases. *Cell.* **117**: 561–574.

Nakagawa, T., Zhu, H., Morishima, N., Li, E., Xu, J., Yankner, B. A. and Yuan, J., 2000. Caspase-12 mediates endoplasmic-reticulum-specific apoptosis and cytotoxicity by amyloid-beta. *Nature.* **403**: 98–103.

Nakagawa, T. and Yuan, J., 2000. Cross-talk between two cysteine protease families. Activation of caspase-12 by calpain in apoptosis. *Journal of Cell Biology.* **150**: 887–894.

Ren, G., Wang, K., Chang, R., Su, Y., Wang, J., Su, J. and Han, B., 2011. Simultaneous administration of fluoride and selenite regulates proliferation and apoptosis in murine osteoblast-like MC3T3-E1 cells by altering osteoprotegerin. *Biological Trace Element Research.* **144**: 1437–1448.

Salvesen, G. S. and Dixit, V. M., 1997. Caspases: intracellular signaling by proteolysis. *Cell.* **91**: 443–446.

Slee, E. A., Adrain, C. and Martin, S. J., 2001. Executioner caspase-3, -6, and -7 perform distinct, non-redundant roles during the demolition phase of apoptosis. *The Journal of Biological Chemistry.* **276**: 7320–7326.

Song, J. S., Lee, H. Y., Lee, E., Hwang, H. J. and Kim, J. H., 2002. Cytotoxicity and apoptosis induction of sodium fluoride in human promyelocytic leukemia (HL-60) cells. *Environmental Toxicology and Pharmacology.* **11**: 85–91.

Song, J. S., Park, Y. D., Hyun, J. W. and Kim, J. H., 2005. Induction of apoptosis by sodium fluorosilicate treatment in human osteogenic sarcoma (HOS) cells. *Anticancer Research.* **25**: 391–395.

Sun, Z., Niu, R., Wang, B., Jiao, Z., Wang, J., Zhang, J. and Wang, S., 2011. Fluoride-induced apoptosis and gene expression profiling in mice sperm in vivo. *Archives of Toxicology.* **85**: 1441–1452.

Thornberry, N. A. and Lazebnik, Y., 1998. Caspases: enemies within. *Science.* **281**: 1312–1316.

Wang, H., Zhou, B., Cao, J., Gu, X., Cao, C. and Wang, J., 2009. Effects of dietary protein and calcium on thymus apoptosis induced by fluoride in female rats (Wistar rats). *Environmental Toxicology.* **24**: 218–224.

Yan, X., Feng, C., Chen, Q., Li, W., Wang, H., Lv, L., Smith, G. W. and Wang, J., 2009. Effects of sodium fluoride treatment in vitro on cell proliferation, apoptosis and caspase-3 and caspase-9 mRNA expression by neonatal rat osteoblasts. *Archives of Toxicology.* **83**: 451–458.

Yu, R. A., Xia, T., Wang, A. G. and Chen, X. M., 2006. Effects of selenium and zinc on renal oxidative stress and apoptosis induced by fluoride in rats. *Biomedical and Environmental Sciences.* **19**: 439–444.

Yuan, J., Shaham, S., Ledoux, S., Ellis, H. M. and Horvitz, H. R., 1993. The C. elegans cell death gene ced-3 encodes a protein similar to mammalian interleukin-1 beta-converting enzyme. *Cell.* **75**: 641–652.

Zhan, X. A., Wang, M., Xu, Z. R., Li, W. F. and Li, J. X., 2006. Evaluation of caspase-dependent apoptosis during fluoride-induced liver lesion in pigs. *Archives of Toxicology.* **80**: 74–80.

Subject Index

Illustrations and figures are in **bold**. Tables are in *italics*.